EUPHORBIACEAE (Part 2)

SUSAN CARTER & A.R.-SMITH*

This part contains the tribe Euphorbieae,.and also concludes with the treatments of those five genera (73. *Bischofia*, 74. *Uapaca*, 75. *Antidesma, Hymenocardia* (Hymenocardiaceae) & *Microdesmis* (Pandaceae)) about which there has been considerable dispute as to their familial status.

Tribe **Euphorbieae**

Boiss. in DC., Prodr. 15(2): 4(1862)

Flowers monoecious, with numerous ♂ flowers surrounding a solitary ♀ flower and all enclosed within a cup-like involucre; each ♂ flower reduced to a single pedicellate stamen without a perianth; ♀ flower consisting of a pedicellate tricarpellary pistil with perianth absent or reduced to a rim or 3 lobes. Milky latex is always present.

A well-defined tribe which includes, besides the very large genus of *Euphorbia* itself, 10 other small genera of tropical and subtropical regions; of these 6 are endemic in tropical Africa, 3 occurring in East Africa (*Elaeophorbia, Synadenium* and *Monadenium*). *Pedilanthus*, from tropical America, is also used as an ornamental.

KEY TO GENERA

1. Cyathia actinomorphic 2
   Cyathia zygomorphic 4
2. Gland continuous, rim-like, with no gap to one side   **70. Synadenium**
   Gland single, lateral, or glands 2–8, distinct 3
3. Fruit not or only slightly fleshy, dehiscent . . . . **68. Euphorbia**
   Fruit thick, fleshy, indehiscent, drupaceous . . . . . **69. Elaeophorbia**
4. Gland a continuous rim except for a gap at one side,
   surrounding the upper part of the cyathium . . . **71. Monadenium**
   Glands completely enclosed by the cyathium . . . . **72. Pedilanthus**

68. EUPHORBIA

L., Sp. Pl.: 450 (1753) & Gen. Pl., ed. 5: 208 (1754); Boiss. in DC., Prodr. 15(2): 78 (1862); G.P. 3(1): 258 (1800); N.E. Br. in F.T.A. 6(1): 470 (1911); Pax in E. & P. Pf., ed. 2, 19C: 208 (1931)

Annual, biennial or perennial herbs, shrubs or trees, sometimes succulent and unarmed or spiny, with a milky usually caustic latex, monoecious (elsewhere rarely dioecious). Roots fibrous, or thick and fleshy, sometimes tuberous. Leaves opposite, alternate or verticillate, often stipulate, subtended by spiny outgrowths on some succulent species. Inflorescence with cyathia in simple, dichotomous or umbellate terminal or axillary cymes; bracts paired, often leaf-like, sometimes brightly coloured and showy. Involucres with (1–)4–5(–8) glands around the rim alternating with 5 fringed lobes. Male flowers in 5 groups separated by fringed membranes, bracteolate. Female flower

* Euphorbieae by Susan Carter, other genera by A.R.-Smith.

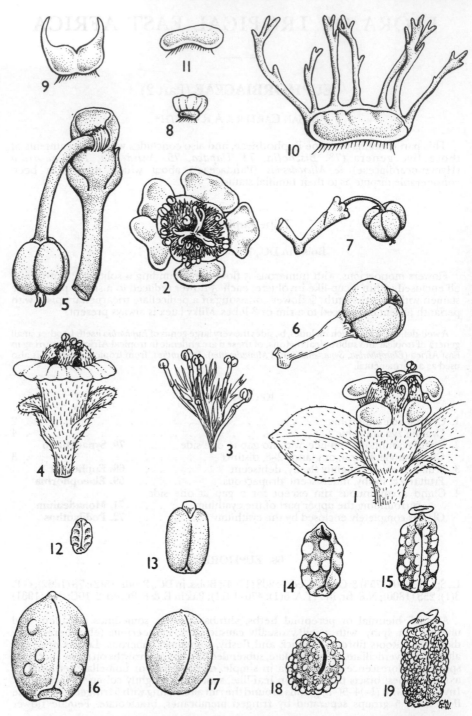

FIG. 76. Key characters. Cyathia of 1, *Euphorbia*, with glands distinct, 2, surface view, 3, stamens and bracteoles, 4, *Synadenium*, with glandular rim complete, 5, *Monadenium*, with glandular rim broken at one side. Capsules 6, sessile or subsessile, 7, exserted. Involucral glands of 8, *Euphorbia* subgen. *Chamaesyce*, 9, subgen. *Esula*, 10, subgen. *Trichadenia*, 11, subgen. *Euphorbia*. Seeds of 12, *Euphorbia* subgen. *Chamaesyce*,-13, subgen. *Esula*, 14, subgen. *Eremophyton* sect. *Pseudacalypha*, 15, sect. *Eremophyton*, 16, subgen. *Trichadenia*, 17, subgen. *Lyciopsis*, 18, subgen. *Euphorbia*, 19, *Monadenium*. Drawn by Christine Grey-Wilson.

subsessile or with the pedicel usually elongating and reflexed in fruit but straightening before dehiscence; perianth reduced to a rim below the ovary, seldom 3-lobed; ovary 3-locular, with 1 pendulous ovule in each locule; styles 3, partly united. Fruit a capsule, sometimes fleshy but becoming woody at maturity, loculicidally and septicidally dehiscent (a regma). Seeds with or without a caruncle.

Euphorbia is one of the largest genera of flowering plants, with possibly over 2000 species and a world-wide distribution in temperate and tropical zones. About 1300 species are herbaceous. The other shrubby, tree and succulent species are confined almost entirely to the tropics and subtropics. With relatively few exceptions the succulent species, numbering over 500, occur in drier regions of the African continent (including Madagascar), well adapted to survive in often extreme xerophytic conditions. Their growth habit is extremely diverse, from small herbs to shrubs and large trees.

A number of species possessing showy leaves or bracts are of horticultural value, those introduced into East Africa including the well-known *E. pulcherrima* Klotzsch (Poinsettia) from Central America and *E. milii* Des Moulins from Madagascar (see notes on Cultivated Species, p. 413). All the succulent species are of interest to specialist collectors, many being at risk in the wild and listed as endangered. The lightweight wood of some tree species has occasionally been used commercially for such items as matchsticks and banana crates.

The latex of all species is abundant and usually caustic, sometimes exceedingly so. Because of this and its irritant qualities, it is often a component, in Africa, of arrow and fish poisons, and is used widely in native medicines. As such it is known occasionally to promote some forms of cancer and consequently its chemical properties are of importance in cancer research. Attempts have been made in the past to use the latex as a rubber substitute, but resulting compounds proved to be unstable. Currently some species are being investigated for their hydrocarbon content as a source of fuel.

Because of the unifying structure of the cyathium, classification of this large and otherwise diverse aggregate of species has always been difficult and no system proposed so far has proved entirely satisfactory. Some authors have separated *Euphorbia* into a number of genera, with reasons usually based upon slight peculiarities of the cyathial structure. As treatment is often confined to limited geographical areas, such concepts invariably break down when applied to related species from other regions. The arrangement employed here loosely follows Pax's adaptation (1931) of Boissier's work (1862), with modifications to better accommodate a number of distinct groups recognized within East Africa. These are placed in subgenera and sections in a sequence which is not strictly phylogenetic but shows increasing specialization towards succulence, and a synopsis of which follows.

SYNOPSIS OF SUBGENERA AND SECTIONS

Subgen. **Chamaesyce** *Raf.* in Amer. Month. Mag. 2:119 (1817). Principally of New World origin and regarded by some authors as an easily defined genus with several distinguishing and often advanced characters. Its main stem is reduced, the entire plant at maturity consisting of an expanded dichotomously branching pseudumbellate inflorescence, with the floral bracts assuming the appearance and function of normal leaves. Stipules are obvious, the 4 involucral glands bear petaloid appendages on the outer margin and the seeds are ecarunculate.
  Sect. **Hypericifoliae** *Pojero*, Fl. Sicula 2(2): 327 (1907). Includes species with 10 or more cyathia grouped or congested into often capitate cymes, and of which there are 3 introduced species in East Africa. Species 1–3.
  Sect. **Chamaesyce**. Includes species with solitary cyathia or a few clustered in leafy cymes. Species 4–22.

Subgen. **Poinsettia** (*Grah.*) *House* in Bull. N. York St. Mus. 254: 472 (1924). Syn. — *Poinsettia* Grah. in Edinb. New Philos. Journ. 20: 412 (1836). Also of New World origin and often regarded as a separate genus. Large floral bracts and a distinctive single funnel-shaped involucral gland. 2 herbaceous species have been introduced and become naturalized in East Africa. Species 23, 24.

Subgen. **Esula** *Pers.*, Syn. Pl.: 14 (1806). Generally (not always) accepted as the most primitive, although its species possess some characters more advanced than those of many others in Africa, such as the development of an herbaceous habit, loss of stipules and a condensed pseudumbellate inflorescence. It has diversified as a group especially in temperate zones to become possibly the subgenus with the largest number of species. All indigenous East African species belong to sect. *Esula*, with leafy bracts, 4 involucral glands and carunculate seeds. Species 25–35.

Subgen. **Eremophyton** (*Boiss.*) *Wheeler* in Amer. Midl. Nat. 30: 483 (1943). Syn. — sect. *Eremophyton* Boiss. in DC., Prodr. 15 (2): 70 (1862). Contains an unsatisfactory grouping of 2, or possibly 3 sections, of annual or short-lived perennial herbaceous species, with a woody or somewhat succulent stem, rarely woody shrubs, petiolate leaves, stipules modified as glands or subulate filaments, leafy bracts and usually 4 involucral glands.

Sect. **Pseudacalypha** *Boiss.* in DC., Prodr. 15 (2): 98 (1862). Used here in its broadest sense to include species with terminal and/or axillary simple or umbellate cymes, and ecarunculate conical seeds with horizontal grooves and ridges. Species 36–43.

Some authors prefer to use the section in its narrowest sense, as represented by the typical species *E. acalyphoides*, to include only those species with axillary cymes which dichotomize just once or twice, and involucral glands hairy on the upper surface (Species 36–41). Other species are then placed in sect. *Holstianae* Pax & K. Hoffm. in V.E. 3(2): 148 (1921), characterized by axillary and terminal 3-rayed umbels branching dichotomously many times and glabrous involucral glands (Species 42, 43).

Sect. **Eremophyton**. In East Africa typified by *E. agowensis*, includes woody stemmed herbaceous species with axillary and terminal inflorescences of dichotomously branching 3-rayed umbels, elongated capsules and flattened seeds with a conspicuous cap-like caruncle. Its relationship with the previous section is somewhat tenuous, linking factors existing in the inflorescence structure and in habit, which is especially evident in the close similarity between *E. pirottae* in this section and *E. longituberculosa*. Species 44–47.

Subgen. **Trichadenia** (*Pax*) *S. Carter* in K.B. 40: 816 (1985). Syn. — sect. *Trichadenia* Pax in V.E. 3(2): 152 (1921). Includes species with dichotomously branching 3–many-rayed pseudumbels, leafy bracts, stipules modified as glands or subulate filaments, 4 or 5 crenulate, lobed or large pectinate involucral glands and relatively large capsules with ecarunculate seeds. They are trees or shrubs, with woody or semi-succulent branches, or herbs which have developed a fleshy rootstock. Species 48–58.

Subgen. **Lyciopsis** (*Boiss.*) *Wheeler* in Amer. Mill. Nat. 30: 483 (1943). Syn. — sect. *Lyciopsis* Boiss. in DC., Prodr. 15(2): 97 (1862). Consists of 3 well-defined sections of woody shrubs or herbs with a thick woody rootstock, with conspicuous glandular stipules, terminal cymes with up to 3(–7) lateral branches and cyathia with 5 glands.

Sect. **Somalica** (*S. Carter*) *S. Carter, comb. nov.* Syn. — subgen. *Trichadenia* (Pax) S. Carter sect. *Somalica* S. Carter in K.B. 40: 817 (1985). Accommodates softly woody semi-succulent stemmed shrubs, with usually small scarious deciduous bracts, crenulate or usually pectinate involucral glands and large often ornamented capsules, with compressed ecarunculate seeds. Habit, glandular stipules, inflorescence features and seeds are more characteristic of subgen. *Lyciopsis* than they are of subgen. *Trichadenia* where it was first placed. This section occurs predominantly in Somalia, with one species in East Africa. Species 59.

Sect. **Lyciopsis**. Includes woody shrubs with leafy or scarious bracts, often saucer-shaped involucral glands, sessile capsules and subglobose seeds. Species 60–68.

Sect. **Espinosae** *Pax & K. Hoffm.* in V.E. 3(2): 149 (1921). Predominantly southern African, with only one species extending into East Africa, *E. espinosa* itself. This is a woody subscandent shrub, with solitary cyathia, scarious bracts, large exserted capsules and seeds with a conspicuous caruncle. Fruit and seeds are typical of subgen. *Tirucalli* which follows in this sequence. Species 69.

Subgen. **Tirucalli** (*Boiss.*) *S. Carter* in K.B. 40: 823 (1985). Syn. — sect. *Tirucalli* Boiss. in DC., Prodr. 15 (2): 94 (1862). Includes in East Africa 2 well-defined sections which have not, however, been formally separated, as distinctions are less apparent in southern Africa where a greater proliferation of species occurs. Characteristic features include cylindrical succulent branches, with small quickly deciduous leaves, prominent leaf-scars, 5 involucral glands, fairly large exserted capsules and subglobose carunculate seeds. *E. tirucalli* itself, with no close relatives in East Africa is a shrubby tree with glandular stipules, a much congested inflorescence of the dichotomously branching rays of a pseudumbel and small scarious bracts. The other species can be grouped and related to *E. mauritanica* L. in South Africa. They are erect or scrambling shrubs, with terminal pseudumbels of 3 to 8 rays and leafy bracts. There are no stipules, but callouses, which become very prominent and appear to be glandular, quickly form around and over the leaf-scars. Species 70–73.

Subgen. **Euphorbia**. The largest subgenus in tropical Africa, and probably developed from a group which became distinct at a relatively early stage, but after the separation of Madagascar which has no endemic species. The tree habit has persisted in some species, while others show advancement in extreme specialization, involving increased succulence of the stems and a reduction in size. Stipules have become modified as prickles mounted on a horny pad (the spine-shield) which surrounds the base of the leaf-scar and bears in addition a pair of spines. The structure of this armature is usually diagnostic. Inflorescences are always axillary, consisting of one or more cymes branching usually once only, a feature which probably represents a reduction of a more complex pseudumbellate arrangement. Involucral glands number 5 and are always entire. Capsules may be sessile or exserted, a feature indicative of species relationships. Seeds are subglobose, smooth or tuberculate and ecarunculate.

Attempts by Pax and a few other authors to distinguish sections within this subgenus have so far proved unsatisfactory, and none are used here. All the known species of spiny *Euphorbia* (probably about 200) would have to be taken into account before this could be successfully accomplished. There are 86 species from East Africa described in this Flora, but several more taxa exist which I know only from incomplete material or photographs. Species 74–159.

Subgen. **Lacanthis** (*Raf.*) *M.G. Gilbert* in K.B. 42: 238 (1987). Syn. —*Lacanthis* Raf., Fl. Tell. 2: 94 (1837). Represented in East Africa only by *E. brunellii*. Gilbert considers this, together with several other Ethiopian species to be related to a much larger group of species which have proliferated in Madagascar. They all have succulent or semi-succulent stems, stipules which have become feathery or developed as crests of soft spines, and usually colourful bracts subtending the cyathia. The stipules are absent in *E. brunellii* but the floral bracts are relatively large and the succulent stem has developed as an underground fleshy caudex. Species 160.

CULTIVATED SPECIES

E. cotinifolia L., Sp. Pl.: 453 (1753). This popular, small shrubby tree, originating from Central America produces vast numbers of seedlings and naturalizes easily. It grows to ± 3 m. high and is prized for its ornamental foliage, its leaves being purplish brown and broadly ovate with long petioles. The inflorescences are relatively inconspicuous, produced as small, few-branched cymes consisting of tiny cyathia bearing purplish glands with creamy lobed petaloid appendages. Its latex is apparently particularly noxious. Reference specimen — Kenya, Machakos District, Donyo Sabuk, 15 Apr. 1963, *Bally* 12666!

E. leucocephala Lotsy in Coult. Bot. Gaz. 20: 350 (1895). A densely branching shrub to ± 2m. high, introduced from Central America as an ornamental. Its small lanceolate leaves ± 3 cm. long are unexceptional, but in flower the branch apices are smothered with clusters of tiny cyathia bearing inconspicuous petaloid glands and subtended by showy creamy-white oblanceolate bracts ± 1 cm. long. Reference specimen — Kenya, Nairobi, Museum [Ainsworth] Hill, 2 June 1961, *Verdcourt* 3183!

E. fulgens Klotzsch, Allg. Gartenz. 2: 26 (1834); Jex-Blake, Gardening in E. Africa, ed. 4: 113 (1957). A small shrub with arching branches, and cyathia ± 1 cm. in diameter bearing yellow glands with large bright red petaloid appendages. A native of Central America it is difficult to propagate and not widely grown. No E. African specimens seen.

E. pulcherrima Klotzsch, Allg. Gartenz. 2: 27 (1834); U.O.P.Z.: 255 (1949); Jex Blake, Gardening in E. Africa, ed. 4: 113 (1957); F.F.N.R.: 199 (1962). Usually called *Poinsettia*, this species from Central America is commonly grown throughout East Africa as an ornamental shrub. It can grow to ± 4 m. high and is characterized by the large (to + 15 cm. long) brilliant red, occasionally creamy yellow, bracts of the inflorescence. Reference specimen — Tanzania, Lushoto District, Chakechake, 17 May 1969, *Ngoundai* 326!

E. alluaudii Drake in Bull. Mus. Hist. Nat. Paris, sér. 2, 9: 43 (1903). Syn. — *E. leucodendron* Drake, l.c. 9: 46 (1903). A densely branched leafless tree to ± 4 m. high, with succulent yellowish-green cylindrical branches ± 1.5 cm. thick and tiny yellow cyathia loosely clustered towards the ends of the branchlets. Introduced from Madagascar and occasionally seen in cultivation, where it is most usually known as *E. leucodendron*. This species, however, was described by Drake in a footnote with no specimens cited, and was considered to be probably synonymous by Friedmann & Cremers in Adansonia 16: 256 (1976). Reference specimens — Kenya, Nairobi, 27 Nov. 1971, *Mwangangi* 1870!; Tanzania, Lushoto District, Amani, 5 Nov. 1936, *Greenway* 4725!

E. stenoclada Baill. in Bull. Soc. Linn. Paris 1: 672 (1887). Syn. — *E. insulae-europae* Pax in E.J. 43: 224 (1909); T.T.C.L.: 210 (1949). Introduced from Madagascar as an ornamental, this species eventually forms a large shrub or tree to ± 6 m. high. Its leafless succulent pale olive-green branches bear alternating spines 1–3 cm. long formed from modified branchlets. In flower these terminate in tightly packed clusters of tiny crimson cyathia. Reference specimen — Tanzania, Lushoto District, Amani, Sigi Chini, 9 Dec. 1940, *Greenway* 6072!

E. milii Des Moulins var. *splendens* (Hook.) Ursch & Leandri in Mem. Inst. Sc. Mad. B, 5: 148 (1954); & as *E. splendens* Hook. in T.T.C.L.: 213 (1949); U.O.P.Z.: 255 (1949); F.F.N.R.: 198 (1962); F.P.U.: 112 (1962). *E. milii* is a very spiny ± succulent stemmed species native to Madagascar, with a purplish-brown bark, dark green lanceolate leaves 1–5 cm. long, and paired subcircular brilliant red, or occasionally yellow petaloid bracts below the sessile cyathia. There are a number of varieties of which var *splendens*, usually known as the Christ Thorn, is most popularly cultivated. It is semi-prostrate and quickly forms a low dense hedge which can be easily trimmed without harming the plant. Var. *hislopii* (N.E.Br.) Ursch & Leandri, loc. cit., another popular form is more sturdy with an erect habit. Reference specimens: var *splendens* — Tanzania, Lushoto District, Amani, 26 Aug. 1929, *Greenway* 1702!; var. *hislopii* — Kenya, Nairobi, Hilton Hotel, 26 Jan. 1971, *Wendelberger* U167!

E. hedyotoides N.E. Br. in F.T.A. 6(1): 515 (1911); Léandri in Adansonia 2: 220 (1962). Type: Tanzania, Lushoto District, cult. Mombo, *Braun* in *Herb. Amani* 1680 (B, holo.†, K, drawing of holo.!, EA, iso.!). In his description of a small woody almost leafless shrub, with leaf-scars fasciculated at the thickened apices of the branches, N.E. Brown did not mention that the specimen he used was from a cultivated plant. It originated from Madagascar and was being tested as a possible source of rubber, but apart from Zimmerman's collection made two years later, also from a cultivated plant, the species has not since been recorded from East Africa. Reference specimen — Tanzania, Lushoto District, Amani, 25 Oct. 1910, *Zimmerman* 3218!

KEY TO SUBGENERA AND SECTIONS

1. Hysteranthous geophytes with leaves in a rosette at ground
   level 2
   Herbs, shrubs or trees 3
2. Cyathia solitary . . . . 8. subgen. **Euphorbia**
   (159, *E. monadenioides*), p.531
   Cyathia in dichotomous cymes . . . . 9. subgen. **Lacanthis**
   (160, *E. brunellii*), p.531
3. Herbs with apparently opposite leaves (leaf-like bracts);
   involucral glands entire, with petaloid appendages 4
   Herbs, shrubs or trees with alternate leaves; involucral
   glands entire, lobed, horned or with finger-like
   processes 5
4. Cyathia 10 or more in capitate or occasionally loose cymes,
   with small bracts . . . . 1a. subgen.
   Chamaesyce sect. **Hypericifoliae**, p.415
   Cyathia solitary or up to 5 in congested leafy cymes   1b. subgen.
   Chamaesyce sect. **Chamaesyce** p.418
5. Involucral gland 1, funnel-shaped . . . . 2. subgen. **Poinsettia**,
   p.430
   Involucral glands 2–8 6
6. Branches succulent, with horny pads or longitudinal
   ridges bearing spines (sometimes minute) at least on
   young growth . . . . 8. subgen. **Euphorbia**,
   p.475
   Branches herbaceous, woody or if succulent then
   spineless 7
7. Branches succulent or semi-succulent, strictly cylindrical
   to 1 cm. thick; leaves small (to 4 × 1 cm.), quickly
   deciduous . . . . 7. subgen. **Tirucalli**,
   p.471
   Branches herbaceous or woody, or if semi-succulent then
   not strictly cylindrical and leaves large (10 × 2cm. or
   more) and persistent 8
8. Cyathial bracts scarious, shorter than the involucres 9
   Cyathial bracts leafy, longer than the involucres 11
9. Glands with finger-like processes; shrubs with semi-
   succulent branches . . . . 6a. subgen. **Lyciopsis**
   sect. **Somalica**, p.459
   Glands entire; woody shrubs 10
10. Capsules sessile; seeds ecarunculate . . . . 6b. subgen. **Lyciopsis**
    sect. **Lyciopsis**, p.461
    Capsules exserted; seeds carunculate . . . . 6c. subgen. **Lyciopsis**
    sect. **Espinosae**, p.470
11. Seeds carunculate 12
    Seeds ecarunculate 13
12. Stipules absent; cyathial bracts deltoid; seeds ovoid   3. subgen. **Esula**, p.432
    Stipules glandular, or if apparently absent then cyathial
    bracts similar to the leaves rarely deltoid; seeds
    compressed dorsi-ventrally with a pointed apex   4b. subgen.
    Eremophyton sect. **Eremophyton**, p.448
13. Trees or shrubs with succulent or semi-succulent stems
    and branches, or perennial herbs with a large fleshy
    rootstock producing annual stems; glands with
    marginal processes, horned, lobed or crenulate, rarely
    entire . . . . 5. subgen. **Trichadenia**,
    p.451

Annual or short-lived perennial herbs, without a fleshy
    rootstock, rarely woody shrubs; glands entire 14
14. Involucral glands 2 or 4; seeds conical or ovoid with
    pointed apex, usually ornamented   . . . . . .   4a. subgen.
                             Eremophyton sect. **Pseudacalypha**, p.442
    Involucral glands 5; seeds smooth, subglobose or rarely
    angular . . . .   6b. subgen. **Lyciopsis**
                               sect. **Lyciopsis**, p.461

1. Subgen. **Chamaesyce**

Annual or perennial herbs, erect or prostrate, branching from near the base. Leaves (bracts) opposite, the base obliquely rounded to subcordate; stipules present. Cyathia terminal and axillary or cymose and often capitate. Involucres bisexual with 4 glands, rarely unisexual (♂) with 5 glands; glands with an entire or lobed petaloid appendage on the outer margin. Stamens just exserted, with subsessile anthers. Perianth of the ♀ flower reduced to a rim below the ovary; styles joined at the base only. Seeds ± 4-angled, without a caruncle.

1a. Sect. **Hypericifoliae**

Annual herbs. Cyathia clustered 10 or more together in terminal and axillary usually pedunculate capitate cymes. Seeds oblong-conical.

1. Capsules always completely glabrous   . . . . . .   *3. E. hyssopifolia*
    Capsules pubescent, at least when young 2
2. Pubescence of white hairs interspersed with long yellow
    hairs . . . .   *1. E. hirta*
    Pubescence of white hairs only   . . . .   *2. E. indica*

1. E. hirta *L.*, Sp. Pl.: 454 (1753); N.E. Br. in F.T.A. 6(1): 496 (1911) & in Fl. Cap. 5(2): 249 (1915); W.F.K.: 36 (1948); U.O.P.Z.: 254 (1949); F.P.S.2: 71 (1952); E.P.A.: 449 (1958); F.W.T.A., ed. 2, 1: 419 (1958); F.P.U.: 112 (1962); Hadidi in B.J.B.B. 43: 86 (1973); U.K.W.F.: 221 (1974); S. Carter in K.B. 39: 643 (1984). Types: India, *Linnean Herbarium* 630: 5, 6, 7 (LINN, syn., K, photos.!)

Annual herb, prostrate to ascending, with branches to 50 cm. long, the whole plant pilose, including the inflorescence and capsule, with minute white adpressed hairs interspersed by yellow spreading segmented hairs ± 1.5 mm. long principally on the branches and especially on younger growth. Leaves ovate, 1–4 × 0.5–2 cm., base very obliquely rounded, apex subacute, margin finely toothed, upper surface sometimes almost glabrous, often blotched with purple especially in the region of the midrib; petiole to 3.5 mm. long; stipules linear, rarely laciniate on lush specimens, to 2.5 mm. long. Cyathia in dense capitate, terminal and axillary cymes to 15 mm. in diameter on peduncles to 15(–20) mm. long, occasionally subtended by 1–2 small leafy bracts ± 1 cm. long; cyathial peduncles to 1 mm. long; bracts deltoid, deeply laciniate to 1 mm. long. Cyathia ± 0.8 × 0.8 mm. with cup-shaped involucres, usually tinged with purple; glands 4, minute, elliptical, green or purplish, with minute entire white to pink appendages; lobes triangular, fimbriate. Male flowers: bracteoles linear, fimbriate; stamens 1 mm. long. Female flower: ovary shortly pedicellate; styles 0.4 mm. long, spreading, bifid almost to the base, with thickened apices. Capsule just exserted on a pedicel 1 mm. long, acutely 3-lobed, with truncate base, 1 × 1.25 mm., pilose with short yellow adpressed hairs. Seeds oblong-conical, 0.8 × 0.4 mm., pinkish brown, with slight transverse wrinkles.

UGANDA. W. Nile District: Maracha rest camp, 3 Aug. 1953, *Chancellor* 109!; Teso District: Serere, 25 Oct. 1955, *Langdale-Brown* 1596!; Mengo District: Old Entebbe, 28 Jan. 1956, *Harker* 137!
TANZANIA. Musoma District: Seronera, 17 Apr. 1962, *Greenway* 10603!; Mbulu District: Lake Manyara National Park, Main Gate, Marera R., 9 Mar. 1964, *Greenway & Kanuri* 11328!; Uzaramo District: Kisarawe, 4 Mar. 1964, *Semsei* 3652!; Zanzibar I., Massazine, 24 Sept. 1959, *Faulkner* 2368!
DISTR. U 1–4; K 1–7; T 1–8; Z; P; a very common pantropical weed native to Central America
HAB. Cultivated land, roadsides and waste places; 0–2000 m.

FIG. 77. *EUPHORBIA INDICA* — 1, habit, × ⅘; 2, cyathium, × 8; 3, seeds, × 18. *E. HYSSOPIFOLIA* — 4, flowering branch, × 4; 5, seeds, × 18. 1, 3, from *Tanner* 2907; 2, from *Tweedie* 2253; 4, 5, from *Harris* 1093. Drawn by Christine Grey-Wilson.

SYN. [*E. pilulifera* sensu Boiss. in DC., Prodr. 15(2): 21 (1862), *non* L. et al., see N.E. Br. in F.T.A. 6(1): 496 (1911)]
  E. pilulifera Boiss. var. *procumbens* Boiss. in DC., Prodr. 15(2): 21 (1862); F.P.N.A. 1: 476 (1948). Numerous syntypes from tropical America and Asia (G, syn., K, photos.!)
  E. hirta L. var. *procumbens* (Boiss.) N.E.Br. in F.T.A. 6(1): 497 (1911)

VAR. Mostly a prostrate herb (N.E. Brown's variety) growing on bare ground and strongly coloured with reddish purple, but it can be erect and entirely green when occurring amongst other vegetation in damp shady situations. All gradations between the two extremes can be found, with no justification for the erection of varieties.

2. **E. indica** *Lam.*, Encycl. Méth. Bot. 2: 423 (1786); Boiss. in DC., Prodr. 15(2): 22 (1862); F.P.S.2: 71 (1952); Raju & Rao in Ind. J. Bot. 2: 202 (1979). Type: East Indies, *Sonnerat* (P, holo., K, photo.!)

Annual herb, spreading or erect with branches to 50(–100) cm. long, the whole plant including the capsule at least when young sparsely pilose, more so on young growth, and often purplish tinged. Leaves ovate, to 3(–5) × 1.5(–2.5) cm., base obliquely rounded, apex rounded, margin obscurely toothed, upper surface sometimes almost glabrous, glaucous especially the lower surface; petiole to 3 mm. long; stipules broadly triangular, laciniate, to 1.5 mm. long. Cyathia in terminal and axillary capitate cymes to 1.5 cm. in diameter on peduncles to 3 cm. long, subtended by a pair of small leafy bracts; cyathial peduncles to 2 mm. long; bracts linear, to 2.5 mm. long. Cyathia ± 1 × 1 mm., with cup-shaped involucres; glands 4, minute, rounded, green, with appendages varying in size to 1 mm. in diameter, white; lobes acutely triangular, 0.5 mm. long. Male flowers: bracteoles linear; stamens 1.25 mm. long. Female flower: ovary shortly pedicellate; styles 0.5 mm. long, suberect, bifid almost to the base. Capsule exserted on a reflexed pedicel 1.5 mm. long, acutely 3-lobed, 1.5 × 2 mm. Seeds oblong-conical, 1 × 0.75 mm., reddish brown, with obscure transverse ridges. Fig. 77/1–3.

UGANDA. Karamoja District: Lochoi, 24 May 1940, *A.S. Thomas* 3519!; Teso District: Serere, Nov.–Dec. 1931, *Chandler* 87!; Mengo District: Kampala, Jan. 1936, *Chandler* 1540!
KENYA. Kilifi District: Malindi, Sabaki R., Oct. 1965, *Tweedie* 3138!; Tana River District: 48 km. S. of Garsen, 23 Sept. 1961, *Polhill & Paulo* 536! & 2 km. S. of Ngao, 1 Mar. 1977, *Hooper & Townsend* 1122!
TANZANIA. Pangani District: Mwera, Mseka, Mtaru, 7 June 1956, *Tanner* 2907!; Buha District: Gombe Stream Reserve, Rutanga valley, 19 Feb. 1964, *Pirozynski* 408!; Ufipa District: Rukwa, 3 May 1935, *Michelmore* 1145!; Zanzibar I., Mwera, 16 Nov. 1930, *Vaughan* 168!
DISTR. U 1–4; K 1, 3–5, 7; T 3, 4, 6, 8; Z; P; a common weed introduced originally from India
HAB. In damp grassland on seasonally waterlogged, usually black cotton soils, or near permanent water; 0–1330 m.

SYN. *E. indica* Lam. var. *angustifolia* Boiss. in DC., Prodr. 15(2): 22 (1862). Types: Sudan, Kordofan, *Kotschy* 154 (K, isosyn.!) & Ethiopia, *Schimper* 1632 (syn., whereabouts unknown) & Java, *Blume* (syn., whereabouts unknown)
  E. indica Lam. var. *pubescens* Pax in E.J. 19: 117 (1894). Types: Tanzania, Pangani, *Stuhlmann* 308 & 841 (B, syn.†)
  [*E. hypericifolia* sensu N.E. Br. in F.T.A. 6(1): 498 (1911) & in Fl. Cap. 5(2): 248 (1915); F.P.N.A. 1:476 (1948); E.P.A.: 449 (1958); Hadidi in B.J.B.B. 43: 87 (1973), *non* L.]

VAR. Leaf-size and pubescence varies according to environmental conditions, plants with more hairy, spreading branches bearing smaller leaves occurring in drier situations. Such variability easily encompasses the above varieties.

NOTE. N.E. Brown mistakenly considered this Old World species to be synonymous with *E. hypericifolia* L., a name which has been used in African taxonomy ever since. *E. hypericifolia* is, however, a widespread species of the New World tropics and subtropics, apparently not, so far, introduced into East Africa. It is characterised by a slightly smaller capsule which is always completely glabrous.

3. **E. hyssopifolia** *L.*, Syst. Nat. 10: 1048 (1759); F.W.T.A., ed. 2, 1: 419 (1958). Type: Jamaica, *Brown* in *Linnean Herbarium* 630: 9 (LINN, holo., K, photo.!)

Annual herb, semi-prostrate to erect, with branches to 40 cm. long, glabrous except for usually a few scattered hairs on branches, leaves and young growth. Leaves lanceolate-ovate, to 2.5 × 1 cm. but usually much smaller, base obliquely rounded, apex rounded, margin minutely toothed; petiole 1 mm. long; stipules broadly triangular, to 0.75 mm. long, fimbriate. Cyathia in terminal and axillary cymes, with peduncles elongating up to 10 mm. within each cyme; bracts enlarging from 1 mm. long to eventually resemble the leaves. Cyathia ± 0.8 × 0.8 mm., with cup-shaped involucres; glands 4, green tinged red, minute, rounded, with minute creamy appendages; lobes acutely triangular, less than 0.2 mm. long. Male flowers: bracteoles fimbriate; stamens 1 mm. long. Female flower: ovary shortly pedicellate; styles 0.4 mm. long, suberect, bifid almost to the base. Capsule exserted on a reflexed pedicel 1.25 mm. long, acutely 3-lobed, with base truncate, 1.5 × 1.5 mm., glabrous. Seeds oblong-conical, 1 × 0.5 mm., reddish black, with 3 transverse ridges. Fig. 77/4, 5, p. 416.

TANZANIA.   Kilosa District: Mikumi Game Lodge, 15 Nov. 1970, *Batty* 1101!; Uzaramo District: Dar es Salaam, University College, 15 Oct. 1967, *Harris* 1093!; Rungwe District: 2 km. beyond Kiwira R. on Mbeya–Tukuyu road, 17 Mar. 1975, *Hooper & Townsend* 850!
DISTR.   T 6, 7; recently introduced weed of central America, also found extensively in West Africa
HAB.   Disturbed ground in grassland and stony, sandy soils; 0–1500 m.

SYN.   *E. brasiliensis* Lam., Encycl. Méth. Bot. 2: 423 (1786). Type: Brazil, *Commerson* (P, holo., K, photo.!)

1b. Sect. **Chamaesyce**

Annual or perennial herbs. Cyathia terminal and pseudoaxillary (one axillary bud develops into a strong branch, reducing the terminal cyathium to an apparently axillary position), solitary or up to 5 together in congested leafy cymes. Seeds conical.

1. Annuals with fibrous roots 2
    Perennials with thick woody or tuberous roots 18
2. Seeds smooth 3
    Seeds variously pitted, wrinkled or grooved 4
3. Leaves subcircular; stipules broadly triangular . . .   *8. E. serpens*
    Leaves obovate; stipules linear . . . .   *16. E. lissosperma*
4. Branches all distinctly hairy, at least with some scattered
      hairs evenly distributed 5
    Branches glabrous, rarely with a few hairs on young
      growth or scattered on leaves and capsules 16
5. Leaves entire 6
    Leaves toothed, sometimes obscurely so 8
6. Capsules sparsely pilose with long spreading hairs   *14. E. granulata*
    Capsules densely pilose with short adpressed hairs 7
7. Leaves ovate-oblong; gland-appendages minute, red   *4. E. kilwana*
    Leaves lanceolate; gland-appendages obvious, white, 2
      larger than the others . . . .   *6. E. lupatensis*
8. Capsule subsessile, included in the involucre . . . .   *7. E. thymifolia*
    Capsule exserted from the involucre 9
9. Capsule densely pilose . . . .   *5. E. pilosissima*
    Capsule glabrous, or pilose but not densely so 10
10. Capsule constricted about the middle, with a fleshy
      swelling between the sutures . . . .   *12. E. allocarpa*
    Capsule regularly 3-lobed 11
11. Capsule with a row of hairs along each suture, otherwise
      glabrous . . . .   *9. E. prostrata*
    Capsule glabrous or sparsely pilose 12
12. Upper surface of the prostrate branches pilose, the lower
      surface glabrous 13
    Branches pilose on both surfaces 14
13. Gland-appendages large, to 0.6 × 1.5 mm.; capsule 1.6 mm.
      in diameter . . . .   *11. E. mossambicensis*
    Gland-appendages minute; capsule 1.2 mm. in diameter   *10. E. diminuta*
14. Leaves very distinctly toothed . . . .   *15. E. serratifolia*
    Leaves obscurely toothed .15
15. Branches fairly densely pilose with long adpressed hairs;
      gland-appendages large, to 0.6 × 1.5 mm. . . . .   *13. E. fischeri*
    Branches very sparsely pilose with long spreading hairs;
      gland-appendages very small . . . .   *14. E. granulata*

16. Leaves entire; plant glabrous except for a minute
    pubescence at the stem-base . . . .   *18. E. arabica*
    Leaves toothed, sometimes at the apex only; plant
    glabrous, or rarely with a few long scattered hairs on
    branches, leaves or capsules 17
17. Plants erect, glabrous or rarely with a few hairs confined to
    basal branches and leaves; stipules no more than 1 mm.
    long, acutely triangular but not divided into linear
    teeth . . . .   *19. E. polycnemoides*
    Plants prostrate, rarely suberect, glabrous or rarely with a
    few hairs on leaves or capsules; stipules ± 1.5 mm. long
    deeply divided into 3–5 linear teeth . . . . . .   *17. E. inaequilatera*
18. Plants pilose, at least with some hairs on the branches
    and/or capsules .19
    Plants completely glabrous 20
19. Whole plant pilose, or at least with some hairs on stems,
    leaves or capsules; leaves ovate to ovate-lanceolate with
    apiculate apex . . . .   *20. E. zambesiana*
    Branches pilose; leaves glabrous, broadly ovate with
    rounded apex . . . .   *21. E. selousiana*
20. Leaves ± lanceolate, apex apiculate; cyathial peduncle to
    25 mm. long . . . .   *20. E. zambesiana*
    Leaves subcircular, apex rounded; cyathial peduncle not
    more than 4 mm. long . . . .   *22. E. rivae*

4. E. kilwana *N.E. Br.* in F.T.A. 6(1): 507 (1911). Type: Tanzania, Kilwa District,
Kilwa–Singino, *Braun* in *Herb. Amani* 1292 (B, holo. †, K, fragment with drawing!, EA, iso.!)

Annual semi-prostrate herb, with branches to 25 cm. long, often red tinged, pilose with
short curved adpressed hairs. Leaves ovate-oblong, to 25 × 8 mm., base obliquely rounded
to subcordate, apex obtuse to rounded, margin entire or very obscurely toothed, glabrous
or almost so above, thinly pilose beneath, slightly glaucous, paler beneath; petiole to 2
mm. long, pilose; stipules lanceolate, 1 mm. long, pilose. Cyathia ± 5 clustered together in
terminal and axillary cymes, with peduncles to 1 mm. long; bracts similar to, but much
smaller than the leaves. Cyathia ± 1.3 × 1.25 mm., with barrel-shaped pilose involucres;
glands 4, minute, circular, red, with minute red appendages; lobes acutely triangular,
minute. Male flowers: bracteoles filamentous; stamens 1.5 mm. long. Female flower:
ovary pedicellate; styles 0.5 mm. long, erect, bifid almost to the base. Capsule exserted on
a reflexed pilose pedicel 2 mm. long, acutely 3-lobed, with truncate base, 2 × 2.5 mm.,
densely pilose with very short, curved, adpressed hairs. Seeds ovoid, 1.2 mm. × 0.75 mm.,
pinkish brown, with 3 or 4 obscure transverse ridges.

TANZANIA. Tanga District: Tanga–Pangani road, Machui, 12 Jan. 1956, *Faulkner* 1793!; Kilosa
District: Msada, 29 June 1973, *Greenway & Kanuri* 15289!; Kilwa District: Selous Game Reserve, 13
July 1968, *Rodgers & Agnew* in MRC 376!
DISTR. T 3, 4, 6, 8; not known elsewhere
HAB. Seasonally wet grassland, in clay; 0–600 m.

SYN. *E. convolvuloides* Benth. var. *integrifolia* Pax in E.J. 43: 85 (1909). Type as for *E. kilwana*

5. E. pilosissima *S. Carter* in K.B. 39: 643, fig. 1 A–D (1984). Type: Tanzania, Dodoma
District, Itigi–Chunya road, *Greenway & Polhill* 11533 (K, holo.!, EA, iso.)

Annual herb erect to 45 cm., the whole plant pilose with long spreading white hairs ± 1
mm. long. Leaves obovate, to 18 × 8 mm., base very obliquely rounded, apex obtuse,
margin deeply serrate with teeth 0.5 mm. long; petiole 1 mm. long; stipules free, linear,
1.25 mm. long. Cyathia clustered in terminal and axillary cymes of 2–5 cyathia, subsessile;
bracts similar to but smaller than the leaves. Cyathia 1.5 × 1.25 mm., with barrel-shaped
involucres; glands 4, erect, transversely elliptic, 0.5 mm. wide, red, with minute red
3–5-lobed appendages; involucral lobes acutely triangular, 0.5 mm. long. Male flowers
very few (± 5): bracteoles filamentous; stamens 1.75 mm. long. Female flower: ovary
pedicellate; styles red, 0.75 mm. long, erect, bifid almost to the base. Capsule just exserted
on a reflexed pedicel 2 mm. long, 3-lobed, with truncate base, 2 mm. long and in diameter,
densely pilose with hairs sometimes tinged red. Seeds oblong-conical, 1.2 × 0.75 mm.,
pinkish brown, obscurely transversely wrinkled.

TANZANIA. Dodoma District: 11 km. from Itigi on Chunya road, 12 Apr. 1964, *Greenway & Polhill* 11533!
DISTR. T 5; known only from this collection
HAB. Sandy soil in valley of impeded drainage; 1310 m.

NOTE. This collection represents a distinct species related to *E. kilwana* N.E. Br., from which it differs by its overall covering of long, soft hairs, the larger lobed appendages of the involucral glands, and most noticeably by the very distinct serrations of the obovate leaves. More material is needed before variation from the cited specimen can be ascertained. The collectors noted that the plant was common in the valley where it was found.

6. E. lupatensis *N.E. Br.* in F.T.A. 6(1): 514 (1911). Type: Mozambique, Lower Zambesi, Lupata, *Kirk* (K, holo.!)

Sparsely branched annual erect to ± 30 cm. high, the whole plant usually red tinged, pilose, sometimes sparsely so, with short curved hairs. Leaves lanceolate, to 35 × 5 mm., base obliquely rounded to subcordate, apex acute, margin entire; petiole 1 mm. long; stipules free, linear, 1 mm. long. Cyathia clustered 2–4 together in terminal and axillary cymes, with peduncles ± 1 mm. long; bracts similar to but much smaller than the leaves. Cyathia 2 × 1.5 mm., with barrel-shaped involucres; glands 4, elliptic, ± 0.5 mm. wide, with conspicuous appendages which are cream with a somewhat lobed margin, 2 subcircular ± 1 mm. in diameter, the other 2 obliquely attached, ± 2 × 1.25 mm.; involucral lobes triangular, 0.5 mm. long, pilose. Male flowers: bracteoles lanceolate, pilose; stamens 2 mm. long. Female flower: ovary pedicellate; styles 1.25–1.5 mm. long, suberect, bifid for ⅓ their length. Capsule exserted on a reflexed pedicel 1.5 mm. long, 3–lobed, 1.5 × 1.75 mm., densely pilose. Seeds oblong-conical, 1.1 × 0.6 mm., pinkish brown, with 3–4 distinct transverse ridges and grooves.

TANZANIA. Tunduru District: Tunduru–Masasi road, 5 Mar. 1963, *Richards* 17752!; Masasi District: Masasi–Muiti, 25 Apr. 1945, *Schlieben* 6417!
DISTR. T 8; scattered distribution elsewhere, from Mozambique, Malawi, Zambia and Zimbabwe
HAB. Open places by roadsides and open woodland, in stony, sandy soils; 300–900m.

NOTE. The two specimens cited above appear to be the only collections of this species so far made in Tanzania. A third specimen, *Richards* 17891 from Masasi District, 64 km. from Masasi on the Tunduru road, possesses distinctly serrate, rather more ovate leaves to 18 × 6 mm., but appears otherwise to be identical. More material of this is needed before a decision can be made upon its taxonomic status.

7. E. thymifolia *L.*, Sp. Pl. 1: 454 (1753); Boiss. in DC., Prodr. 15(2): 47 (1862); F.W.T.A., ed. 2, 1: 421 (1958). Type: India, *Linnean Herbarium* 630: 10 (LINN, holo., K, photo.!)

Prostrate densely branching annual herb, with branches to 25 cm. long, glabrous on the undersurface, the upper surface densely pilose with fairly long curved adpressed hairs, the whole plant tinged reddish brown. Leaves ovate, to 8 × 4 mm., base obliquely subcordate, apex obtuse, margin shallowly toothed, upper surface glabrous, the lower surface with long scattered hairs; petiole 0.5 mm. long; stipules free, linear, to 1.25 mm. long, often deeply 2–3-toothed, pilose. Cyathia solitary, subsessile, terminal and pseudoaxillary on congested leafy shoots, ± 0.5 × 0.5 mm., with funnel-shaped involucres, pilose; glands 4, minute, subcircular, red, with often almost invisible red appendages; lobes minute, triangular, ciliate. Male flowers very few (5 or less): bracteoles reduced to 1 or 2 threads; stamens 0.8 mm. long. Female flower: ovary subsessile; styles 0.6 mm. long, erect, bifid to halfway. Capsule subsessile (pedicel 0.2 mm. long), splitting the involucre during development, 3-lobed, with truncate base, 1 × 1 mm., pilose with short adpressed hairs. Seeds conical, sharply 4-angled, 0.6 × 0.4 mm., reddish brown, with shallow transverse ridges and grooves.

TANZANIA. Mpanda, 15 July 1968, *Sanane* 228 partly! (the majority of plants on the Kew specimen are *E. prostrata*); Songea, near Government Rest House, 26 Feb. 1956, *Milne-Redhead & Taylor* 8750!
DISTR. T 4, 8; West Africa eastwards to Central African Republic, Zaire and Zambia, with one specimen seen from Mozambique; a pantropical weed apparently only recently spreading into East Africa
HAB. On disturbed ground in sandy soil, 1050–1150 m.

SYN. *E. afzelii* N.E. Br. in F.T.A. 6(1): 506 (1911). Type: Sierra Leone, *Afzelius* (B, holo. †, K, fragment and drawing!)

8. E. serpens *Kunth,* Nov. Gen. & Sp. 2: 52 (1817); N.E. Br. in F.T.A. 6(1): 511 (1911). Type: Venezuela, *Humboldt & Bonpland* (P, holo.)

Annual much-branched herb, prostrate, with branches to 10(–20) cm. long, the whole plant completely glabrous. Leaves subcircular, 1–3 mm. in diameter, base obliquely cordate, margin entire; petiole 0.2–0.5 mm. long; stipules united at the swollen nodes, triangular, ± 0.5 × 1 mm., with toothed margin. Cyathia solitary on peduncles to 2 mm. long, terminal and pseudoaxillary on short leafy shoots, ± 0.7 × 0.7 mm., with cup-shaped involucres; glands 4, minute, transversely elliptic, with small white shallowly lobed petaloid appendages; involucral lobes minute, margin fringed. Male flowers: bracteoles laciniate; stamens 0.7 mm. long. Female flower: perianth evident as a 3-lobed rim below the shortly pedicellate ovary; styles 0.2 mm. long, spreading, bifid almost to the base. Capsule exserted on a reflexed pedicel to 1.8 mm. long, obtusely 3-lobed, with truncate base, ± 1.5 × 1.8 mm., yellowish green. Seeds oblong-conical, 1 × 0.6 mm., greyish pink, smooth.

KENYA. Teita District: Voi, 14 Apr. 1969, *Bally* 13250!
DISTR. **K** 7; a pantropical weed but in E. Africa known only from this collection
HAB. Not known, probably by the roadside; 560 m.

SYN. *E. minutiflora* N.E. Br. in F.T.A. 6(1): 1036 (1913). Type: Zimbabwe, Victoria Falls, *Schwartz* in *Bolus* 13027 (K, fragment of holo.!)

9. E. prostrata *Ait.* in Hort. Kew. 2: 139 (1789); Boiss. in DC., Prodr. 15(2): 47 (1862); N.E. Br. in F.T.A. 6(1): 510 (1911) & in Fl. Cap. 5(2): 245 (1915); F.P.N.A. 1: 477 (1948); E.P.A.: 455 (1958); F.W.T.A., ed. 2, 1: 421 (1958); Hadidi in B.J.B.B. 43: 98 (1973); U.K.W.F.: 222 (1974). Type: West Indies, cult. *Miller* in England 1758 (BM, holo.!)

Prostrate much-branched annual herb, with branches to 20 cm. long, glabrous on the underside, pilose above with short curled hairs, the whole plant often tinged purplish. Leaves ovate, to 8 × 5 mm., base obliquely rounded, apex rounded, margin obscurely toothed, upper surface glabrous, lower surface sparsely pilose towards the apex; petiole to 1 mm. long; stipules pilose, free on the upper surface of the branch, triangular, 0.5 mm. long, joined on the lower surface to form a broad triangle to 1 mm. long with 2 unequal teeth. Cyathia solitary on peduncles 1.25 mm. long, terminal and pseudoaxillary on short leafy shoots, 1 × 0.6 mm., with barrel-shaped involucres; glands 4, minute, red, with minute white or pink appendages; lobes minute, triangular, pilose. Male flowers few: bracteoles hair-like; stamens 1 mm. long. Female flower: ovary pedicellate; styles 0.2 mm. long, spreading, bifid to the base. Capsule exserted on a reflexed pilose pedicel 1.5 mm. long, acutely 3-lobed, with truncate base, 1.25 × 1.25 mm., the base and purple-tinged sutures beset with long spreading hairs. Seeds oblong-conical, acutely 4-angled, 1 × 0.5 mm., grey-brown, with numerous distinct transverse ridges and grooves. Fig. 78/2, p. 422.

UGANDA. Toro District: Fort Portal, 20 Nov. 1931, *Hazel* 15!; Teso District: Serere, Dec. 1931, *Chandler* 215!; Mengo District: Kampala, Makerere University Hill, 9 Apr. 1970, *Lye* 5190!
KENYA. Machakos District: Makindu, Hunter's Lodge, 5 Apr. 1971, *Bally & Carter* 14135!; S. Kavirondo District: Kisii, Sept. 1933, *Napier* 2951 in *C.M.* 5263!; Kilifi District: Jilori, 26 Nov. 1961, *Polhill & Paulo* 854!
TANZANIA. Lushoto District: Mkomazi, July 1955, *Semsei* 2340!; Ulanga District: Mahenge, 3 Apr. 1932, *Schlieben* 2041!; Mbeya, 14 May 1956, *Milne-Redhead & Taylor* 10240!; Zanzibar I., Mazizini, 10 Aug. 1963, *Faulkner* 3252!
DISTR. **U** 2–4; **K** 1, 4, 5, 7; **T** 1–4, 6–8; **Z**; **P**; introduced from the West Indies, a common weed throughout the tropics and subtropics
HAB. Disturbed ground in gardens, cultivated land and by roadsides especially in dry sandy soil; 0–2040 m.

10. E. diminuta *S. Carter* in K.B. 39: 646 fig. 1 E–H (1984). Type: Kenya, Northern Frontier Province, 32 km. E. of Wajir, *Gillett* 21319 (K, holo.!, EA, iso.)

Branching prostrate annual herb, with branches to 13 cm. long, usually much less (± 5 cm.), sparsely hairy on the upper surface, glabrous beneath. Leaves obovate, to 8 × 4 mm., base very obliquely rounded, apex rounded, margin distinctly toothed especially round the apex, usually glabrous, occasionally sparsely hairy on both surfaces; petiole to 0.3 mm. long; stipules free, filamentous, 0.3 mm. long. Cyathia solitary on peduncles to 0.4 mm. long, terminal and pseudoaxillary on short leafy shoots, 0.8 × 0.8 mm., with cup-shaped involucres; glands 4, minute, transversely elliptic, with minute deeply 2–3-lobed pink or white appendages; involucral lobes minute, triangular, ciliate. Male flowers few:

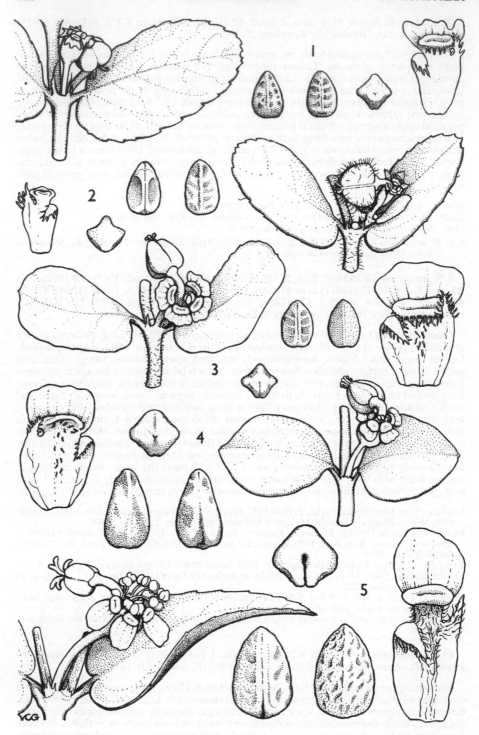

FIG. 78. *EUPHORBIA* subgen. *CHAMAESYCE* sect. *CHAMAESYCE*. Cyathium and leaves (bracts), × 6, involucral gland and lobes, × 16, seeds, × 12 of **1**, *E. inaequilatera*, **2**, *E. prostrata*, **3**, *E. mossambicensis*, **4**, *E. rivae*, **5**, *E. zambesiana*. 1, from *Richards 24905*; 2, from *Glover & Samuel 2993*; 3, from *Richards 16037*; 4, from *Bally 7659*; 5, from *Verdcourt 2806A*. Drawn by Victoria Gordon-Friis.

bracteoles filamentous; stamens 0.8 mm. long. Female flower: ovary pedicellate; styles 0.2 mm. long, erect, bifid almost to the base. Capsule exserted on a reflexed pedicel 2 mm. long, acutely 3-lobed, with truncate base, 1.2 × 1.2 mm., glabrous. Seeds conical, 0.8 × 0.5 mm., pinkish brown, with shallow wrinkles.

KENYA. Northern Frontier Province: 19 km. from Garissa on Dadaab road, 11 May 1974, *Gillett & Gachathi* 20597!; Turkana District: 40 km. SW. of Lodwar, 12 May 1953, *Padwa* 141!; Tana River district: Thika–Garissa road, 4 km. E. of Namorumat Drift, 10 June 1974, *R.B. & A.J. Faden* 74/778!
DISTR. K 1, 2, 7; not known elsewhere
HAB. On sandy soil in open *Acacia-Commiphora* bushland, usually in shade; 210–550 m.

NOTE. The disjunct distributions of the Garissa–Wajir area and Turkana suggest that this very insignificant little species should also occur in dry areas between.

11. E. mossambicensis (*Klotzsch & Garcke*) *Boiss.* in DC., Prodr. 15(2): 36 (1862); N.E. Br. in F.T.A. 6(1): 509 (1911), both as '*E. mozambicensis*'; S. Carter in K.B. 39: 644 (1984). Type: Mozambique, Rios de Sena, *Peters* 33 (B, holo. †, K, fragment!)

Much-branched prostrate annual herb, with branches to 35 cm. long, the upper surface covered with short, adpressed hairs, the lower surface glabrous. Leaves obovate, to 14 × 8 mm., base very obliquely rounded, apex rounded and entire, obscurely toothed or minutely apiculate, glabrous or rarely with a few scattered hairs around the margin; petiole to 0.5 mm. long; stipules free, linear to 0.4 mm. long, sometimes 2–3-toothed at the broader base. Cyathia solitary on peduncles 0.5–1 mm. long, terminal and pseudoaxillary on short leafy shoots, ± 1.25 × 1.25 mm., with cup-shaped involucres; glands 4, transversely elliptic, to almost 1 mm. wide but usually much less, reddish, with distinct white or pinkish lobed appendages, to 0.6 × 1.5 mm.; involucral lobes minute, triangular with ciliate margins. Male flowers: bracteoles laciniate; stamens 1.25 mm. long. Female flower: ovary pedicellate; styles minute (0.3 mm. long), erect, bifid to halfway. Capsule exserted on a reflexed pedicel to 3 mm. long, acutely 3-lobed with truncate base, 1.6 × 1.6 mm., usually glabrous, occasionally with a few long spreading hairs. Seeds ovate-conical, sharply 4-angled, 0.9 × 0.5 mm., pinkish brown, with a few very obscure ridges and grooves. Fig. 78/3.

TANZANIA. Shinyanga, 1938, *Koritschoner* 2118!; Mpanda District: Katabi Plain, 16 Feb. 1971, *Sanane* 1572!; Iringa District: Mtera, 31 Jan. 1962, *Polhill & Paulo* 1299!
DISTR. T 1, 3, 4, 7, 8; Mozambique, Malawi, Zambia and Zimbabwe
HAB. On sandy, gritty soils in open bushland and grassland; 500–1220 m.

SYN. *Anisophyllum mossambicense* Klotzsch & Garcke in Klotzsch, Nat. Pfl. Tric.: 30 (1860)
     Euphorbia mozambicensis (Klotzsch & Garcke) Boiss. var. *nyasica* N.E. Br. in F.T.A 6(1): 510 (1911). Types: Malawi, Nyika Plateau, Nymkowa, *McClounie* 169 & Mt. Malosa, *Whyte* (both K, syn.!)

VAR. The slightly larger size and occasional increased pubescence of N.E. Brown's var. *nyasica* is easily encompassed within the variation shown by the wide distribution of the species, and is influenced by local environmental conditions.

12. E. allocarpa *S. Carter* in K.B. 39: 644, fig. 1/J–L (1984). Type: Tanzania, Kilosa District, Ruaha Gorge, *Bally & Carter* 16418 (K, holo.!, EA, iso.!)

Prostrate branching annual, with branches ± 15 cm. long covered in curved adpressed hairs which are absent on the under surface of at least the lower half of the branches. Leaves subsessile, obovate, to 10 × 6 mm., base very obliquely rounded, apex rounded, margin obscurely denticulate around the apex, glabrous, the lower surface tinged red; stipules free, broadly triangular, 0.4 × 0.4 mm., deeply toothed. Cyathia solitary on peduncles 1 mm. long, terminal and pseudoaxillary on leafy shoots, 1.5 × 1.5 mm., with cup-shaped involucres; glands 4, transversely elliptic, ± 0.8 mm. wide, with 4–6-lobed white appendages, 0.4 mm. wide; involucral lobes minute, triangular, ciliate. Male flowers: bracteoles laciniate; stamens 1.5 mm. long. Female flower: ovary pedicellate; styles 0.3 mm. long, erect, bifid to halfway with thickened apices. Capsule exserted on a reflexed pedicel 3.5 mm. long, 3-lobed, 1.8 × 1.8 mm., somewhat constricted about the middle, the lower half with a row of very short adpressed hairs along each side of the sutures and a fleshy hollow swelling between them. Seeds contained in the lower half of the capsule, conical, 0.8 × 0.6 mm., pinkish brown, with a few very obscure transverse ridges and grooves.

TANZANIA.   Kilosa District: Ruaha Gorge, 35 km. E. of Mbuyuni, 12 Feb. 1968, *Robertson* 875! & 3 Feb.
1974, *Bally & Carter* 16418!
DISTR.   **T** 6; known only from this one locality
HAB.   Exposed stony slopes and sandy soil with deciduous bushland, in shade; 600 m.

NOTE.   Known so far only from this one locality, this otherwise insignificant species should be easily
recognised from its strangely distorted capsule.

13. E. fischeri *Pax* in E.J. 19: 117 (1894). Type: Tanzania, Dodoma District, Ugogo,
Unyangwira [Nganguina], *Stuhlmann* 386 (B, syn. †) & Saranda [Salanda], *Fischer* 32
(B, syn. †, part in K!)

Much-branched, prostrate, annual herb with branches to 25 cm. long covered with
long, spreading white hairs. Leaves obovate to 10 × 5 mm., base very obliquely rounded,
apex rounded and very obscurely toothed, glabrous or with a few scattered hairs on both
surfaces; petiole to 0.5 mm. long; stipules filamentous, 0.4 mm. long. Cyathia solitary on
peduncles to 0.75 mm. long, terminal and pseudoaxillary, crowded on leafy shoots, 1.25 ×
1.25 mm., with cup-shaped involucres; glands 4, transversely elliptic, ± 0.6 mm. wide, with
large white, lobed appendages ± 0.6 × 1.5 mm.; involucral lobes minute, triangular, ciliate.
Male flowers: bracteoles filamentous; stamens 1.25 mm. long. Female flower: ovary
pedicellate; styles 0.75 mm. long, spreading, bifid to nearly halfway, with obviously
thickened apices. Capsule exserted on a reflexed pedicel to 2.5 mm. long, acutely 3-lobed,
1.25 × 1.25 mm., with a few long scattered hairs or occasionally glabrous. Seeds oblong-
conical, 0.8 × 0.4 mm., pinkish brown, with numerous shallow wrinkles.

TANZANIA.   Dodoma District: 30 km. E. of Manyoni, 23 Apr. 1964, *Greenway & Polhill* 11713!; Iringa
District: Ruaha National Park by ferry over Great Ruaha R., 19 Jan. 1966, *Richards* 21003! & near
River opposite Lunda, 17 May 1968, *Renvoize* 2201!
DISTR.   **T** 5, 7; not known elsewhere
HAB.   In open *Brachystegia* woodland on sandy soil; 800–1430 m.

SYN.   *E. mozambicensis* (Klotzsch & Garcke) Boiss. var. *fischeri* (Pax) N.E. Br. in F.T.A. 6(1): 510 (1911)

NOTE.   The completely different pubescence of the branches, much longer, deeply bifid styles, and
slightly smaller, more obviously wrinkled seeds serve to separate this taxon as a species distinct
from *E. mossambicensis*. It appears to be confined to a small area in central and southern Tanzania.

14. E. granulata *Forssk.*, Fl. Aegypt.-Arab.: 94 (1775); N.E. Br. in F.T.A. 6(1): 502 (1911);
F.P.S. 2: 72 (1952); E.P.A.: 447 (1958). Type: Yemen, *Forsskål* (C, holo., K, photo.!)

Prostrate annual herb, with branches to 15 cm. long, the whole plant puberulous with
short straight hairs, or more sparsely covered with long spreading hairs and the upper
surface of the leaves glabrous. Leaves obovate or oblong-ovate, to 8 × 4.5 mm., base
obliquely rounded to subcordate, apex rounded, margin entire or toothed; petiole to 1
mm. long; stipules to 1.5 mm. long, usually deeply divided into 2–4 linear teeth. Cyathia
solitary on peduncles to 0.5 mm. long, terminal and pseudoaxillary on short leafy shoots,
scarcely 1 × 1 mm., with cup-shaped involucres; glands 4, minute, transversely elliptic, with
very small obscurely lobed pink or white appendages; involucral lobes minute, triangular,
ciliate. Male flowers few: bracteoles filamentous; stamens 1 mm. long. Female flower:
ovary pedicellate; styles minute (0.2 mm. long), spreading, bifid to halfway. Capsule
exserted on a reflexed pedicel 1.5 mm. long, acutely 3-lobed, 1.5 × 1.5 mm. Seeds
oblong-conical, acutely 4-angled, 1 × 0.5 mm., pinkish brown, with numerous shallow
transverse wrinkles.

var. **glabrata** *(Gay) Boiss.* in DC., Prodr. 15(2): 34 (1862); N.E. Br. in F.T.A. 6(1): 503 (1911); F.P.S.
2:72 (1952); E.P.A.: 448 (1958). Types: Oman, Muscat, *Aucher* 5304 (K, syn.!) & Arabia, *Schimper* 754 (K,
syn.!) & *Botta* (P, 2 syn.) & Sudan, Kordofan, *Kotschy* 69 (P, syn., K, syn. partly!)

Pubescence sparse, with long spreading hairs. Leaves obovate to 6 × 4 mm., base obliquely
rounded, margin entire; upper surface glabrous; stipules to 1 mm. long, with 2–3 linear teeth. Gland
appendages pink or white.

KENYA.   Northern Frontier Province: Lokori, 29 July 1968, *Mwangangi & Gwynne* 1046! & Suguta
valley, 29 May 1970, *Mathew* 6467!; Turkana District: Lodwar, 13 May 1953, *Padwa* 156!
DISTR.   **K** 1, 2; Arabian Peninsula, Egypt and NE. Africa
HAB.   In exposed sandy gritty soils; 300–800 m.

SYN.   *E. forskalii* Gay var. *glabrata* Gay in Webb & Berth., Phyt. Canar. 3(3): 243 (1847)

NOTE. A. Radcliffe-Smith, in Fl. Iraq 4: 335 (1980), considered var. *glabrata* to be an extreme form of a variable species, placing it in synonymy. However, in Kenya, the varietal distinction is constant, NW. Kenya being the most southerly point of its distribution. The typical variety extends over the same area as far south as the Kordofan region of the Sudan and northern Somalia and eastwards from Arabia to Pakistan. It differs by being more densely pilose, including both surfaces of the leaves, and with shorter hairs.

var. **dentata** *N.E. Br.* in F.T.A. 6(1): 503 (1911); E.P.A.: 448 (1958). Type: Ethiopia, *Ellenbeck* 724 (B, syn.†) & Kenya, Lake Turkana [Lake Rudolph], *Wellby* (K, syn.!)

As for var. *glabrata*, but more laxly branched; hairs confined to branches, capsules and occasionally the lower surface of the leaves. Leaves oblong-ovate, to 8 × 4.5 mm., base obliquely subcordate, margin toothed at least near the apex; stipules to 1.5 mm. long with 3–4 teeth. Gland appendages a little more noticeable, white.

KENYA. Northern Frontier Province: Lokori, 16 May 1970, *Mathew* 6236!; Masai District: 5 km. from Magadi on Nairobi road, 30 June 1962, *Glover & Samuel* 2993!; Tana River District: Galole, 18 Dec. 1964, *Gillett* 16410!
TANZANIA. Masai District: Lake Chala [Dschalla], 30 Mar. 1952, *Bally* 8112!
DISTR. **K** 1, 6, 7; **T** 2; Ethiopia
HAB. Dry stony soils, often on lava; 60–1000 m.

SYN. [*E. mossambicensis* sensu Agnew, U.K.W.F.: 222 (1974), *non* Boiss.]

NOTE. The few widely scattered gatherings from East Africa suggest this variety is more common than the number of available specimens suggest. Ethiopian specimens are less hairy and possess rather larger leaves with more distinctly toothed margins. They all, including the Ellenbeck syntype, which unfortunately appears no longer to exist, originate from higher altitudes towards the northeast around Addis Ababa, Awash and Harar. They may prove to represent a distinct taxon.

15. E. serratifolia *S. Carter* in K.B. 35: 413, fig. 1/A–D (1980). Type: Tanzania, Kondoa District, Chungai, *Polhill & Paulo* 1163 (K, holo.!, B, BR, EA, LISC, P, PRE, SRGH, iso.)

Annual herb, with spreading branches suberect to 25 cm. high, the whole plant often tinged red, very sparsely covered with long spreading hairs, the upper side of the branches in addition with shorter adpressed hairs. Leaves ovate to 18 × 12 mm., base very obliquely subcordate, apex subacute, margin distinctly toothed, with the teeth apparently gland-tipped; petiole to 2 mm. long; stipules 0.75 mm. long, deeply divided into 2–3 linear teeth. Cyathia solitary, on peduncles to 2.5 mm. long, terminal and pseudoaxillary on leafy shoots, 1 × 1 mm., with cup-shaped involucres; glands 4, transversely elliptic, ± 0.3 mm. wide, red, with obvious, deeply 3–6-lobed pink or red appendages; involucral lobes broadly triangular, margin ciliate. Male flowers: bracteoles linear, deeply divided; stamens 1 mm. long. Female flower: ovary pedicellate; styles 0.3 mm. long, spreading, bifid almost to the base. Capsule exserted on a reflexed pedicel 1.25 mm. long, 3-lobed, with broadly truncate base, 1.5 × 2 mm. Seeds ovoid, 1 × 0.75 mm., reddish brown, with 3 distinct transverse ridges.

TANZANIA. Dodoma District: 37 km. S. of Itigi on Chunya road, 17 Apr. 1964, *Greenway & Polhill* 11605!; Iringa District: Ruaha National Park, Magangwe, 18 May 1968, *Renvoize* 2237! & 8 Mar. 1972, *Bjørnstad* 1452!
DISTR. **T** 5, 7; Malawi
HAB. Open *Brachystegia* woodland on sandy stony soil; 1000–1400 m.

16. E. lissosperma *S. Carter* in K.B. 35: 414, fig. 1/E–H (1980). Type: Tanzania, Mwanza District, Bulingwa, *Tanner* 550 (K, holo.!)

Prostrate or suberect annual herb, with branches to 25 cm. long, longitudinally ridged, the whole plant glabrous or usually at least the lower branches, leaves and capsules sparsely covered with fairly short spreading hairs. Leaves ovate, to 10 × 7 mm., base rounded to subcordate and markedly oblique, apex rounded, margin entire or very obscurely toothed near the apex, hairs (when present) noticeable around the margin; petiole to 1 mm. long; stipules linear to 1 mm. long, occasionally deeply divided into 2–3 teeth. Cyathia solitary on peduncles to 1 mm. long, terminal and pseudoaxillary on short leafy shoots, scarcely 1 × 1 mm., with cup-shaped involucres; glands 4, minute, transversely elliptic, green to purplish, with small entire or shallowly lobed white appendages; involucral lobes minute, triangular, ciliate. Male flowers: bracteoles linear, deeply divided; stamens 0.8 mm. long. Female flower: ovary pedicellate; styles 0.4 mm. long, spreading, bifid almost to the base. Capsule exserted on a reflexed pedicel 2.5 mm. long, acutely 3-lobed, 1.5 × 1.75 mm., with the ridges often purple tinged. Seeds ovoid, 1 × 0.6 mm., reddish brown, perfectly smooth.

UGANDA. Karamoja District: Meriss Camp, June 1930, *Liebenberg* 180!
KENYA. S. Nyeri District: Mwea Rice Irrigation Scheme, Research Station, 3 June 1976, *Kahurananga & Kibui* 2801!; Nairobi National Park, 26 May 1961, *Verdcourt* 3172!; Machakos District: Thika–Garissa road near Kongonde, 7 June 1974, *R.B. & A.J. Faden* 74/735!
TANZANIA. Mwanza District: Mbarika, Lubili, 22 May 1953, *Tanner* 1528!; Musoma District: Ikoma, Mugumu, 7 Apr. 1959, *Tanner* 4101!; Singida District: 43 km. from Issuna on Singida–Manyoni road, 13 Apr. 1964, *Greenway & Polhill* 11547!
DISTR. U 1; K 4, 5; T 1, 5; Burundi
HAB. Seasonally wet open grassland on black fissuring clay; 800–1750 m.

17. E. inaequilatera *Sond.* in Linnaea 23: 105 (1850); N.E. Br. in Fl. Cap. 5(2): 246 (1915); F.P.N.A.1: 478 (1948); W.K.F.: 37 (1948); F.P.S.2: 72 (1952); E.P.A.: 449 (1958); Hadidi in B.J.B.B. 43: 97 (1973); U.K.W.F.: 221 (1974); S. Carter in K.B. 39: 647 (1984). Type: South Africa, *Gueinzius* 167 (whereabouts unknown)

Annual much-branched herb, prostrate, to ± 50 cm. in diameter, or sometimes decumbent with branches to 30 cm. long, longitudinally ridged, conspicuously so when dry, the whole plant completely glabrous or with a few scattered hairs on the lower leaf-surface and occasionally on the capsules. Leaves ovate to occasionally lanceolate, to 14 × 6 mm., base very obliquely rounded to subcordate, apex obtuse, margin serrate, sometimes obscurely so; petiole to 1.5 mm. long; stipules ± 1.5 mm. long, deeply divided with 3–5 linear points. Cyathia solitary, on peduncles 1 mm. long, terminal and pseudoaxillary on short, leafy shoots, 1 × 1 mm., with cup-shaped involucres; glands 4, minute, transversely elliptic, red, with small lobed pink or white appendages; involucral lobes minute, triangular, margin sharply toothed. Male flowers: bracteoles laciniate; stamens 1.25 mm. long. Female flower: ovary pedicellate; styles 0.5 mm. long, spreading, bifid almost to the base. Capsule exserted on a reflexed pedicel 2 mm. long, acutely 3-lobed, 1.5 × 1.75 mm., with the angles often purple-tinged. Seeds oblong-conical, 1.25 × 0.75 mm., greyish brown, with shallow transverse wrinkles and pits. Fig. 78/1, p. 422.

var. **inaequilatera**

Plants completely glabrous.

UGANDA. Kigezi District: Ruwenzori National Park [Queen Elizabeth Park], Ishasha River Camp, 13 May 1961, *Symes* 708!; Mbale District: Bufumbo, Nov. 1932, *Chandler* 1029!; Mengo District: Entebbe, Aug. 1922, *Maitland* 138!
KENYA. Elgon, eastern slope above Japata Estate, 12 Feb. 1948, *Hedberg* 41!; Nairobi, Jan. 1949, *Bally* 6572!; Masai District: Olorgasailie, 11 Aug. 1951, *Verdcourt* 584!
TANZANIA. Mbulu District: Mt. Hanang, Katesh, 2 May 1962, *Polhill & Paulo* 2289!; Moshi District: Engari Nairobi, 19 June 1944, *Greenway* 6856!; Mbeya, 13 May 1956, *Milne-Redhead & Taylor* 10099!
DISTR. U 1–4; K 1–7; T 1, 2, 4–7; a common weed from the Arabian Peninsula through Somalia, Ethiopia, East Africa, southern tropical Africa to South Africa
HAB. Open patches amongst grass on seasonally wet gravelly, sandy or clay soils; 275–2500 m.

SYN. *E. sanguinea* Boiss. in DC., Prodr. 15(2): 35 (1862); N.E. Br. in F.T.A. 6 (1): 508 (1911); W.K.F.: 37 (1948). Types: N. Yemen, *Schimper* 753 (G, syn., K, photo.!, K, isosyn.!) & Ethiopia, *Schimper* 182 (G, syn., K, isosyn.!) & 1324 (G, syn., K, photo.!)
    E. sanguinea Boiss. var. *natalensis* Boiss. in DC., Prodr. 15(2): 35 (1862), *nom. illegit.* Type as for *E. inaequilatera* Sond.
    E. sanguinea Boiss. var. *intermedia* Boiss. in DC., Prodr. 15 (2): 35 (1862); N.E. Br. in F.T.A. 6(1): 508 (1911). Types: Ethiopia, Shoata, *Schimper* 1133 (K, syn.!) & 2472 (syn., whereabouts unknown)
    E. inaequalis N.E. Br. in F.T.A. 6(1): 512 (1911); E.P.A.: 449 (1958). Type: Somalia, *E. Cole* (K, holo.!)

VAR. Almost always found as a prostrate herb growing on dry, sandy soils and usually red-tinged, this species becomes decumbent or almost erect and without the red colouration, when growing in damper situations amongst denser grass clumps. The var. *intermedia* and *E. inaequalis* are encompassed by this range of variation. The holotype of *E. inaequalis* is a lushly grown specimen typical in all respects of *E. inaequilatera*. At the northern limit of its range plants are often weak with the toothing of the leaf margins obscure or even obsolete.

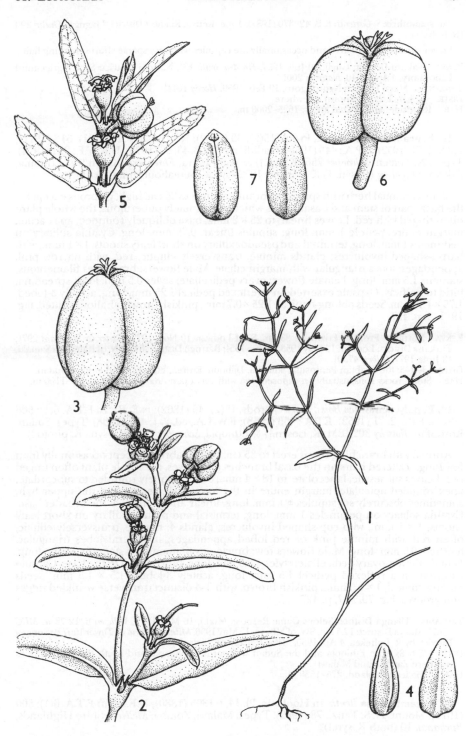

FIG. 79. *EUPHORBIA ARABICA* — **1**, habit, × ⅔; **2**, flowering branch, × 8; **3**, capsule, × 12; **4**, seeds, × 12. *E. POLYCNEMOIDES* — **5**, flowering branch, × 6; **6**, capsule, × 18; **7**, seeds, × 18. 1, from *Gilbert* 4750; 2, from *Tweedie* 1749; 3, 4, from *Carter & Stannard* 297; 5–7, from *Milne-Redhead & Taylor* 9459. Drawn by Christine Grey-Wilson.

var. **spanothrix** S. *Carter* in K.B. 42: 370 (1987). Type: Kenya, Kiambu District, Maguga, *Thulin* 290 (K, holo.!)

Lower surface of the leaves and occasionally the capsules with very sparse short spreading hairs.

KENYA. Nakuru District: Gilgil, 20 July 1957, *Harden-Smith* 18!; Nairobi District: Scott Agricultural Laboratory, 4 Apr. 1940, *Nattrass* 200!
TANZANIA. Masai District: Ngorongoro, 19 Feb. 1959, *Heady* 1641!
DISTR. **K** 3, 4; **T** 2; not known elsewhere
HAB. Roadsides and grazed areas; 1625–2000 m.

18. E. arabica *Boiss*. in DC., Prodr. 15(2): 33 (1862); N.E. Br. in F.T.A. 6(1): 513 (1911); F.P.S. 2: 71 (1952); E.P.A.: 443 (1958); Hadidi in B.J.B.B. 43: 90 (1973); U.K.W.F.: 221 (1974). Types: N. Yemen, *Schimper* 756 (K, syn.!) & Wadi Djara, *Ehrenberg* (K, syn.!) & Ethiopia, Arkiko, *Ehrenberg* (B, syn. †) & *Schimper* 1272 (syn., whereabouts unknown)

Slender annual herb with spreading branches erect to 25 cm. high, glabrous except for the basal part of stem and basal leaves which are minutely puberulous, the whole plant often tinged with red. Leaves linear, to 23 × 2 mm., base obliquely rounded, apex acute, margin entire; petiole 1 mm. long; stipules linear, 0.75 mm. long. Cyathia solitary on peduncles 1 mm. long, terminal and pseudoaxillary on short leafy shoots, 1 × 1 mm., with barrel-shaped involucres; glands minute, transversely elliptic, red, with narrow pink appendages; lobes triangular with margin ciliate. Male flowers: bracteoles filamentous; stamens 1.5 mm. long. Female flower: ovary pedicellate; styles 0.5 mm. long, spreading, bifid to halfway. Capsule exserted on a reflexed pedicel 1.25 mm. long, acutely 3-lobed, 1.75 × 1.75 mm. Seeds oblong-conical, 1.25 × 0.7 mm., pinkish brown, shallowly pitted. Fig. 79/1–4, p. 427.

KENYA. Northern Frontier Province: 33 km. S. of Lodwar, 10 Nov. 1977, *Carter & Stannard* 297!; Turkana District: Lodwar, 13 May 1953, *Padwa* 155!; Baringo District: 6 km. N. of Kampi ya Samaki, 13 June 1977, *Gilbert* 4750!
DISTR. **K** 1–3; the Arabian Peninsula, Somalia, Djibouti, Eritrea, Ethiopia and eastern Sudan
HAB. Stony, rocky soils, usually on exposed lava with very open *Acacia* bushland; 450–1100 m.

19. E. polycnemoides *Boiss*. in DC., Prodr. 15(2): 46 (1862); N.E. Br. in F.T.A. 6(1): 506 (1911); F.P.S. 2: 71 (1952); E.P.A.: 455 (1958); F.W.T.A., ed. 2, 1: 421 (1958). Types: Sudan, Kordofan, *Kotschy* 302 (BM, K, isosyn.!) & Ethiopia, *Schimper* 1500 (G, syn., K, photo.!)

Annual with branching stems erect to 35 cm. high, glabrous except occasionally for a few long, scattered hairs on the basal branches and leaves, the whole plant often tinged red. Leaves subsessile, lanceolate, to 18 × 4 mm., base obliquely rounded to subcordate, apex rounded apiculate, margin entire in the lower half, serrated in the upper half, sometimes obscurely so; stipules ± 1 mm. long, linear and toothed at the broader base. Cyathia solitary on peduncles 1 mm. long, terminal and pseudoaxillary on short leafy shoots, 1 × 1 mm., with cup-shaped involucres; glands 4, minute, transversely elliptic, often red, with minute pink or red lobed appendages; involucral lobes triangular, toothed, 0.5 mm. long. Male flowers few: bracteoles filamentous; stamens 1 mm. long. Female flower: ovary pedicellate; styles ± 0.5 mm. long, bifid almost to the base. Capsule exserted on a reflexed pedicel 1.5 mm. long, acutely 3-lobed, 1.5 × 1.5 mm. Seeds oblong-conical, 1 × 0.5 mm., pinkish brown, with 3–4 distinct transverse wrinkled ridges and grooves. Fig. 79/5–7, p. 427.

TANZANIA. Ulanga District: Selous Game Reserve, Mlahi, 16 May 1977, *Vollesen* 6/12-78 in MRC 4567!; Mbeya District: 17.5 km. SW. of Mbeya, 12 May 1956, *Milne-Redhead & Taylor* 10077!; Songea District: by R. Luhekea, 4 Ap. 1956, *Milne-Redhead & Taylor* 9459!
DISTR. **T** 6–8; from Ethiopia and the Sudan westwards to Nigeria, and south through Zaire to northern Zambia and Malawi
HAB. Wooded grassland; 275–1530 m.

20. E. zambesiana *Benth*. in Hook., Ic. Pl. 14, t. 1305 (1880); N.E.Br. in F.T.A. 6(1): 500 (1911); Mount. Fl. S. Tanz.: 78 (1982). Types: Malawi, Zomba, *Meller* & Shire Highlands, *Buchanan* 10 (both K, syn.!)

Perennial herb with a woody twisted root 1 cm. thick and to 25 cm. or more long, producing several woody underground stems to 3 cm. long, which branch profusely at ground-level; branches densely rebranching, leafy and prostrate to ± 10(–25) cm. long, or more floriferous and erect to ± 5 cm. high, often tinged red. Leaves lanceolate to ovate to 16 × 9 mm., base obliquely subcordate, apex minutely apiculate, the very narrowly cartilaginous margin entire or occasionally minutely toothed, under surface often tinged red; petiole to 2 mm. long; stipules linear or deeply divided into 2–4 linear teeth, ± 1 mm. long. Cyathia solitary on peduncles 1–25 mm. long, terminal and pseudoaxillary, 2 × 2.5 mm., with broadly cup-shaped involucres; glands 4 in ♂ cyathia, or sometimes 5 in cyathia which develop only ♂ flowers, transversely elliptic, ± 1 mm. wide, red with conspicuous white or pink, entire or obscurely 2–3-lobed appendages to 1.5 × 2 mm.; involucral lobes acutely triangular, 0.8 mm. long, margin ciliate. Male flowers many, especially in unisexual cyathia: bracteoles deeply laciniate, apices ciliate; stamens 2 mm. long. Female flower: ovary pedicellate; styles 1 mm. long, erect, bifid to halfway. Capsule exserted on a reflexed pedicel 5.5 mm. long, deeply 3-lobed, 3 × 3 mm. Seeds ovate-conical, 1.75 × 1.25 mm., reddish buff and deeply pitted. Fig. 78/5, p. 422.

var. **zambesiana**

Plant entirely glabrous; leaves lanceolate to ovate, 16 × 4.5 mm.

TANZANIA. Ngara District: Bushubi, Muganza, 2 Dec. 1959, *Tanner* 4579!; Kigoma District: 24 km. from Kigoma on Kasulu road, 14 July 1960, *Verdcourt* 2806A!; Mbeya District: Mkama, 4 Nov. 1912, *Stolz* 1650!
DISTR. T 1, 4, 7; Mozambique, Malawi and Zambia, southern Zaire and Angola, Rwanda and Burundi
HAB. In open wooded grassland; 900–2750 m.
SYN. *E. poggei* Pax in E.J. 19: 118 (1894). Type: Zaire, Lulua R., *Pogge* 121 (B, holo.†)
NOTE. A species which is among the first to appear after burning, it produces short, erect, very floriferous shoots, which later give way to longer, more leafy prostrate branches. Amongst the first shoots are solitary peduncles often bearing unisexual (♂) cyathia with 5 glands. The illustration accompanying the type description is erroneous in that it also shows 5 glands instead of 4 on bisexual cyathia. Examination of the type specimens show no such abnormality.
  Plants from Malawi often posses longer stems and larger leaves (to 30 × 9 mm.) than those found in Tanzania.

var. **villosula** (*Pax*) *N.E. Br.* in F.T.A. 6(1): 501 (1911); S. Carter in K.B. 39: 647 (1984). Types: Tanzania, Bukoba District, Karagwe, Kinuni, *Stuhlmann* 1654 & 1656 (B, syn.†)

Whole plant densely pilose with long, spreading hairs, or at least some hairs present on stems, leaves or capsules; leaves ovate to ovate lanceolate, to 16 × 9 mm.

UGANDA. Masaka District: Nabugabo, Sept. 1932, *Hopkins* in *Tothill* 1154! & NW. side of Lake Nabugabo, 9 Oct. 1953, *Drummond & Hemsley* 4709!
KENYA. Masai District; Lebetero Hills, Jan. 1961, *van Someren* in *E.A.H.* 12279!
TANZANIA. Bukoba District: Ndama, Oct. 1931, *Haarer* 2221!; Kigoma District: 24 km. from Kigoma on Kasulu road, 14 July 1960, *Verdcourt* 2806!; Mbeya, 29 Sept. 1950, *Backland*!
DISTR. U 4; K 6; T 1, 4, 7, 8; Angola, Zambia, southern Zaire, Burundi and Rwanda
HAB. Sandy soil in grass and open bush; 600–2400 m.
SYN. *E. villosula* Pax in E.J. 19: 118 (1894)
  E. angolensis Pax in E.J. 19: 117 (1894). Type: Angola, *Teuscz* in *Mechow* 207 (B, holo.†)
  E. poggei Pax var. *benguelensis* Pax in E.J. 23: 532 (1897). Type: Angola, *Antunes* 91A (whereabouts unknown)
  E. poggei Pax var. *villosa* Pax in Bull. Herb. Boiss. 6: 737 (1898). Type: Angola, *Antunes* 84 (whereabouts unknown)
  E. serpicula Hiern in Cat. Afr. Pl. Welw. 1: 941 (1900). Type: Angola, *Welwitsch* 279 (?BM, iso.)
  E. andongensis Hiern in Cat. Afr. Pl. Welw. 1: 943 (1900). Type: Angola, *Welwitsch* 281 (?BM, iso.)
  E. zambesiana Benth. var. *benguelensis* (Pax) N.E. Br. in F.T.A. 6(1): 501 (1911); W.K.F.: 36 (1948)
  [*E. zambesiana* (as *E. zambesiaca*) sensu Agnew, U.K.W.F.: 221 (1974), *non* Benth. sensu stricto]

VAR. Distribution of this pilose form overlaps that of the typical glabrous variety, but has so far not been collected east of Zambia. However, it extends further north, through western Tanzania into Uganda and towards Lake Victoria with one specimen from SW. Kenya. The degree of hairiness is extremely variable, from all parts of the plant, including the involucre and capsule being densely hairy, to the stems only or sometimes also the leaves (var. *benguelensis*), and/or the capsules with a few sparse hairs, or rarely the capsule alone with some hairs. All these forms occur with each other and with the glabrous form, so no justification can be found for upholding other species or varieties.

21. E. selousiana S. *Carter* in K.B. 42: 369 (1987). Type: Tanzania, Kilwa District, 43 km. SW. of Kingupira, *Rodgers & Vollesen* 2632 (EA, holo.!)

Perennial herb, probably with a thick woody rootstock, giving rise to woody persistent underground stems to ± 5 cm. long which branch at ground level; branches procumbent to ± 25 cm. long, pilose with spreading hairs. Leaves broadly ovate to ± 8 × 7 mm., base rounded and markedly oblique, apex rounded, margin entire or obscurely toothed, leathery, glabrous; petiole 0.5 mm. long; stipules triangular, 0.5 mm. long, margin toothed. Cyathia solitary on peduncles to 1.5 mm. long, terminal and pseudoaxillary, 1.7 × 2 mm., with cup-shaped involucres; glands 4, transversely elliptic, ± 0.4 × 0.7 mm., brownish, with deeply 3–5-lobed white petaloid appendages 2–5 mm. long; involucral lobes triangular, 0.3 mm. long, margin shortly ciliate. Male flowers: bracteoles laciniate, apices ciliate; stamens 1.8 mm. long. Female flower: styles 0.6 mm. long, bifid to halfway, with thickened apices. Capsule exserted on a reflexed pedicel 4.5 mm. long, deeply 3-lobed, ± 2 × 2 mm. Seeds oblong-conical, 1.3 × 0.8 mm., greyish pink, surface wrinkled.

TANZANIA. Kilwa District: Mbwera, 18 June 1932, *Schlieben* 2441! & 43 km. SW. of Kingupira, 5 Aug. 1975, *Rodgers & Vollesen* 2632!
DISTR. T 8; known only from these 2 collections
HAB. Deciduous woodland; 300–320 m.

22. E. rivae *Pax* in Ann. Ist. Bot. Roma 6: 186 (1897); N.E. Br. in F.T.A. 6(1): 505 (1911); E.P.A.: 456 (1958); U.K.W.F.: 221 (1974). Type: Ethiopia, Sidamo Province, *Riva* 94 (B, holo.†, FT, iso., K, drawing!)

Glabrous perennial herb with a thick tuberous rootstock to ± 12 × 4 cm. giving rise to several woody underground stems to ± 5 cm. long, which branch at ground-level; branches prostrate and leafy to 15 cm. long, or occasionally more floriferous and erect to 5 cm. high. Leaves broadly ovate to subcircular, to 17 × 13 mm., base rounded and obscurely oblique, apex rounded, margin entire, undersurface often tinged red; petiole to 1 mm. long; stipules broadly triangular to 0.8 × 0.6 mm., margin deeply toothed, sometimes joined at the base. Cyathia solitary on peduncles to 4 mm. long, terminal and pseudoaxillary, 2 × 2 mm., with cup-shaped involucres; glands 4, transversely elliptic to 1 mm. wide, red with white entire or obscurely 2–3-lobed appendages to 0.75 × 1.5 mm.; involucral lobes triangular, 0.4 mm. long, shortly ciliate. Male flowers: bracteoles laciniate with ciliate apices; stamens 2 mm. long. Female flower: ovary pedicellate; styles 0.75 mm. long, suberect, bifid to halfway, with thickened apices. Capsule exserted on a reflexed pedicel 5 mm. long, deeply 3-lobed, 2.25 × 2.25 mm. Seeds oblong-conical, 1.5 × 0.8 mm., pinkish buff and shallowly pitted. Fig. 78/4, p. 422.

KENYA. Kiambu District: 30 km. N. of Nairobi on Thika road, 1 Dec. 1957, *Verdcourt* 1962!; Nairobi District: below High Commission Headquarters, 22 Apr. 1961, *Polhill* 373!; Embu District: Emberre, 19 Sept. 1932, *M.D. Graham* in A.D. 2190!
DISTR. K 4; SE. Sudan and southern Ethiopia
HAB. Amongst grass in poor, often water-logged soil (black cotton or sandy); 725–1825 m.

SYN. [*E. zambesiana* sensu Cufodontis, E.P.A.: 462 (1958) [*Mooney* 5436], *non.* Benth.]

NOTE. The apparently disjunct distribution of this species from three limited, widely separated areas, seems to indicate that it should also occur elsewhere in northern Kenya.

2. Subgen. **Poinsettia**

Erect annual herbs (in East Africa). Leaves petiolate; stipules modified as obvious glands. Cyathia in densely branching cymes, with basal bracts large and leaf-like, sometimes brightly coloured. Involucres bisexual, with 1 funnel-shaped gland and 5 lobes. Stamens just exserted, with subsessile anthers, and bracteoles included in the involucre. Perianth-rim of the ♀ flower sometimes prominent; styles joined at the base, with bifid apices. Capsule exserted on a reflexed pedicel. Seeds conical, without a caruncle.

Floral bracts ovate (in East Africa), uniformly green; involucral
    glands funnel-shaped, with circular opening; seed surface
    tuberculate . . . . 23. *E. heterophylla*
Floral bracts panduriform, with a red blotch at the base;
    involucral glands funnel-shaped, with elliptical opening;
    seed surface sharply tuberculate . . . . 24. *E. cyathophora*

23. E. heterophylla *L.*, Sp. Pl.: 453 (1753); F.P.S. 2: 76 (1952); F.W.T.A., ed. 2, 1: 421 (1958). Type: tropical America, t. 12 in Pluk., Alm. Bot.: 369 (1696)

Erect annual herb to 1 m. high, with branches glabrous to sparsely pilose towards the apices, often tinged red. Leaves ovate, to 12 × 6 cm., base cuneate, apex obtuse, margin with minute distant gland-tipped teeth, occasionally more coarsely toothed, upper surface glabrous to sparsely pilose around the edges, lower surface pilose with septate hairs especially on the midrib and nerves, glabrescent; petiole to 2(-4) cm. long; glandular stipules fairly large, purplish. Cyathia densely clustered in axillary and terminal cymes forking ± 5 times, with progressively shorter rays from ± 15 cm. to ± 2 mm. long; basal bracts similar to the leaves but paler green, progressively smaller, more lanceolate and subsessile above. Cyathia glabrous, ± 3.5 × 2.5 mm., with goblet-shaped involucres; gland peltate, funnel-shaped, 1 mm. long, the opening circular, 1.2 mm. across, red-rimmed; lobes subcircular, ± 1.3 mm. long, deeply and sharply toothed with margins minutely ciliate. Male flowers: bracteoles few, ligulate, feathery; stamens 4 mm. long. Female flower: ovary pedicellate, glabrous or occasionally with minute scattered hairs, the perianth forming an obvious rim; styles ± 1 mm. long, occasionally minutely puberulous, bifid to halfway. Capsule exserted on a reflexed pedicel to 6 mm. long, deeply 3-lobed, ± 4.5 × 5.5 mm. Seeds conical with acute apex, 2.6 × 2.4 mm., blackish brown, surface bluntly tuberculate.

UGANDA. Mbale District: Budama, Paya, July 1926, *Maitland* 1169!; Teso District: Serere, Dec. 1931, *Chandler* 351!; Mengo District: Luzira, May 1931, *Snowden* 2061!
KENYA. Nairobi, University Way, 20 Mar. 1967, *Mwangangi* 38!; Machakos District: Kibwezi, 16 May 1938, *Bally* in *C.M.* 7705!; N. Kavirondo District: Mlaba Forest, June 1965, *Tweedie* 3057!
TANZANIA. Lushoto District: E. Usambara Mts., Mtindero, 21 Oct. 1940, *Greenway* 6038!; Mpanda District: Illembo, 5 July 1968, *Sanane* 211!; Iringa District: Kidatu, 6 Mar. 1971, *Batty* 1250!
DISTR. U 1–4; K 1, 3–5, 7; T 1–4, 6–8; throughout tropical Africa, a pantropical weed originating from Central America
HAB. A weed of cultivation; 150–3000 m.

SYN. *E. geniculata* Ortega, Hort. Matr. Dec.: 18 (1797); U.K.W.F.: 222 (1974). Type: Cuba, *Hort. Madrid* (? MA)

VAR. Plants occurring in East Africa show little of the variation which characterises this very widespread species, with scarcely any indication of the irregularity in leaf-shape giving the species its name. Elsewhere in tropical Africa, the leaves, especially the upper ones, may be more lanceolate or slightly panduriform in shape, and a pale cream or purplish blotch may sometimes occur at the base.

24. E. cyathophora *Murr.*, Comm. Götting. 7: 81, t. 1 (1786). Type: tropical America, *l.c.* t. 1 (or ? dried material in GOET)

Erect shrubby annual herb to 1 m. high, with glabrous stems and branches. Leaves ovate to markedly panduriform, to 10 × 5 cm., base cuneate, apex obtuse, margin shallowly toothed, upper surface glabrous, lower surface pilose with curved septate hairs, glabrescent; petiole to 1.5 cm. long, pilose; glandular stipules brownish. Cyathia densely clustered in terminal cymes, forking ± 4 times, with progressively shorter rays from ± 5 cm. to 2 mm. long; basal bracts similar to the leaves with a bright red blotch at the base, the upper ones progressively smaller, more lanceolate, sub-sessile and entirely red. Cyathia glabrous, ± 3.5 × 3 mm., with barrel-shaped involucres; gland peltate, funnel-shaped, 1.5 mm. long, the opening transversely elliptical, 2 mm. wide; involucral lobes rounded, 1.5 × 1.5 mm., margin deeply lobed. Male flowers: bracteoles ligulate, apices feathery; stamens 4 mm. long. Female flower: ovary pedicellate, glabrous; styles 2 mm. long, bifid almost to the base. Capsule exserted on a reflexed pedicel to 5.5 mm. long, deeply 3-lobed, 4 × 5 mm. Seeds ovoid-conical, with acute apex, 2.8 × 2.2 mm., blackish brown, surface sharply tuberculate.

KENYA. Nairobi, Feb. 1920, *Battiscombe* 1153!; Mombasa District: Bamburi, 4 Aug. 1970, *Bally* 13929!; Kilifi, Sept. 1932, *Napier* 2300 in *C.M.* 5601!
TANZANIA. Shinyanga, 17 Mar. 1953, *Welch* 183!; Tabora, 27 Aug. 1952, *Bally* 8310!; Uzaramo District: Dar es Salaam, Oyster Bay, 2 June 1969, *Mwasumbi* 10522!; Zanzibar I., Chwaka, 29 Nov. 1963, *Faulkner* 3312!
DISTR. K 4, 7; T 1, 3, 4, 6; Z; pantropical, introduced into a number of places throughout tropical Africa as an ornamental plant and becoming naturalised
HAB. A garden escape around habitation and on waste ground; 0–1800 m.

SYN. [*E. heterophylla* sensu Williams, U.O.P.Z.: 254 (1949); Jex-Blake, Gard. E. Afr., ed. 4: 63 (1957), non L.]

3. Subgen. Esula

Annual or perennial herbs, erect and sometimes shrubby. Leaves sessile or subsessile, usually entire; stipules absent. Cyathia in terminal and often axillary umbellate cymes; bracts deltoid, or leaf-like below the umbel. Involucres bisexual, or the primary central one of umbel, if present, entirely ♂; glands 4, rarely 5–8 on involucres with all ♂ flowers, entire or more often with 2 horns; lobes 5. Stamens with anthers clearly exserted from the involucre, but bracteoles included. Perianth of ♀ flower reduced to a rim below the ovary; styles usually deeply bifid. Capsule exserted on a reflexed pedicel. Seeds ovoid, with a caruncle.

1. Small annual herbs; seeds sculptured 2
   Herbs or shrubs; seeds smooth 3
2. Capsule with fleshy ridges along the angles; seeds with
   rows of deep pits . . . . *25. E. peplus*
   Capsule smooth; seed surface with a close network of
   ridges . . . . *26. E. dracunculoides*
3. Capsules 2-locular . . . . *29. E. brevicornu*
   Capsules 3-locular 4
4. Capsules tuberculate 5
   Capsules smooth 6
5. Herbs with simple or sparsely-branched stems 30(–100)
   cm. long from a woody rootstock . . . . *33. E. depauperata*
   Woody shrubs to 3 m. high . . . . *34. E. ugandensis*
6. Herbs with simple or sparsely-branched stems to 30 cm.
   high 7
   Much-branched shrubby herbs to 4.5 m. high 9
7. Rootstock woody but slender, 3–5 mm. thick . . . . *30. E. wellbyi*
   Rootstock thick and woody, 1–2 cm. in diameter 8
8. Leaves oblanceolate; cyathia 3.5 mm. in diameter; capsule
   4 × 4.5 mm. . . . . *31. E. daviesii*
   Leaves linear to linear-lanceolate; cyathia 5–8 mm. in
   diameter; capsule 5.5 × 6.5 mm. . . . . . . *32. E. cyparissioides*
9. Cymes unbranched or forking once only, in terminal
   umbels of 3–5, with (primary) rays to 3 cm. long;
   involucral glands without horns . . . . . . *35. E. usambarica*
   Cymes forking many times, in terminal and axillary
   umbels of 3–15, with primary rays 5–15 cm. long;
   involucral glands horned 10
10. All the involucral glands without further processes
    between the horns . . . . *27. E. schimperiana*
    Most of the involucral glands with 1–4(–8) processes
    between the horns . . . . *28. E. fuscolanata*

25. E. peplus *L.*, Sp. Pl.: 456 (1753); N.E. Br. in Fl. Cap. 5(2): 255 (1915); F.P.S. 2: 72 (1952); E.P.A.: 454 (1958). Types: Europe, *Linnean Herbarium* 630: 24 (LINN, syn., K, photo.!) & *Hort. Cliff.* 16 (BM, syn.)

Glabrous annual herb to 25 cm. high. Leaves obovate, to 25 × 15 mm., base cuneate, apex rounded, margin entire; petiole of lower leaves to 1 cm. long, shorter above. Cymes axillary and in terminal 3-branched umbels forking many times, with primary rays to 3.5 cm. long; bracts sessile, similar to the leaves but more ovate. Cyathia on peduncles to 1.3 mm. long, 1 × 1 mm., with cup-shaped involucres; glands 4, transversely oblong, 0.5 mm. broad, with 2 horns to 0.8 mm. long; lobes rounded, minute, margins ciliate. Male flowers: bracteoles linear, tips minutely ciliate; stamens 1 mm. long. Female flower: styles 0.3 mm. long, joined at the base, spreading, bifid for ⅔. Capsule exserted on a pedicel to 3 mm. long, deeply 3-lobed, with truncate base, 2 × 2 mm., with longitudinal fleshy ridges each side of the sutures. Seeds oblong-ovoid, 1.5 × 1 mm., reddish brown becoming grey, with longitudinal rows of deep pits and a smooth white caruncle 0.3 mm. across.

TANZANIA. Ufipa District: Malonje Farm, 14 Mar. 1957, *Richards* 8735!; Mbeya, 22 Feb. 1975, *Hellqvist* 9!

DISTR. T 4, 7; originating from Europe and E. Asia, a weed of world-wide temperate and subtropical zones including South Africa, and fairly recently introduced into Zimbabwe and Sudan

HAB.   Weed of cultivation and disturbed ground, in shade; 1700–2100 m.

26. E. dracunculoides *Lam.*, Encycl. Méth. Bot. 2: 428 (1786); N.E. Br. in F.T.A. 6(1): 515 (1911); F.P.S. 2: 75 (1952). Type: Mauritius, *Commerson* 545 (P-LA, holo., K, photo.!)

Glaucous glabrous annual herb to 25 cm. high. Leaves sessile, linear to linear-lanceolate, to 6 × 0.5 cm., entire. Cymes axillary and in terminal 3-branched umbels forking many times, with primary rays to 5 cm. long; bracts leaf-like. Cyathia subsessile, 1.5 × 1.5 mm., with cup-shaped involucres; glands 4, transversely oblong, 0.8 mm. broad with 2 horns to 0.8 mm. long; lobes rounded, margin ciliate. Male flowers: bracteoles linear, apices ciliate; stamens 2 mm. long. Female flower: styles 1 mm. long, joined at the base, spreading, bifid to almost halfway. Capsule exserted on a pedicel to 4 mm. long, deeply 3-lobed, 3.5 × 4 mm. Seeds ovoid, 2.5 × 1.7 mm., blackish brown, with a close network of paler ridges, and a cream caruncle 1 mm. across.

TANZANIA.   Uzaramo District: Dar es Salaam, Government garden, 1 Oct. 1901, *Stuhlmann* 97!
DISTR.   T 6; Sudan (Red Sea Coast) and Djibouti; S. Spain, N. Africa, Arabia, India and Mauritius
HAB.   Presumably a garden weed; near sea-level

NOTE.   Known in East Africa from this one collection, obviously introduced. Various subspecies occur in S. Spain and N. Africa.

27. E. schimperiana *Scheele* in Linnaea 17: 344 (1843); Boiss. in DC., Prodr. 15(2): 155 (1862); N.E. Br. in F.T.A. 6 (1): 533 (1911); F.P.N.A. 1: 482 (1948); W.F.K.: 37 (1948); F.P.S. 2: 75 (1952); E.P.A.: 457 (1958); F.W.T.A., ed. 2, 1: 421 (1958); U.K.W.F.: 222 (1974); S. Carter in K.B. 40: 812 (1985). Type: N. Yemen, *Schimper* 897 (K, iso.!)

Much-branched annual or short-lived perennial herb, erect to 2 m., completely glabrous, or hairy with hairs present at least on the stem below the leaves or on the capsule, the stem or occasionally the whole plant sometimes tinged reddish purple. Leaves sessile, ovate-lanceolate to lanceolate, to 15 × 2 cm., base cuneate, apex apiculate, margin entire, glabrous or sparsely hairy on the lower surface, usually crowded and leaving prominent scars. Cymes axillary and in terminal 3–15-branched umbels forking many times with primary rays to 15 cm. long; bracts sessile, deltoid, occasionally ± 3-lobed, 1–4 × 1–2 cm , apex apiculate, margin entire, but the bracts below the umbel ± similar to the leaves. Cyathia subsessile or on peduncles to 3 mm. long, 2 × 2 mm., with cup-shaped involucres, glabrous or occasionally sparsely hairy; glands 4, spreading, transversely elliptic to reniform, 1 × 1.5–2 mm., 2-horned, the horns 0.5–1.5 mm. long, green becoming brownish red; involucral lobes subquadrate, 0.5 mm. long, apex shallowly 2-lobed, ciliate. Male flowers; bracteoles linear, apices ciliate; stamens to 4.5 mm. long, glabrous or occasionally hairy below the articulation. Female flowers: ovary glabrous or pilose; styles erect to 2.5 mm. long, joined at the base for 0.5 mm., spreading, bifid for up to 1 mm. Capsule exserted on a reflexed pedicel to 5.5 mm. long, glabrous or pilose, deeply 3-lobed, with truncate base, to 4 × 4.5 mm., glabrous or sparsely pilose with long spreading hairs. Seeds oblong, slightly compressed, 2–2.5 × 1.5–2 mm., smooth, shiny black but eventually grey speckled with brown; caruncle 0.5 mm. across, wrinkled, yellow-brown.

KEY TO INFRASPECIFIC VARIANTS

1. Stems glabrous except sometimes for a few hairs in the leaf
   axils . . . . a. var. **schimperiana**
   Stems pubescent at least below the junction of the leaves
   with the stem 2
2. Capsule glabrous . . . . b. var. **pubescens**
   Capsule pilose . . . . c. var. **velutina**

a. var. **schimperiana**

Plants glabrous except sometimes for a few short hairs in the axils of the upper leaves. Fig. 80/1–4, p. 434.

UGANDA.   Karamoja District: near Kotido, July 1958, *J. Wilson* 484!; Kigezi District: Bufumbira, Bukimbiri, Sept. 1946, *Purseglove* 2108!; Mbale District: Elgon, Mt. Nkokonjeru, 20 Dec. 1924, *Snowden* 948!

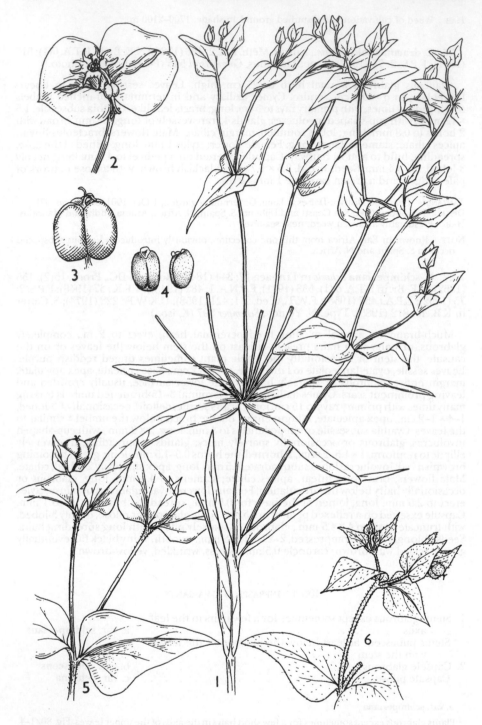

FIG. 80. *EUPHORBIA SCHIMPERIANA* var. *SCHIMPERIANA* — 1, flowering branch, × ²⁄₃; 2, cyathium, × 2; 3, capsule, × 1; 4, seeds, × 4 Var. *PUBESCENS* — 5, fruiting branch, × 2. Var. *VELUTINA* — 6, fruiting branch, × 2. 1, from *Drummond & Hemsley* 2016; 2, from *Nutt* 1328; 3, 4, from *Bally* 2490; 5, from *A.S. Thomas* 4378; 6, from *Richards* 24684. Drawn by Christine Grey-Wilson.

KENYA. Trans-Nzoia District: Kitale, Kirk's Bridge, 18 May 1943, *Bally* 2490!; N. Nyeri District: Mt. Kenya National Park, Naro Moru track, 10 Apr. 1977, *Hooper & Townsend* 1684!; Masai District: Keshemoruo, 13 July 1961, *Glover et al.* 2153!

TANZANIA. Masai District: Ngorongoro Crater above Laroda, 12 Sept. 1932, *B.D. Burtt* 4319!; Lushoto District: W. Usambara Mts., Lushoto–Malindi road, saddle near Magamba Peak, 29 May 1953, *Drummond & Hemsley* 2816!; Njombe, 10 July 1956, *Milne-Redhead & Taylor* 10798!

DISTR. U 1–4; K1–6; T 2–4, 6, 7; the Arabian Peninsula, N. Somalia and Ethiopia; westwards to Zaire and Cameroun; and south throughout Malawi and into Zimbabwe

HAB. In grassland, forest edges and clearings and a weed of land cleared for cultivation; 1100–3350 m.

SYN. *E. dilatata* A. Rich., Tent. Fl. Abyss. 2: 240 (1851); E.P.A.: 445 (1958). Type: Ethiopia, Mt. Bachit, *Schimper* 543 (BM, K, iso.!)

E. monticola A. Rich., Tent. Fl. Abyss. 2: 242 (1851). Type: Ethiopia, Djeledjeranne, *Schimper* 1706 (K, iso.!)

E. ampla Hook.f. in J.L.S. 6: 20 (1861); N.E. Br. in F.T.A. 6(1): 532 (1911); W.F.K.: 37 (1948). Type: Equatorial Guinea, Fernando Po. *Mann* 614 (K, holo.!)

E. ampla Hook.f. var. *tenuior* Hook.f. in J.L.S. 7: 215 (1864). Types: Cameroun, *Mann* 1265 & 2006 (K, syn.!)

E. longecornuta Pax in Hochgebirgsfl. Trop. Afr.: 287 (1892); N.E. Br. in F.T.A. 6(1): 535 (1911); F.P.N.A. 1: 481 (1948); W.F.K.: 37 (1948); E.P.A.: 451 (1958). Type: Ethiopia, R. Reb. near Gerra, *Schimper* (BM, K, iso.!)

E. kilimandscharica Pax in Hochgebirgsfl. Trop. Afr.: 287 (1892). Type: Tanzania, Kilimanjaro, *Meyer* 124 (B, holo. †)

E. hochstetteriana Pax in E.J. 19: 123 (1894), *nom. illegit.*, type as for *E. schimperiana* Scheele

E. preussii Pax in E.J. 19: 123 (1894). Type: Equatorial Guinea, *Preuss* 636 (BM, K, iso.!)

E. stuhlmannii Pax in E.J. 23: 535 (1897). Type: Tanzania, Uluguru Mts., *Stuhlmann* 221 (B, holo.†)

E. lehmbachii Pax in E.J. 28: 27 (1899). Type: Equatorial Guinea, *Lehmbach* 14 (B, holo.†)

b. var. **pubescens** (*N.E. Br.*) S. *Carter* in K.B. 40: 813 (1985). Type: Uganda, Toro District, Ruwenzori, Kivata, *Scott-Elliot* 7573 (K, holo.!)

Stem pubescent with long crisped hairs, sometimes present only below the junction of the leaves with the stem. Lower surface of young leaves often sparsely pilose. Cyathia and capsules glabrous. Fig. 80/5.

UGANDA. Acholi District: Chua, Agoro, 14 Nov. 1945, *A.S. Thomas* 4378!; Toro District: Ruwenzori, Bwera, 1951, *Osmaston* 3751!; Kigezi District: Kachwekano Farm, Apr. 1949, *Purseglove* 2722!

KENYA. Elgeyo District: Cherangani Hills, Kaibwibich, 3–11 Aug. 1968, *Thulin & Tidigs* 142!; Nakuru District: E. Mau Forest Reserve, 28 Aug. 1949, *Geesteranus* 5944!; Masai District: N. slope of Kibo, SE. of Loitokitok, 25 Feb. 1933, *C.G. Rogers* 493!

TANZANIA. Masai District: Ngorongoro, Empakaai Crater, 3 Sept. 1972, *Frame* 52!; Arusha District: SW. side of Mt. Meru, Olmotonyi, 21 Feb. 1969, *Richards* 24120!; Rungwe District: Poroto Mts., Ngozi, 16 Oct. 1956, *Richards* 6495!

DISTR. U 1, 2; K 3, 4, 6; T 2, 4, 6, 7; Nigeria and Cameroun, S. Sudan, N. Mozambique, N. Malawi and NE. Zambia

HAB. Forest clearings and edges, often in grass; 1525–2970 m.

SYN. *E. longipetiolata* Pax & K. Hoffm. in E.J. 45: 241 (1910). Type: Cameroun, *Ledermann* 5899 (B, holo.†, K, drawing!)

E. longecornuta Pax var. *pubescens* N.E. Br. in F.T.A. 6(1): 535 (1911); F.P.N.A. 1: 482 (1948)

c. var. **velutina** *N.E. Br.* in F.T.A. 6(1): 534 (1911); S. Carter in K.B. 40: 813 (1985). Types: Tanzania, Moshi District, Kilimanjaro, Marangu, *Volkens* 587 (B, syn. †, K, portion of syn.! & BM, isosyn.!) & Kibosho, *Volkens* 1665 & Moshi, *Merker* 603 & Mt. Meru, *Merker* 590 (all B, syn.†)

All parts of the plant, at least on young growth, pilose with long crisped brownish hairs. Ovary and capsule always pilose. Fig. 80/6.

UGANDA. Acholi District: Zututuro, 6 June 1963, *Kertland*!; Mengo District: Kiwafu, Sept. 1914, *Dummer* 42! & Sisa, Sept. 1936, *Chandler* 1936!

KENYA. Kiambu District: Kabete, 1930, *Mettam* 293!; Meru District: Meru Forest, Mariani, 25 July 1943, *J. Bally* 1 in *Bally* 3201!; Kericho District: Sotik, Kibajet Estate, 12 Sept. 1949, *Bally* 7475!

TANZANIA. Arusha District: Arusha National Park, track through 'Right of Way', 12 Nov. 1969, *Richards* 24684!; Moshi District: Kilimanjaro, S. slope between Umbwe R. and Weru Weru R., 1 Sept. 1932, *Greenway* 3224!; Songea District: Ngwambo, 10.5 km. N. of Miyau, 22 May 1956, *Milne-Redhead & Taylor* 10283!

DISTR. U 1, 4; K 4, 5; T 2, 8; Malawi and Zimbabwe

HAB. In grassland in very open forest, often near water or swampy ground; 1175–2150 m.

SYN. *E. velutina* Pax in P.O.A. C: 242 (1895), *non* Greene (1886) & *non* K. Schum. (1889). Type: Tanzania, Kilimanjaro, Marangu, *Volkens* 587 (B, holo. †, K, fragment!, BM, iso.!)

E. buchananii Pax in E.J. 28: 27 (1901). Type: Malawi, Blantyre, *Buchanan* 7058 (whereabouts unknown)

E. schimperiana Scheele var. *buchananii* (Pax) N.E. Br. in F.T.A. 6(1): 534 (1911)

VAR. A widespread species in which all combinations of variable characters can be found throughout its distribution. Depending upon local environmental conditions of shade and moisture, the size of the plants and of their leaves, bracts and cyathia all vary considerably. Cyathial gland size and the length of the horns can vary on a single plant, as also can the size of capsules and seeds. Whether the staminal pedicels are glabrous or pubescent and the length of the capsular pedicels likewise hold no significance. Pubescence of the stems and capsules however, are somewhat more constant characters and are thus used to differentiate the above varieties. Their distribution ranges fall within that of the typical variety, with all three sometimes occurring at the same locality.

28. E. fuscolanata *Gilli* in Ann. Nat. Mus. Wien 74.: 438 (1970). Type: Tanzania, Njombe District, Madunda, *Gilli* 274 (W, holo.!)

Much-branched shrub 1 m. high, probably annual; stem pubescent with crisped brown hairs. Leaves sessile, lanceolate, to 8.5 × 1 cm., base cuneate, apex apiculate, entire, glabrous. Cymes axillary and in terminal 5-branched umbels forking many times, with primary rays to 8 cm. long; bracts sessile, deltoid, ± 1.5 × 1.5 cm., margin entire, apex apiculate, the bracts of the umbel ± similar to the leaves. Cyathia subsessile, ± 3 × 4.5 mm. excluding the gland-processes, with cup-shaped involucres, glabrous; glands 4, spreading, transversely rectangular, ± 1.2 × 1.5–2.5 mm., 2 horned and with the margin between entire, or irregularly crenulate, or more usually with 1–4(–8) finger-like processes similar to the horns, horns and processes to 1.5 mm. long; lobes subquadrate, ± 1.2 × 1 mm., apex deeply and irregularly toothed, margin ciliate towards the base. Male flowers: bracteoles linear, apex minutely ciliolate; stamens 4.5 mm. long. Female flower: ovary glabrous; styles 3 mm. long, joined for 0.8 mm., deeply bifid. Capsule exserted on a reflexed pedicel 3 mm. long, obtusely 3-lobed, ± 5 × 5 mm. Seeds ovoid, 2.5 × 2 mm., smooth, becoming black, with a caruncle 1 mm. across.

TANZANIA. Njombe District: Livingstone Mts., Madunda, 29 July 1958, *Gilli* 274!
DISTR. **T** 7; known only from the type collection

NOTE. In all except details of the cyathial structure, this collection is virtually indistinguishable from robust specimens of *E. schimperiana* var. *pubescens*. However, until further material shows it to be perhaps an abnormal form, it must be considered as a distinct species, distinguished by the larger cyathia, with almost glabrous deeply toothed lobes, and especially by the extremely irregular shape of the glands.

29. E. brevicornu *Pax* in E.J. 43: 88 (1909); S. Carter in K.B. 40: 813 (1985). Type: Kenya, Masai District, Mau Escarpment, *Baker* (whereabouts unknown)

Annual or probably short-lived shrubby perennial herb to 2 m. high, glabrous except for crisped hairs below the junction of the leaves and bracts with the branches and usually on the stem and young leaves. Leaves densely crowded and leaving prominent scars, sometimes tinged red on the lower surface, sessile, lanceolate to 8 × 1 cm., base cuneate, apex obtuse and minutely apiculate, margin entire. Cymes lax, axillary and in terminal 3–15-branched umbels forking many times, with primary rays to 10 cm. long, all parts of the cyme often tinged red; bracts sessile, broadly ovate to 24 × 12 mm., apex acute, margin entire. Cyathia on peduncles to 1 mm. long, 1.75 × 2 mm., with cup-shaped involucres; glands 4, spreading, broadly crescent-shaped, to 1.25 mm. broad, with curved horns 0.5 mm. long; involucral lobes subquadrate with 2-lobed apex, 0.5 mm. long, margin ciliate. Male flowers: bracteoles linear, apex ciliate; stamens 2 mm. long. Female flowers: ovary 2-locular; styles 2, 1.5 mm. long, joined at the base, bifid for ⅓. Capsule exserted on a reflexed pedicel to 4.5 mm. long, 2-lobed, 2.25 × 2.75 mm. Seeds ovoid, 1.8 × 1.4 mm., smooth, reddish black, with a caruncle 1 mm. across, wrinkled, yellow.

UGANDA. Mbale District: Bulambuli, 12 Nov. 1933, *Tothill* 2345! & Benet Sabei, 14 Dec. 1938, *A.S. Thomas* 2644! & Elgon, Sebei, Suam Ridge, 23 Dec. 1954, *Dale* U.850!
KENYA. Trans-Nzoia District: Elgon, Mbere R. valley, S. of Koitcut, 24 Dec. 1967, *Gillett* 18429!; N. Nyeri District: Mt. Kenya, 5 Apr. 1975, *Hepper, Field & Townsend* 1867!; Masai District: Enesambulai valley, 2 Nov. 1969, *Greenway & Kanuri* 13862!
DISTR. **U** 3; **K** 2–4, 6; not known elsewhere
HAB. In open forest, in shade or damp situations; 2000–3250 m.

SYN. [*E. repetita* sensu N.E. Br. in F.T.A. 6(1): 536 (1911) pro parte; W.F.K.: 37 (1948); U.K.W.F.: 22 (1974), *non* A. Rich.]
    E. euryops Bullock in K.B. 1932: 492 (1932) & Hook., Ic. Pl. 32, t. 3193 (1933); W.F.K.: 37 (1948). Type: Kenya, Trans-Nzoia District, Elgon, *Lugard* 380 (K, holo.!)

NOTE. *E. repetita* A. Rich. is confined to central Ethiopia, distinguished from *E. brevicornu* by being completely glabrous except for the shorter, pilose pedicel of the similarly 2-locular capsule, and by the longer horns on the glands and the more oblong seeds, which are grey and shallowly pitted. The inflorescence and leaves also appear to have less of a tendency to colour red.

30. E. wellbyi *N.E. Br.* in F.T.A. 6(1): 541 (1911); A.V.P.: 126 (1957); E.P.A.: 462 (1958); U.K.W.F.: 222 (1974); S. Carter in K.B. 40: 814 (1985). Type: Ethiopia, Shoa Province, *Wellby* (K, holo.!)

Annual or probably short-lived perennial herb branching from near the base, to ± 30 cm. high, seldom more, glabrous or the stem pilose with short curved hairs, the whole plant sometimes tinged red. Leaves densely crowded, sessile, oblanceolate, to 25 × 7 mm., apex rounded, margin entire. Cymes occasionally axillary, usually in terminal 3-5-branched umbels, with primary rays to 4 cm. long, forking ± 5 times; bracts sessile, deltoid, to 1.5 × 1.5 cm., seldom larger. Cyathia on peduncles to 1 mm. long, 1.8 × 2 mm., with cup-shaped involucres; glands 4, spreading, transversely oblong, 1 mm. broad, with 2 horns to 0.8 mm. long; involucral lobes ovate with 2-lobed apex, 0.8 mm. long, margin minutely ciliate. Male flowers: bracteoles linear, ciliate; stamens 2.5 mm. long. Female flower: styles 1.5 mm. long, joined at the base for ⅓, erect, bifid to almost halfway. Capsule exserted on a glabrous or pilose pedicel to 3.5 mm. long, deeply 3-lobed, 2.7 × 3.2 mm. Seeds ovoid, 1.5 × 1.2 mm., smooth, reddish black, with a cream, wrinkled caruncle, 0.6 mm. across.

var. **wellbyi**

Stem hairy, at least on young sterile growth and below the inflorescence; capsular pedicel sometimes hairy.

KENYA. N. Nyeri District: Aberdare Range, near W. part of Nyeri track, 15 July 1948, *Hedberg* 1582!
TANZANIA. Masai District: Ngorongoro Crater rim, 9 Sept. 1966, *V.C. Gilbert* F. 3!
DISTR. **K** 4; **T** 2; Ethiopia
HAB. Montane moorland; 2375-3600 m.

var. **glabra** *S. Carter* in K.B. 40: 814 (1985). Type: Uganda, Mbale District, Elgon, W. slope above Butadiri along track via Mudangi, *Hedberg* 4550 (K, holo.!)

Stem and pedicel of the capsule always completely glabrous.

UGANDA. Mbale District: Elgon, 22 Oct. 1916, *Snowden* 462! & Mudangi Camp, 1930, *Liebenberg* 1625! & by Sasa stream, 23 Mar. 1951, *G. Wood* 150!
KENYA. Trans-Nzoia District: Elgon, E. slope above Tweedie's saw-mill, 2 Mar. 1948, *Hedberg* 214!; Elgeyo District: Arror valley, 26 Aug. 1969, *Mabberley & McCall* 254!; N. Nyeri District: Gikururu R., 1.5 km. W. of Giandogoro Gate, 14 Oct. 1970, *Mabberley* 333!
TANZANIA. Masai District: Elanairobi Volcano, 20 Sept. 1932, *B.D. Burtt* 4173!
DISTR. **U** 3; **K** 3, 4; **T** 2; Ethiopia
HAB. Amongst grass in heathland above the forest line or in clearings at the forest edge, usually in swampy ground; 2900-4000 m.

VAR. Usually no more than 30 cm. high in open heathland, but occasionally reaching 1 m. amongst higher vegetation at the edge of forest clearings.

31. E. daviesii *E.A. Bruce* in K.B. 1940: 51 (1940). Type: Tanzania, Njombe District, Kitulo [Elton] Plateau, *R.M. Davies* E.17 (K, holo.!)

Glabrous perennial herb with a woody rootstock ± 1 cm. thick, producing numerous usually simple annual stems, erect to 30 cm. Leaves sessile, slightly fleshy, oblanceolate, to 2.5 × 1 cm., apex acute to ± obtuse, apiculate, margin entire. Cymes occasionally axillary, usually in terminal 3-7-branched umbels, with primary rays to 3 cm. long, forking ± 4 times; bracts sessile, often tinged purplish, ± deltoid, to 1.5 × 1.5 cm., base subcordate, apex obtuse to rounded, apiculate. Cyathia sessile, 3 × 3.5 mm., with cup-shaped involucres; glands 4, rarely 5 on cyathia developing all ♂ flowers, spreading, transversely oblong, 1.5 × 2-2.5 mm., outer margin entire to shallowly sinuate, with 2 minute horns, yellow-orange; involucral lobes rounded, 1 mm. long, apex shallowly 2-lobed, margin ciliate. Male flowers many: bracteoles ligulate, apices ciliate; stamens 4 mm. long. Female flower: styles

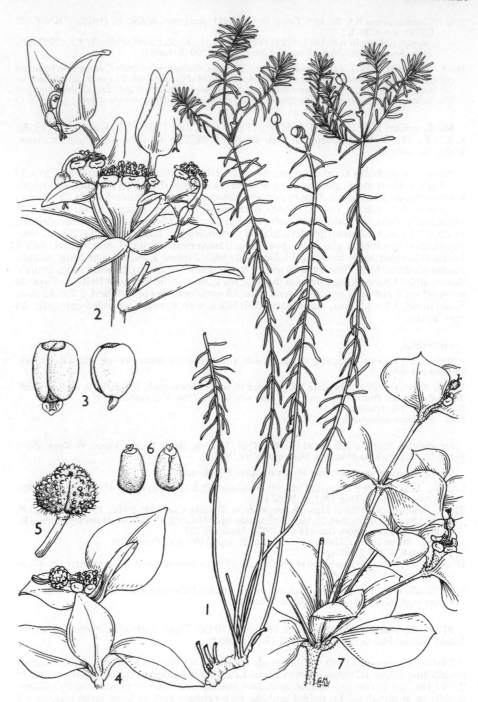

FIG. 81. *EUPHORBIA CYPARISSIOIDES* — **1**, habit, × ²⁄₅; **2**, flowering branch, × 2; **3**, seeds, × 4. *E. DEPAUPERATA* var. *DEPAUPERATA* — **4**, flowering branch, × ²⁄₅; **5**, capsule, × 2; **6**, seeds; × 2. Var. *TRACHYCARPA* — **7**, flowering branch, × ²⁄₅. 1, 3, from *Milne-Redhead & Taylor* 8032; 2, from *Richards* 12991; 4–6, from *Milne-Redhead & Taylor* 8190, 7, from *Hooper & Townsend* 797. Drawn by Christine Grey-Wilson.

1.5 mm. long, joined at the base, spreading, bifid for ⅓. Capsule exserted on a pedicel to 6 mm. long, 3-lobed, 4 × 4.5 mm. Seeds ovoid, 3 × 2–4 mm., smooth, grey speckled with brown; caruncle cream, 0.5 mm. across.

TANZANIA. Rungwe District: Kikondo, 21 Oct. 1956, *Richards* 6687!; Njombe District: Lupembe, Ruhudji R., 29 Aug. 1931, *Schlieben* 1157A! & Kitulo [Elton] Plateau, by Ndumbi R., 1 Dec. 1963, *Richards* 18507!
DISTR. T 7; N. Malawi
HAB. In grass amongst rocks and in dry sandy soil, usually appearing after burning; 1700–2700 m.
SYN. *E. imbricata* E.A. Bruce in K.B. 1933: 468 (1933), *nom. illegit., non* Vahl. Type as for *E. daviesii*

32. E. cyparissioides *Pax* in E.J. 19: 123 (1894); N.E. Br. in F.T.A. 6(1): 542 (1911); F.P.S. 2: 75 (1952); U.K.W.F.: 223 (1974); S. Carter in K.B. 40: 814 (1985). Types: Sudan, Niamniam, *Schweinfurth* 3979 & Dar Fertit, *Schweinfurth* Ser. III, 149 (K, isosyn.!)

Glabrous perennial herb, with a woody rootstock 1–2 cm. thick, producing densely tufted simple or sparsely branched annual stems, erect to 30 cm. seldom more. Leaves sessile, spreading to reflexed, linear-lanceolate to 30 × 5 mm., base cuneate, apex rounded and strongly apiculate, margins entire and ± revolute, densely crowded on the upper part of sterile growths, sparser amd often shorter on flowering stems. Cymes occasionally axillary, usually clustered in terminal 3–7-branched umbels, with primary rays to 2 cm. long, forking usually once only or sometimes branching to produce further umbels; bracts similar to the leaves but more ovate-lanceolate, ± 12 × 6 mm. Cyathia usually subsessile, but sometimes on peduncles to 5 mm. long, especially those at the centre of the umbels which sometimes develop only ♂ flowers, ± 3 × 5 mm., with broadly cup-shaped involucres, the central cyathia larger to 8 mm. in diameter; glands 4–7 or occasionally more and sometimes proliferating amongst the ♂ flowers of a "double" involucre, transversely oblong, 1–2 × 1.5–3 mm., margin very irregular, rounded, shortly 2-horned, or crenulate and ± convoluted, bright yellow; involucral lobes rounded, 1–1.5 mm. long, margin lobed or toothed, finely ciliate. Male flowers very many: bracteoles linear, apex ciliate; stamens 3.5–4.5 mm. long. Female flower: styles 1.8 mm. long, joined for ⅓, with spreading thickened bifid apices. Capsule exserted on a pedicel to 1.5 cm. long, deeply 3-lobed, 5.5 × 6.5 mm., surface roughened. Seeds ovoid, slightly compressed, 3.5 × 3 mm., apex rounded, smooth, blackish with faint brown speckles; caruncle 1.5 mm. across, wrinkled, orange. Fig. 81/1–3.

UGANDA. Acholi District: SE. Imatongs, Lomwaga Mt., 5 Apr. 1945, *Greenway & Hummel* 7277!
KENYA. Trans-Nzoia District: Kitale, Jan. 1965, *Tweedie* 2967!; Nandi, Jan. 1894, *Scott-Elliot* 6864!; Laikipia Escarpment, *Gardner* in F.D. 3515!
TANZANIA. Ufipa District: Nsanga Forest, 7 Aug. 1960, *Richards* 12991!; Mbeya Peak Forest Reserve, 3 July 1962, *Mgaza* 525!; Songea District: Lumecha Bridge, 21 km. N. of Songea, 3 Jan. 1956, *Milne-Redhead & Taylor* 8032!
DISTR. U 1; K 3; T 1, 4, 6–8; south central Ethiopia and Sudan, Zaire and Cameroun, Angola, Zambia, Zimbabwe, Malawi and Mozambique
HAB. In grassland and open woodland on well-drained soils appearing especially after burning; 930–2700 m.
SYN. *E. huillensis* Pax in E.J. 28: 27 (1899). Type: Angola, Huila, *Dekindt* 1029 (B, holo.†)
E. ericifolia Pax in E.J. 33: 288 (1903). Type: Tanzania, Uluguru to Usanga, *Prittwitz & Gaffron* 97 (K, iso.!)
E. dejecta N.E. Br. in F.T.A. 6(1): 541 (1911). Type: Malawi, Tanganyika Plateau, *Whyte* (K, holo.!)
E. cyparissioides Pax var. *minor* N.E. Br. in F.T.A. 6(1): 542 (1911). Various syntypes including Malawi, *Whyte* & Zambia, *Allen* 354 & *Kassner* 2125 (BM, syn.!) & Angola, *Johnston* (all K, syn.!)
VAR. Like most widespread species this is extremely variable. The short stems which typically appear after burning can sometimes persist and elongate. Leaves are mostly needle-like and 1–1.5 cm. long, but can be scarcely 5 mm. long and strongly reflexed, or spreading and much larger and broader. Size of the cyathium varies considerably, and especially the number, size and shape of the glands.
NOTE. The species was described from its northernmost limit of distribution, in Sudan, but is especially common in SE. Tanzania in Ufipa and Mbeya Districts, and in Malawi, Northern Province and Zambia, Mbala District.

33. E. depauperata *A. Rich.*, Tent., Fl. Abyss. 2: 241 (1851); Boiss. in DC., Prodr. 15(2): 119 (1862); N.E. Br. in F.T.A. 6(1): 537 (1911); W.F.K.: 37 (1948); F.P.S. 2: 73 (1952); E.P.A.: 445 (1958); F.W.T.A., ed. 2,1: 421 (1958); U.K.W.F.: 223 (1974); S. Carter in K.B. 40: 815 (1985). Types: Ethiopia, Mt. Sholoda, *Schimper* 336 (BM, K, isosyn.!) & 1532 (K, isosyn.!) & *Quartin Dillon* (P, syn.)

Perennial herb, with a thick woody rootstock producing numerous simple or sparsely branched annual stems, erect or spreading and ± decumbent, 30–100(–150) cm. long, glabrous or pilose with long spreading hairs. Leaves subsessile, linear-lanceolate, or ovate on short flowering stems (10–15 cm. high) produced after burning and often tinged with red, to 4(–5) × 1(–1.5) cm., base abruptly cuneate to rounded, apex shortly apiculate, margin narrowly cartilaginous, sometimes ± revolute, and often minutely toothed, glabrous or with some long spreading hairs on the lower surface especially along the midrib, rarely with a few on the upper surface; petiole to 1 mm. long, flattened, with a short tuft of hairs in the axil. Cymes axillary and in (3–)5(–6)-branched umbels around a terminal ♂ cyathium, with primary rays to 7 cm. long, forking ± 3 times; bracts sessile, deltoid to subcircular, ± 1.5 × 1.5 cm., longer below the umbel, apex acute to rounded, base with a tuft of hairs on the upper surface and sometimes a few hairs on the lower surface. Cyathia sessile, glabrous, or pilose with spreading hairs, sometimes sparsely so, ± 3.5 × 4.5 mm., with broadly cup-shaped involucres; glands 4–6, spreading, circular to transversely oblong, 1.5–3 mm. broad, yellow; lobes rounded, 1 mm. long, densely ciliate on the inner surface. Male flowers: bracteoles fan-shaped, densely ciliate; stamens 4.5 mm. long. Female flower: ovary tuberculate, usually densely so, glabrous or pilose; styles 3 mm. long, joined to nearly halfway, erect, spreading, bifid for ⅓. Capsule exserted on a pedicel to 6 mm. long, shallowly 3-lobed, ± 5 × 6.5 mm., densely to sparsely covered with large fleshy tubercles, occasionally almost smooth, glabrous or fairly sparsely pilose with long spreading hairs, usually tinged reddish. Seeds ovoid, slightly compressed, ± 2.5 × 2 mm., smooth, greyish brown; caruncle 1 mm. across, cream.

KEY TO INFRASPECIFIC VARIANTS

1. Apices of sterile shoots pilose . . . . c. var. **trachycarpa**
   Apices of sterile shoots glabrous 2
2. Tubercles on ovary and capsule obvious . . . . . a. var. **depauperata**
   Tubercles on ovary and capsule obscure . . . . . . b. var. **laevicarpa**

a. var. **depauperata**

Stem glabrous or sparsely pilose, with the apices of young sterile shoots glabrous. Involucres glabrous or pilose. Ovary and capsule tuberculate, glabrous or pilose. Fig. 81/4–6, p. 438.

UGANDA. Karamoja District: Napak, June 1950, *Eggeling* 5908! & Mt. Moroto, 6 Sept. 1956, *Bally* 10772! & June 1963, *J. Wilson* 1456!
KENYA. Elgeyo District: N. Cherangani Hills, Chepkotet, 12 Aug. 1968, *Thulin & Tidigs* 224!; Nakuru District: E. Mau Forest Reserve, camp 10, 31 Aug. 1949, *Geesteranus* 5991!; Naivasaha District: Kinangop, Chania R., Sasumua Dam, 18 Jan. 1953, *Verdcourt* 875!
TANZANIA. Moshi District: Kilimanjaro, track to Shira Plateau, 12 Feb. 1969, *Richards* 24001!; Ufipa District: Malonji Plateau, Mamya Mt., 27 Oct. 1965, *Richards* 20629!; Rungwe District: Kikondo, 20 Oct. 1956, *Richards* 6656!
DISTR. U 1; K 1–4, 6; T 2–5, 7, 8; widespread from Sierra Leone in the west, eastwards to Burundi, and the Imatong Mts., in Sudan, northwards through Ethiopia to Eritrea, and southwards through Malawi, Zambia and Zimbabwe
HAB. Sandy, rocky soils in grassland, often in forest clearings; 1220–3350 m.

SYN. *E. shirensis* Bak.f. in Trans. Linn. Soc., ser. 2, 4: 38 (1894). Type: Malawi, Mt. Mulanje, *Whyte* (K, holo.!)
E. lepidocarpa Pax in E.J. 33: 387 (1903). Type: Ethiopia, Harar Province, *Ellenbeck* 590 (B, holo.†)
E. depauperata A. Rich. subsp. *aprica* Pax in E.J. 39: 631 (1907); N.E. Br. in F.T.A. 6(1): 538 (1911) as var. *aprica;* E.P.A.: 445 (1958). Type: Ethiopia, Shoa Province, *Rosen* (B, holo.†)
E. multiradiata Pax & K. Hoffm. in E.J. 45: 240 (1910). Type: Cameroun, *Ledermann* 2006 (B, holo.†)
E. depauperata A. Rich. var. *pubiflora* N.E. Br. in F.T.A. 6(1): 538 (1911). Types: Malawi, Northern Province, *Whyte* (K, 2 syn.!)

b. var. **laevicarpa** *Friis & Vollesen* in K.B. 37: 470 (1982). Type: Sudan, Imatong Mts., *Friis & Vollesen* 206 (C, holo., K, iso.!)

Stems and apices of young sterile shoots glabrous, and rarely a few hairs present on some leaves. Involucres glabrous. Ovary and capsule almost smooth, with only faint indications of a few small tubercles, pilose.

UGANDA. Acholi District: Imatong Mts., Apr. 1938, *Eggeling* 3000! & Apr. 1943, *Purseglove* 1388!
DISTR. U 1; Sudan, Imatong Mts., not known elsewhere
HAB. Montane grassland; 1830–2740 m.

NOTE. The citation by Friis and Vollesen of *Eggeling* 3608 as the typical variety is clearly in error as the specimen possesses pilose capsules with only faint indications of tubercles, exactly matching the other specimens and the drawing in their publication.

c. var. **trachycarpa** (*Pax*) *S. Carter* in K.B. 40: 815 (1985). Type: Tanzania, Njombe District, E. Ukinga, Lager Kidoko, *Prittwitz & Gaffron* 174 (B, holo.†)

Stems and apices of young sterile shoots ± densely pilose, flowering stems usually more sparsely pilose. Involucres pilose or occasionally glabrous. Ovary and capsule densely tuberculate, glabrous or pilose. Fig. 81/7, p. 438.

TANZANIA. Mbeya District: Chimala Escarpment, Kitakala Mission, 15 Dec. 1963, *Richards* 18568! & 15 km. from Mbeya on Chunya road, 16 Mar. 1975, *Hooper & Townsend* 797!; Iringa District: Sao Hill, Nov. 1959, *Watermeyer* 168!
DISTR. T 2, 4, 7, 8; N. Malawi and N. Zambia
HAB. Grassland; 1050–2440 m.

SYN. *E. trachycarpa* Pax in E.J. 33: 288 (1903)

VAR. This is another widespread species which exhibits a great deal of variation, particularly in size and in pubescence. It is especially common in southern Tanzania and northern Malawi, where it often appears after burning, flowering when only a few centimetres high. The stem can persist and elongate, remaining erect amongst tall grass, or becoming decumbent and ± creeping. The stems, leaves, involucres and capsule can be glabrous, or one or any combination of these features show varying degrees of pubescence, sometimes in the same gathering. Nevertheless, the young sterile leafy shoots are always completely glabrous or very distinctly pilose, so that varietal distinction can be established on this basis. Also an unusually smooth, almost tubercle-less capsule distinguishes some plants from the Imatong Mountains at varietal level. Var. *pubescens* Pax (in Ann. Ist. Bot. Roma 6: 188 (1897)) from central Ethiopia may possibly merit distinction, with stems and both surfaces of the leaves thickly hairy, but the involucre and capsule glabrous.

34. E. ugandensis *Pax* in E.J. 45: 240 (1910); N.E. Br. in F.T.A. 6(1): 531 (1911); K.T.S.: 202 (1961); U.K.W.F.: 223 (1974). Type: Kenya, Kiambu District, Limuru [Lamuru], *Scheffler* 268 (B, holo. †, BM, K, iso.!)

Perennial shrubby herb to 3 m. high, woody at the base; branches in whorls of 5, glabrous or very sparsely pilose, often tinged red. Leaves subsessile, lanceolate, to 11 × 3 cm., base tapering into a very short flattened petiole, apex acute to shortly apiculate, margin entire, petiole and lower surface of the leaf pilose with long, soft hairs, especially the midrib and sometimes the upper surface of young leaves. Cymes in terminal 5-branched umbels, with primary rays to 7 cm. long, forking usually twice; bracts sessile, deltoid, to 2 × 2 cm., apex obtuse to rounded and minutely apiculate, base and lower margins sparsely pilose. Central cyathium of the umbel on a peduncle 1.5 mm. long, the rest sessile, 3 × 3.8 mm., with funnel-shaped involucres, glabrous; glands 4–5, transversely oblong, and ± 2-lipped, ± 1.5 mm. broad, yellow-green; involucral lobes subquadrate, 1 mm. long, apex shallowly bilobed, ciliate on the inner surface. Male flowers: bracteoles spathulate, deeply toothed, densely ciliate; stamens 4 mm. long. Female flower: ovary tuberculate, glabrous; styles 1.8 mm. long joined at the base, spreading, bifid for ⅓, with thickened apices. Capsule exserted on a pedicel to 6 mm. long, 3-lobed, 4 × 5 mm., almost smooth with faint indications of a few tubercles, capsule wall very thick. Seeds ovoid, 3.2 × 2.5 mm., grey, obscurely wrinkled; caruncle whitish, 0.5 mm. across.

UGANDA. Mbale District: Elgon, Bulambuli, 11 Nov. 1933, *Tothill* 2275!
KENYA. Naivasha District: S. Kinangop, Elephant Mt., 5 June 1968, *Mwangangi* 1013!; N. Nyeri District: Mt. Kenya, Sirimon Track, 6 Apr. 1975, *Hepper, Field & Townsend* 4896!; Masai District: Enesambulai Valley, 2 June 1969, *Greenway & Kanuri* 13644!
TANZANIA. Mbeya District: NW. Kitulo [Elton] Plateau, 11 Nov. 1931, *R.M. Davies* E.52!; Rungwe District: Bundali Hills, near Tewe, 7 Nov. 1966, *Gillett* 17592! & Poroto Mts., near Kikondo Hill, 29 May 1980, *Hooper & Townsend* 1737!
DISTR. U 3; K 3, 4, 6; T 7; not known elsewhere
HAB. In forest clearings, often in damp situations, and in the bamboo zone; 1980–3350 m.

35. E. usambarica *Pax* in E.J. 19: 122 (1894); N.E. Br. in F.T.A. 6(1): 538 (1911); T.T.C.L.: 212 (1949); K.T.S.: 202 (1961). Type: Tanzania, Usambara Mts., *Holst* 660 (B, holo. †)

Glabrous shrub to 4.5 m. high, with long thin often subpendent branches, rebranching mostly dichotomously. Leaves subsessile or shortly petiolate, oblanceolate or obovate, to 12 × 3 cm., base cuneate tapering gradually to the base of the petiole, apex acute and markedly apiculate, margin entire, dark green and slightly fleshy. Cymes in terminal

3–5-branched umbels, with (primary) rays to 3 cm. long, sometimes forking once only; bracts subcircular, 6–10 mm. long, apex obtuse to rounded, markedly apiculate. Central cyathium of the umbel on a peduncle to 2 mm. long, the rest sessile to subsessile, 3 × 8 mm., with cup-shaped involucres; glands 4, or 5 on cyathia developing only ♂ flowers (usually the central one), subcircular to transversely elliptic, 1.5–2.5 mm. broad, outer margin rounded and entire to truncate and shallowly crenulate; involucral lobes subquadrate, 1.2 mm. long, apex deeply 2-lobed, lower margins shortly ciliate. Male flowers: bracteoles fan-shaped, deeply divided, feathery; stamens 4 mm. long. Female: styles 2.5 mm. long, joined at the base, with spreading very distinctly bifid slightly thickened apices. Capsule exserted on a pedicel 6(–10) mm. long, deeply 3-lobed, 6 × 8 mm. Seeds ovoid, 3.4 × 3 mm., grey, surface lightly ridged and wrinkled; caruncle 0.7 mm. across, cream.

subsp. **usambarica**

Leaves sessile, oblanceolate, to 12 × 2 cm. Cymes unbranched.

KENYA. Teita District: Teita Hills, 6 Feb. 1953, *Bally* 8776! & Mbololo Hill, 17 Nov. 1969, *Bally* 13599!
TANZANIA. Lushoto District: W. Usambara Mts., Mazumbai Tea Estate, 27 Apr. 1975, *Hepper & Field* 5190!; Morogoro District: Morogoro–Lupanga Peak track, 16 Aug. 1951, *Greenway & Eggeling* 8609!; Iringa District: Dabaga Highlands, Kibengu, 13 Feb. 1962, *Polhill & Paulo* 1458!
DISTR. **K** 7; **T** 2, 3, 6, 7; N. Malawi
HAB. Understorey of fairly open montane forest, and in riverine forest at lower altitudes; (45–)1000–2400 m.

SYN. *E. membranacea* Pax in P.O.A. C: 242 (1895). Type: Tanzania, Usambara Mts., Lutindi, *Holst* 3318 (B, holo.†, K, iso.!)

subsp. **elliptica** *Pax* in E.J. 34: 374 (1904). Type: Tanzania, W. Usambara Mts., near Mbalu, *Engler* 1439 (B, holo.†)

Leaves petiolate, obovate, to 6 × 3 cm.; petiole to 5 mm. long. Cymes forking once.

TANZANIA. Lushoto District: W. Usambara Mts., Mkusi, 23 Feb. 1947, *Greenway* 7930! & Shagayu Forest, May 1953, *Procter* 213! & N. of Matondwe Hill at head of Kwai valley, 12 June 1953, *Drummond & Hemsley* 2908!
DISTR. **T** 3; known only from W. Usambara Mts.
HAB. Understorey shrub of upland forest; 1600–2000 m.

4. Subgen. **Eremophyton**

Erect branching annual or perennial herbs, with stems sometimes woody or ± fleshy. Leaves petiolate; stipules modified as glands or sometimes vestigial. Cymes axillary or 3(–5) in terminal umbels, branching dichotomously. Bracts usually similar to the leaves, or occasionally ± deltoid. Cyathia bisexual, with 4 (rarely 2 or 5) entire involucral glands. Stamens just exserted from the involucre. Capsule subglobose or oblong. Seeds 4-angled (sometimes obscurely so), with dorsally pointed apex, variously ornamented in horizontal ridges and grooves, with or without a caruncle.

4a. Sect. **Pseudacalypha**

Involucral glands 4, entire, spreading, the surface sometimes hairy, or occasionally 2 and ± tubular. Capsule usually subglobose and shortly exserted from the involucre. Seeds conical, smooth or sculptured, without a caruncle.

1. Involucral glands 2, ± tubular . . . . *41. E. longituberculosa*
   Involucral glands 4, spreading 2
2. Glands hairy on the upper surface 3
   Glands glabrous on the upper surface 4
3. Leaves ovate to ovate-lanceolate; seed-ridges smooth . . . . *42. E. acalyphoides*
   Leaves linear; seed-ridges minutely wrinkled . . . . . *43. E. perangustifolia*
4. Whole plant pilose with long spreading hairs; styles erect, to 2.5 mm. long . . . . *40. E. crotonoides*
   Plant glabrous or only sparsely pilose; styles spreading, to 1.5 mm. long 5

5. Seeds smooth, shiny black . . . .   *37. E. lutosa*
   Seeds tuberculate, dull greyish brown to black 6
6. Styles shortly bifid; seeds sharply angled, surface warty and
   angles crested . . . .   *39. E. lophiosperma*
   Styles entire; seeds obscurely angled, surface with small
   tubercles 7
7. Capsule with short adpressed hairs . . . .   *36. E. systyloides*
   Capsule with long spreading hairs . . . .   *38. E. benthamii*

36. E. systyloides *Pax* in E.J. 19: 121 (1894); N.E. Br. in F.T.A. 6(1): 520 (1911); W.F.K.: 37 (1948); E.P.A.: 459 (1958); U.K.W.F.: 222 (1974). Type: Zanzibar I., Mkokotoni [Kokotoni], *Hildebrandt* 1041 (B, holo.†, K, Z, iso.!)

Annual herb to 1.5 m. high, stem often woody at the base, the whole plant sparsely covered with long spreading hairs. Leaves lanceolate to ovate-lanceolate, 5–11 × 1.3–2 cm., base cuneate, apex acute, margin serrate with small gland-tipped teeth, midrib prominent on the lower surface; petiole to 2.5 cm. long; glandular stipules purplish. Cymes axillary and in terminal 3-branched umbels, with primary rays to 15 cm. long; bracts subsessile. Cyathia sessile, 2.5 × 3.5 mm., with cup-shaped involucres, pilose; glands 4, transversely elliptic, 1 × 1.5 mm., green; lobes rounded, 0.7 mm, wide, finely ciliate. Male flowers: bracteoles fan-shaped, deeply laciniate with feathery tips; stamens 2.5 mm. long. Female flower: ovary densely hairy; styles ± 1 mm. long, joined at the base, spreading with somewhat flattened grooved apices. Capsule exserted on a slightly curved pedicel 2 mm. long, subglobose with truncate base and shallow grooves along the sutures, 4 × 5 mm., pubescent with short adpressed hairs. Seeds conical, very obscurely 4-angled, 3 × 2.5 mm., with acute apex, greyish black, with small tubercles in irregular horizontal lines. Fig. 82/4–6, p. 444.

UGANDA. Karamoja District: Kotido, 3 June 1940, *A.S. Thomas* 3694!; Mengo District: Kawanda, *Hazel* 382! & Kipayo, 25 Aug. 1913, *Dummer* 185!
KENYA. Machakos District: Kibwezi Plains, foot of Chyulu Hills, May 1938, *Bally* 85 (703) in *C.M.* 7704!
TANZANIA. Handeni District: Kideleko, 30 June 1965, *Archbold* 462!; Pangani District: Sekera [Sangare] Forest, 26 June 1965, *Faulkner* 3589!; Morogoro District: Kidatu, 11 Apr. 1970, *Batty* 1025!; Zanzibar I., Mkokotoni [Kokotoni], Oct. 1873, *Hildebrandt* 1041!
DISTR. U 1, 2, 4; K 4; T 2, 3, 5, 6, 8; Z; single specimens seen from Mozambique (Zambesia), Malawi (Shire R.) and Zambia (Gwembe valley)
HAB. On sandy soils amongst grass in open woodland and disturbed ground; 0–1525 m.

SYN. *E. holstii* Pax in E.J. 19: 121 (1894). Type: Tanzania, Usambara Mts., *Holst* 530 (B, holo.†)
E. volkensii Pax in P.O.A. C: 242 (1895). Type: Tanzania, Kilimanjaro, Marangu, *Volkens* 638a (B, holo.†)*

NOTE. This species appears to be very scattered in distribution beyond a concentration in the coastal regions of NE. Tanzania, where it is common from the Usambaras south to the Ulugurus.

37. E. lutosa *S. Carter* in K.B. 35: 416, fig. 2/A–C (1980). Type: Tanzania, Dodoma District, 11 km. from Itigi on Chunya road, *Greenway & Polhill* 11532 (K, holo.!)

Annual herb to 60 cm. high, glabrous except for short curved adpressed hairs on young growth. Leaves broadly ovate to 4.5 × 2.5 cm., base cuneate, apex acute, margin closely serrate with small glandular teeth, midrib prominent to winged on the lower surface; petiole to 2 cm. long, somewhat flattened; glandular stipules purplish. Cymes axillary and in terminal 3-branched umbels, with primary rays to 9 cm. long; bracts similar to the leaves. Cyathia subsessile, 2.5 × 3.5 mm., with cup-shaped shortly pubescent involucres; glands 4, transversely elliptic, 1 × 1.75 mm., dull yellow; lobes subquadrate, 1 mm. long, margin ciliate. Male flowers: bracteoles fan-shaped, with ciliate apices; stamens 2.5 mm. long. Female flower: ovary densely puberulous; styles 0.5 mm. long, spreading, joined at the base, shortly bifid. Capsule shortly exserted on a puberulous pedicel 2 mm. long, subglobose, with shallow longitudinal grooves along the sutures, base truncate, 5 × 5.5 mm., puberulous with short adpressed hairs. Seeds ovoid, very obscurely 4-angled, 3.25 × 2.25 mm., apex obtuse, smooth, shiny black, with a white granular patch below the apex.

* *Volkens* 638, the same species from the same locality is in BM.

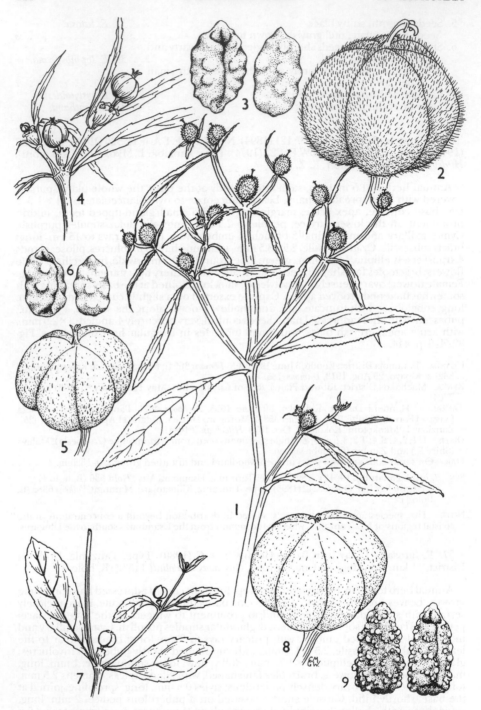

FIG. 82. *EUPHORBIA CROTONOIDES* — **1**, fruiting branch, × ⅔; **2**, capsule, × 6; **3**, seeds, × 6. *E. SYSTYLOIDES* — **4**, part of fruiting branch, × 3; **5**, capsule, × 6; **6**, seeds, × 6. *E. LOPHIOSPERMA* — **7**, part of fruiting branch, × ⅔; **8**, capsule, × 6; **9**, seeds, × 6. 1, from *Polhill & Paulo* 1941; 2, 3, from *Beesley* 118; 4–6, from *Faulkner* 3583; 7–9 from *Gillett* 20180. Drawn by Christine Grey-Wilson.

TANZANIA. Shinyanga, Apr.–May 1932, *Bax* 103!; Singida District: 19 km. from Singida on Itigi road, 28 Mar. 1965, *Richards* 20012!; Dodoma District: Great North Road, 25 km. N. of Dodoma, 21 Apr. 1962, *Polhill & Paulo* 2111!

DISTR. **T** 1, 5; not known elsewhere
HAB. Black clay soil amongst grass in scattered *Acacia* bushland; 100–1500 m

38. E. benthamii *Hiern* in Cat. Afr. Pl. Welw. 1: 943 (1900); N.E. Br. in F.T.A. 6(1): 40 (1911). Type: Angola, *Welwitsch* 283 (BM, holo.!, K, iso.!)

Annual herb to 1 m. high, stem often woody at the base, the whole plant sparsely covered with long spreading white hairs. Leaves ovate-lanceolate to lanceolate, to 10 × 3.5 cm., base cuneate, apex acute, margin subentire to serrate with minute gland-tipped teeth, midrib prominent on the lower surface; petiole to 5 cm. long; glandular stipules purplish. Cymes axillary and in terminal 3-branched umbels, with primary rays to 12 cm. long; bracts subsessile, usually reflexed. Cyathia sessile, 2.5 × 4 mm., with cup-shaped involucres, pilose; glands 4, transversely elliptic, 1 × 1.5 mm., green, becoming reddish brown; lobes rounded, 0.7 mm. in diameter, ciliate. Male flowers: bracteoles fan-shaped, feathery; stamens 2.5 mm. long. Female flower: ovary densely hairy; styles 1 mm. long, joined at the base, spreading, thickened, with grooved apices. Capsule barely exserted on a pilose pedicel 2 mm. long, subglobose, with truncate base, 4 × 5 mm., pilose with long spreading hairs. Seeds conical, 4-angled, 3 × 2 mm. apex acute, brownish black, surface irregularly tubercled in 3–4 distinct (rarely obscure) horizontal ridges.

TANZANIA. Iringa District: Iheme, 23 Feb. 1962, *Polhill & Paulo* 1589!; Songea District: between Peramiho and Wuwawesi [Wuwawezi] R., 21 Feb. 1956, *Milne-Redhead & Taylor* 8847! & Luhimba R., 7 May 1956, *Milne-Redhead & Taylor* 8847a!
DISTR. **T** 7, 8; Angola, Zimbabwe, Zambia and Malawi
HAB. Red sandy soil in grass amongst mixed dry scrub; 960–1750 mm.

NOTE. The fruiting specimen, *Milne-Redhead & Taylor* 8847a, differs from typical *E. benthamii* by its almost completely smooth seeds. Since the inflorescences of the other two specimens seen are immature, the Tanzanian collections may prove to represent a distinct taxon when more material has been obtained.

39. E. lophiosperma *S. Carter* in K.B. 39: 651 (1984). Type: Kenya, Machakos District, Kibwezi, *Scheffler* 77 (K, holo.!, BM, iso.!)

Annual herb to 1 m. high, stem woody at the base; branches longitudinally ridged, glabrous or with a few scattered spreading hairs. Leaves glabrous or with a few scattered hairs on both surfaces, ovate to 7.5 × 4 cm., base cuneate, apex acute, margin serrate, sometimes markedly so with gland-tipped teeth, midrib winged on the lower surface; petiole flattened, to 3 cm. long. Cymes axillary and in terminal 3-branched umbels, with primary rays to 10 cm. long; bracts subsessile, similar to the leaves. Cyathia sessile or very shortly pedunculate, 2.5 × 3.5 mm., with cup-shaped involucres and a few short hairs below the glands; glands 4, transversely elliptic, 1.5 mm. wide, yellow becoming brown; lobes rounded, 0.7 mm. wide, with finely ciliate margins. Male flowers: bracteoles fan-shaped, deeply laciniate, apices ciliate; stamens 3 mm. long. Female flower: ovary shortly pubescent; styles 1.5 mm. long, joined for ⅓, apices spreading, shortly bifid. Capsule exserted on a slightly curved pilose pedicel ± 3.5 mm. long, subglobose with truncate base, 5 × 6 mm., thinly pilose with slightly spreading hairs. Seeds conical, sharply 4-angled, 3.8 × 2.8 mm., apex acute, greyish brown, surface warty forming two horizontal ridges and crested angles. Fig. 82/7–9.

UGANDA. Karamoja District: Chosan, 12 June 1959, *Symes* 552! & Moroto, Kasuneri Estate, July 1972, *J. Wilson* 2147!; Mengo District: Mawokota, 1905, *Dawe* 238!
KENYA. Nairobi, 9 Aug. 1971, *Mwangangi & Msafiri* 1603!; Machakos District: Kibwezi, 5 Jan. 1911, *Scheffler* 365!; Kitui District: 2 km. E. of Mwingi on Nairobi–Garissa road, 8 May 1974, *Gillett & Gachathi* 20485!
TANZANIA. Mwanza District: Ilemera, Rumara, 8 Apr. 1953, *Tanner* 1359!; Masai District: Eluanata, 3 May 1965, *Leippert* 7521!; Dodoma District: Itigi–Singida road, 26 Mar. 1965, *Richards* 19922!
DISTR. **U** 1, 4; **K** 1, 4, 7; **T** 1–3, 5; southern Ethiopia
HAB. Usually on sandy soils in grass amongst scattered trees and shrubs; (0–)600–1700 m.

SYN. *E. systyloides* Pax var. *pedunculata* N.E. Br. in F.T.A. 6(1): 520 (1911). Types: Uganda, Mengo District, Mawokota, *Dawe* 238 & Kenya, Machakos District, Kibwezi, *Scheffler* 77 (both K, syn.!)
 E. systyloides Pax var. *lata* N.E. Br. in F.T.A. 6(1): 521 (1911). Types: Tanzania, Masai District, Sonjo–Sale, *Merker* 583 & Moshi, *Merker* 605 (both B, syn.†)

NOTE. This species represents an intermediate between *E. crotonoides* Boiss. and *E. systyloides* Pax, but can be readily distinguished from both by its sculptured sharply angled seeds, broadly ovate leaves, and usually distinctly bifid style apices. A few specimens from southern Ethiopia possess narrower leaves and may represent a local variation. Two specimens from the Kenya coast (Mombasa District, Likoni, *Ossent* 234 & Kilifi District, Kibarani, *J.W. Jeffery* K841) are immature and are included here only tentatively, on the basis of their spreading styles with bifid tips.

E. platypoda Pax in Jahres. Schles. Ges. 89, Abt. 2b: 2 (1911) may belong here also, the description matching well except for the carunculate seed. Despite this Pax related the species to the ecarunculate *E. volkensii* Pax (syn. of *E. systyloides* Pax). As the type (Tanzania, Moshi, *Winkler* 4022, B, holo.†) no longer exists, its true identity must remain in doubt.

40. E. crotonoides *Boiss.* in DC., Prodr. 15(2): 98 (1862); N.E. Br. in F.T.A. 6(1): 518 (1911); W.K.F.: 37 (1948); F.P.S. 2: 73 (1952); E.P.A.: 444 (1958); U.K.W.F.: 222 (1974); S. Carter in K.B. 39: 650 (1984). Type: Sudan, Kordofan, *Kotschy* 419 (K, iso.!)

Annual much-branched somewhat fleshy herb to 50(–100) cm. high, stem often woody below; branches and upper part of the stem longitudinally ridged to distinctly winged, the whole plant covered in long spreading white hairs. Leaves reflexed, linear-lanceolate to ovate, 3–14 × 0.5–6 cm., base tapering into a 2-winged petiole, apex acute, margin subentire to markedly serrate, with gland-tipped often red-tinged teeth, midrib winged on the lower surface; petiole to 5 cm. long; glandular stipules dark red. Cymes axillary and in terminal 3-branched umbels forking many times, with primary rays to 12 cm. long; bracts narrower than the leaves, with petioles to 5 mm. long. Cyathia subsessile, 2.5 × 3.5 mm., with cup-shaped involucres; glands 4, transversely elliptic, 0.8 × 1.5 mm., yellow, turning brown, then red; lobes quadrangular, 1.75 mm. wide, margin ciliate. Male flowers: bracteoles deeply laciniate, apices ciliate; staments 3 mm. long. Female flower: ovary densely pilose; styles 2–2.5 mm. long, puberulous, joined to almost halfway, erect with apices spreading, thickened. Capsule exserted on a slightly curved pilose pedicel to 3 mm. long, subglobose, 6.5–7 mm. in diameter, with truncate base and shallow longitudinal grooves along the sutures, pilose with long spreading hairs. Seeds ovoid or conical, 4-angled, ± 4.5 × 2.5 mm., apex rounded or acute, brown to grey or reddish black, surface obscurely or distinctly tuberculate in irregular horizontal lines.

subsp. **crotonoides**

Herb to 50(–100) cm. high. Leaves lanceolate to ovate, 7–14 × 1.5–6 cm., margin markedly serrate. Seeds conical, distinctly 4-angled, 4.5 × 2.5 mm., apex acute and distinctly ventrally constricted, surface tuberculate. Fig. 82/1–3, p. 444.

UGANDA. Acholi & Karamoja Districts: Rom and Meriss, June 1930, *Liebenberg* 339 (Field no. 160)!; Karamoja District: 69 km. S. of Moroto, 13 Sept. 1956, *Bally & Hardy* 10814!; Teso District: Serere, Aug. 1932, *Chandler* 871!

KENYA. Northern Frontier Province: Moyale, 1 Sept. 1952, *Gillett* 13770!; Nakuru District: shores of Lake Elmenteita, 17 Aug. 1948, *Bogdan* 1870!; Masai District: between Sultan Hamud and Amboseli, 30 Dec. 1971, *Kokwaro* 2925!

TANZANIA. Arusha District: Ngare Nanyuki, Dec. 1965, *Beesley* 188!; Mpanda District: Lake Katavi, 11 Feb. 1962, *Richards* 16059!; Njombe District: Wangingombe, 29 Mar. 1962, *Polhill & Paulo* 1941!

DISTR. U 1, 3; K 1, 3, 4, 6, 7; T 1–5, 7; eastern Sudan and Ethiopia, northern Malawi and Zambia south to South Africa (Transvaal) and west to southern Angola and Namibia

HAB. Usually on sandy stony often disturbed soils amongst grass in open woodland or scattered bushland; 350–2050 m.

SYN. *E. holstii* Pax var. *hebecarpa* Pax in E.J. 34: 374 (1904). Types: Kenya, Nakuru, *Engler* 2019 & Masailand, *Merker* (both B, syn.†)
E. systyloides Pax var. *hebecarpa* (Pax) N.E. Br. in F.T.A. 6(1): 521 (1911); E.P.A.: 459 (1958)

VAR. In the northern half of its distribution the leaves of this widely spread species are usually all narrowly lanceolate. In Tanzania southwards growth is usually more lush, producing larger ovate-lanceolate leaves, although those on the flowering branches (bracts) are more lanceolate. The size of the glandular teeth on the leaf margins varies considerably throughout the species range, sometimes reaching 1.5 mm. in length on well-grown specimens. The density of the pubescence also varies, as does the colour of the seeds from brown to black.

subsp. **narokensis** *S. Carter* in K.B. 39: 651 (1984). Type: Kenya, Masai District, Narok, *Glover et al.* 2270 (K, holo.!, EA, iso.)

Stunted herb with a woody base, to 45 cm. high. Leaves narrowly lanceolate, to 5 × 0.5 cm., margin entire to very obscurely toothed. Seeds ovoid, obscurely 4-angled, 3.5 × 2.8 mm., apex rounded, surface very shallowly tuberculate.

KENYA. Masai District: Narok, Entasekera Escarpment, 12 July 1961, *Glover et al.* 2270! & Oltarakwai, 18 Aug. 1961, *Glover et al.* 2528!
DISTR. K 6; known only from these 2 collections
HAB. Grassy hillside and forest edge; 2500 m.

41. E. longituberculosa *Boiss* in DC., Prodr. 15(2): 85 (1862); N.E. Br. in F.T.A. 6(1): 558 (1911); E.P.A.: 451 (1958); S. Carter in K.B. 40: 810 (1985). Type: Ethiopia, Eritrea, *Schimper* 2307 (BM, K, iso.!)

Glabrous perennial herb to 30 cm. high, with a thick tap-root merging into a swollen ± succulent tuberculate or occasionally smooth stem to 8 cm. high and 4 cm. thick; branches radiating from the stem apex to 8 cm. long, tuberculate or smooth and ± succulent, sometimes with further thinner branches at the apices in groups of 3–6. Leaves subpetiolate to petiolate from the apices of the tubercles, linear-lanceolate to 50 × 8 mm., base cuneate, apex acute and minutely apiculate, margin entire; petiole to 1 cm. long; glandular stipules minutely filamentous and apparent only on young growth. Cymes clustered 2–6 at the apices of the branches, with primary rays to 6 cm. long, forking up to 5 times often trichotomously, semi-persistent; bracts sessile, the uppermost ovate ± 10 × 5 mm., base rounded, apex acute, the lower ones progressively larger and more similar to the leaves. Cyathia on peduncles 1–1.5 mm. long, ± 3 × 2.5 mm., with funnel-shaped involucres; glands 2, subquadrate, ± 2.5 × 2.5 mm., the outer margin shallowly crenulate, the lateral margins curled inwards and joined at the base to form an erect tube; lobes ± 1.8 × 1 mm., the upper margin deeply and finely toothed to almost halfway. Male flowers few: bracteoles ligulate, deeply divided; stamens to 6 mm. long. Female flower: styles ± 2.7 mm. long, joined almost to the minutely bifid tips. Capsule exserted on a pedicel to 9 mm. long, deeply 3-lobed with truncate base, ± 4.5 × 6 mm. Seeds obtusely trigonous-conical with truncate base, acute apex and 2 shallow horizontal constrictions, ± 3 × 2.5 mm., black to greyish brown with a speckled roughened surface.

KENYA. Northern Frontier Province (no further details), cult. Nairobi Nov. 1934, *Gardner* in *F.D.* 3354! & 5 km. S. of Mandera, 1971, *Classen* in *Bally* 14149! & Malka Dakacha, W. of Mandera, 3 Mar. 1974, *Bally & Carter* 16588, photos only!
DISTR. K 1; Ethiopia, Somalia and S. Yemen
HAB. In bare sandy rocky soils usually near watercourses; 300 m.

SYN. *Tithymalus braunii* Schweinf. in J.B. 1: 295 (Oct. 1863). Type: Ethiopia, Eritrea, *Schimper* 221 (B, holo.†, K. drawing!)

42. E. acalyphoides *Boiss.* in DC., Prodr. 15(2): 98 (1862) & Euph. Ic., t. 49 (1866); Hiern in Cat. Afr. Pl. Welw. 1: 950 (1900); N.E. Br. in F.T.A. 6(1): 517 (1911); F.P.S.2: 69 (1952); E.P.A.: 442 (1958); U.K.W.F.: 222 (1974); S. Carter in K.B. 39: 648 (1984). Types: Sudan, Kordofan, *Kotschy* 88 (K, isosyn.!) & Ethiopia, Eritrea, *Ehrenberg* (B, syn.†) & *Schimper* 1753 (BM, K, isosyn.!) & Tigre Province, *Schimper* 2294 (BM, K, isosyn.!)

Annual herb to 45 cm. high with spreading branches, the whole plant pilose including the flowering parts, with white spreading hairs; stems sometimes slightly swollen and succulent, stem and branches longitudinally grooved. Leaves ovate to ovate-lanceolate, to 4 × 2 cm., base cuneate, apex obtuse to rounded, margin entire, upper surface often glabrous; petiole to 25 mm. long; stipules vestigial. Cymes axillary, 1–2-forked; bracts subsessile, subcircular, to 5 mm. in diameter. Cyathia on peduncles to 8 mm. long, 1.75 × 3 mm. long, with funnel-shaped involucres; glands 4, transversely elliptic, 1 × 1.5 mm., reddish yellow, hairy on both surfaces; lobes circular, 0.75 mm. in diameter, densely ciliate. Male flowers: bracteoles very few, ligulate, apex ciliate; stamens 2.25 mm. long. Female flower: ovary densely pubescent; styles 2 mm. long, joined at the base, distinctly bifid at the apex, pubescent. Capsule exserted on a curved pilose pedicel 1.75 mm. long, 3-lobed, with truncate base, 3.5 × 4.5 mm. Seeds conical, with acute apex and 2 horizontal constrictions, 2.5 × 2 mm., black to blackish brown.

subsp. **acalyphoides**

Stem not noticeably swollen and succulent; mature capsule minutely papillose; seeds blackish brown with the surface minutely wrinkled.

KENYA. Meru District: 58 km. E. of junction with Marsabit road on Isiolo–Wajir road, 8 Dec. 1977, *Stannard & Gilbert* 818!; Masai District: S. of Olorgasailie, 22 Apr. 1960, *Verdcourt* 2757!; Teita District: Tsavo East National Park, 7 km. W. of Voi Gate, 19 Dec. 1966, *Greenway & Kanuri* 12798!

TANZANIA. Pare District: Kisiwani, 25 June 1942, *Greenway* 6493!
DISTR. **K** 1–4, 6, 7; **T** 2, 3; Arabia, eastern Sudan, Ethiopia and southern Somalia, with a disjunct population in Angola
HAB. Sandy soils in open *Acacia* bushland; 330–1160 m.

SYN. *E. acalyphoides* Boiss. var. *arabica* Boiss. in DC., Prodr. 15(2): 98 (1862). Type: Arabia, *Ehrenberg* (B, holo.†, K, iso.!)
   E. fodhliana Defl. in Bull. Soc. Bot. Fr. 43: 230 (1896). Type: Yemen, *Deflers* 494 (P, holo.!)
   E. incurva N.E. Br. in F.T.A. 6(1): 552 (1911); T.T.C.L.: 211 (1949). Type: Tanzania, Masai District, Sonjo–Sale plains, 18 Apr. 1904, *Merker* 579 (B, holo.†, K, drawing!)

NOTE. Subsp. *cicatricosa* S. Carter in K.B. 39: 648 (1984) occurs in arid regions, with very sparse vegetation, of Somalia and south-east Ethiopia, but so far has not been recorded from Kenya. It is a smaller plant, usually less than 30 cm. high, with horizontally spreading branches from a somewhat swollen stem bearing prominent leaf-scars. The leaves are ovate-lanceolate with a slightly crisped margin. It is more puberulous with finer hairs, the mature capsule is puberulous but not papillose, and the seeds are slightly smaller, black with a smooth surface.

43. E. perangustifolia S. *Carter* in K.B. 39: 649 (1984). Type: Kenya, Northern Frontier Province, near Ramu, *Gilbert & Thulin* 1423 (K, holo.!, EA, U, iso.)

Annual herb ± 30 cm. high, with spreading branches, the whole plant puberulous with short spreading hairs; stem and branches longitudinally grooved. Leaves linear to 30 × 3 mm., margin entire; petiole to 8 mm. long; stipules vestigial. Cymes axillary or terminal, forking once; bracts subsessile, ovate, to 3 mm. long. Cyathia on peduncles to 1.5 mm. long, 1.75 × 2.5 mm., with funnel-shaped involucres; glands 4, quadrangular, 1 × 1 mm., reddish, hairy on both surfaces; lobes obovate, ciliate, 1 mm. long. Male flowers: bracteoles linear, ciliate; stamens 2 mm. long. Female flower: ovary densely puberulous; styles 1.25 mm. long, joined at the base, minutely bifid, puberulous. Capsule exserted on a slightly curved pedicel 2.25 mm. long, 3-lobed, with truncate base, 3 × 3.5 mm., surface minutely papillose, and pubescent with short curved hairs. Seeds conical, apex acute, obscurely 4-angled, with 2 deep horizontal constrictions, 2 × 1.4 mm., black with minutely wrinkled surface.

KENYA. Northern Frontier Province: Ramu–Banissa road, 10 km. W. of junction with El Wak road, 4 May 1978, *Gilbert & Thulin* 1423!
DISTR. **K** 1; SE. Ethiopia
HAB. *Commiphora, Boswellia, Acacia* bushland on stony soil over limestone; 480–520 m.

4b. Sect. **Eremophyton**

Involucral glands 4(–5), spreading, entire. Capsule oblong, well exserted from the involucre on a reflexed pedicel. Seeds usually compressed dorsiventrally, oblong and wrinkled in vertical ridges and grooves, with a cap-like transverse caruncle.

1. Leaves toothed . . . . *45. E. agowensis*
   Leaves entire 2
2. Herbs to 30 cm. high, with semi-succulent stem; leaves
      lanceolate . . . . *44. E. pirottae*
   Woody-based herbs or shrubs; leaves obovate 3
3. Shrubby herbs to 75 cm. high, cyme-peduncles minutely
      hispid . . . . *46. E. trichiocyma*
   Woody shrub to 1.5(–3) m. high, with shiny red bark; cyme-
      peduncles glabrous . . . . *47. E. polyantha*

44. E. pirottae *Terrac.* in Ann. Ist. Bot. Roma 5: 97 (1894); N.E.Br. in F.T.A. 6(1): 559 (1911); E.P.A.: 454 (1958). Type: Eritrea, Midir I., *Terracciano* 204 (FT, holo.!)

Sparingly branched annual herb to 30 cm. high, stem slightly swollen at the base and semi-succulent to 1 cm. thick, the whole plant ± sparsely pilose, with spreading white hairs to 1 mm. long, glabrescent at maturity. Leaves ovate-lanceolate to linear-lanceolate, to 8 × 1 cm., base cuneate, apex rounded, margin entire, midrib prominent on the lower surface; petiole to 1 cm. long; stipules absent or glandular and rudimentary. Cymes axillary and in terminal 3-branched umbels, with primary rays to 6 cm. long; bracts similar to the leaves. Cyathia on peduncles 1 mm. long, 2.2 × 2.5 mm., with cup-shaped involucres, sparsely pilose; glands 4, spreading, subcircular, 1 mm. wide, bright reddish green; lobes subquadrate, 1 mm. long, margin toothed, ciliate. Male flowers: bracteoles linear, deeply toothed, ciliate; stamens 2 mm. long. Female flower: ovary pilose; styles 1.4 mm. long, joined at the base, bifid to halfway. Capsule exserted on a reflexed pilose pedicel 6.5 mm. long, oblong-trigonous, 4.8 × 4 mm., sparsely pilose. Seeds oblong, obscurely 4-angled, 4 × 2 mm., greyish brown, lightly wrinkled, with a yellow caruncle 1.6 mm. wide.

KENYA. Northern Frontier Province: 11 km. S. of Kangetet, 21 May 1970, *Mathew* 6322!; Meru District: Isiolo–Wajir road, 15 km. W. of Kula Mawe Police Post, 8 Dec. 1977, *Stannard & Gilbert* 817!; Machakos District: 17.5 km. E. of Mtito Andei, 18 Jan. 1961, *Greenway* 9764!
TANZANIA. Pare District: Kiruru, May 1928, *Haarer* 1438!
DISTR. **K** 1, 4, 6; **T** 3; Somalia, Ethiopia, Saudi Arabia (one collection)
HAB. On sandy soils in open bushland; 450–1650 m.

SYN. *E. gorinii* Chiov., Fl. Somala 2: 401 (1932); E.P.A.: 446 (1958). Type: Somalia, El Ualud, *Gorini* 466 (FT, holo.!)

NOTE. With its very widely scattered distribution, this insignificant species is probably more common than the few specimens seen appear to indicate.

45. E. agowensis *Boiss.* in DC., Prodr. 15(2): 70 (1862); N.E. Br. in F.T.A. 6(1): 540 (1911); F.P.S. 2: 76 (1952); E.P.A.: 443 (1958); U.K.W.F.: 222 (1974). Types: Ethiopia, Agau [Agow], *Schimper* 1414 & 2150 (BM, K, isosyn.!) & Dhalac Is., *Ehrenberg* (B, syn.†)

Annual or short-lived perennial herb with a woody base, 40–100 cm. high, glabrous or usually with the upper branches and young growths shortly pubescent. Leaves ovate to linear-lanceolate, to 8 × 0.6–4 cm., base cuneate, apex rounded, margin indistinctly toothed but often with 1–2 pairs of large gland-tipped teeth at the base, upper surface glabrous, lower surface with scattered hairs and a prominent midrib; lamina tapering into a winged petiole to 5 mm. long; glandular stipules minute. Cymes axillary and in terminal 3-branched umbels with primary rays to 5 cm. long; bracts sessile, often reflexed. Cyathia subsessile, 1.8 × 1.8 mm., with cup-shaped involucres, puberulous; glands 4(–5), transversely elliptic, ± 1 mm. wide, green becoming brown; lobes subquadrate, 0.5 mm. long, margin deeply toothed. Male flowers: bracteoles strap-shaped, deeply divided, ciliate; stamens 2.3 mm. long. Female flower: ovary densely pubescent; styles 1.6 mm. long, joined at the base, erect with spreading apices bifid for ⅛. Capsule exserted on a reflexed pedicel 4 mm. long, oblong-trigonous, 3.5 × 3 mm., pubescent with short adpressed hairs. Seeds oblong, ± flattened, 2.8 × 1.6 mm., wrinkled in 4 longitudinal rows of large tubercles on each side, whitish-grey, with a yellow caruncle 1.2 mm. wide.

var. **agowensis**

Leaves ovate to linear-lanceolate, but the bracts always ovate, to 7 × 4 cm., with marginal teeth at the base rarely present; basal leaves often petiolate. Fig. 83/1–3, p. 450.

UGANDA. Karamoja District: Rupa, Aug. 1954, *J. Wilson* 151!
KENYA. Meru District: Isiolo–Wajir road, 6 km. E. of junction with Marsabit road, 7 Dec. 1977, *Stannard & Gilbert* 805!; Kitui District: Kibwezi–Kitui road, 11 km. after Athi River, 22 Apr. 1969, *Napper & Kanuri* 2040!; Masai District: Olorgasailie, 24 May 1960, *Bally* 12266!
TANZANIA. Masai District: 14.5 km. S. of Longido on Namanga–Arusha road, 2 Jan. 1962, *Polhill & Paulo* 1026! & Masailand, 14.5 km. along the Moshi track, 7 Jan. 1969, *Richards* 23672! & Engaruka road, 24 Feb. 1970, *Richards* 25505!
DISTR. **U** 1; **K** 1, 4, 6; **T** 2; Somalia, Ethiopia, Eritrea, N. Yemen and India
HAB. In grass, usually in the shade of bushes in small thickets; 650–1250 m.

var. **pseudoholstii** *(Pax) Bally & S. Carter* in K.B. 39: 650 (1984). Type: Ethiopia, Boran, *Ellenbeck* 2067 (B, holo.†, K, drawing!)

Leaves and bracts linear-lanceolate, to 8 × 1 cm., often with 1–2 pairs of large gland-tipped marginal teeth at the base, all subsessile. Fig. 83/4, p. 450.

KENYA. Northern Frontier Province: 55 km. S. of Mado Gashi on Garissa road, 12 Dec. 1977, *Stannard & Gilbert* 975! & 30 km. from Ramu on Malka Mari road, 8 May 1978, *Gilbert & Thulin* 1552!; Masai District: Olorgasailie, 24 May 1960, *Bally* 12265!
DISTR. **K** 1, 2, 4, 6, 7; Ethiopia, northern Somalia, Angola, N. Yemen and India
HAB. On sandy soils, often amongst grass in open woodland; 60–1300 m.

SYN. *E. pseudoholstii* Pax in E.J. 33: 287 (1903); N.E. Br. in F.T.A. 6(1): 521 (1911); E.P.A.: 456 (1958)
E. pearsonii N.E. Br. in F.T.A. 6(1): 1038 (1913). Type: Angola, *Pearson* 2071 (K, holo.!)

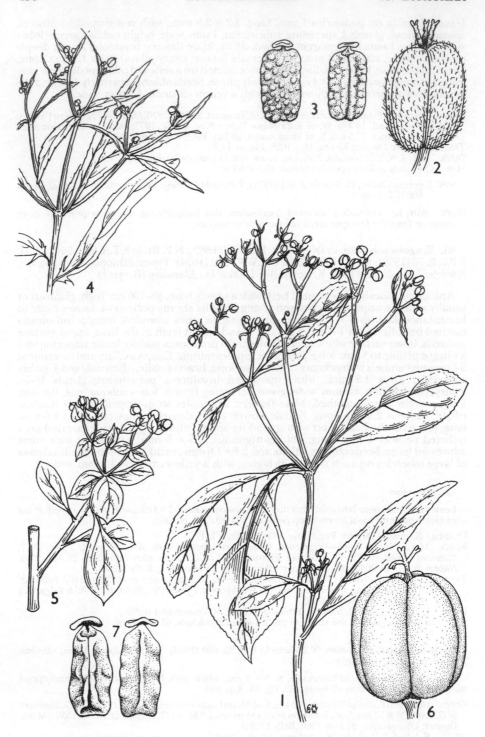

FIG. 83. *EUPHORBIA AGOWENSIS* var. *AGOWENSIS* — 1, fruiting branch, × ⅔; 2, capsule, × 6; 3, seeds, × 6. Var. *PSEUDOHOLSTII* — 4, flowering branch, × ⅔. *E. POLYANTHA* — 5, fruiting branch, × ⅔; 6, capsule, × 6; 7, seeds, × 6. 1–3 from *Polhill & Paulo* 1026; 4, from *Gilbert & Thulin* 1552; 5, 6, from *Verdcourt* 3875. Drawn by Christine Grey-Wilson.

VAR. The narrow-leaved var. *pseudoholstii* appears more frequently under drier habitat conditions than the broader-leaved var. *agowensis,* indicating an environmental influence on leaf-shape. Nevertheless both forms can be found growing together throughout the species distribution, so varietal distinction only is justified.

NOTE. Despite their disjunct distributions, N.E. Brown's distinctions between *E. pearsonii* and *E. pseudoholstii* do not stand up under the examination of more material than the three specimens he had for comparison.

46. E. trichiocyma S. *Carter* in K.B. 40: 809 (1985). Type: Tanzania, Mpwapwa, *Hornby* 642 (K, holo.!, EA, iso.!)

Shrubby, probably short-lived perennial herb to 75 cm. high, woody at the base. Leaves petiolate, ovate to 8 × 4 cm., base cuneate, apex acute, margin entire, young leaves minutely hispid especially on the lower surface, mature leaves with a few hairs towards the base on the lower surface; petiole to 8 mm. long, puberulous; glandular stipules minute. Cymes (detached on the only specimen seen, probably in terminal umbels) with minutely hispid rays, the primary ones apparently ± 1.2 cm. long, forking 3 times; bracts obovate, ± 5 × 4.5 mm., base cuneate, apex apiculate, glabrous. Cyathia sessile, ± 3 × 3.5 mm., with cup-shaped involucres; glands 4–5, transversely elliptic, 0.5 × 1–1.5 mm.; lobes subquadrate, ± 1 × 1 mm., upper margin deeply toothed. Male flowers: bracteoles strap-shaped, apex deeply divided, ciliate; stamens 3 mm. long. Female flower: styles 2 mm. long, joined at the base, apices spreading and distinctly bifid. Capsule (immature) exserted on a minutely hispid pedicel 4 mm. long, deeply 3-lobed, 4 × 5 mm., glabrous. Seed (immature) ovoid with dorsally pointed apex, slightly conpressed, 3.2 × 2.8 mm., surface obscurely wrinkled (probably in longitudinal lines); caruncle 1.2 mm. wide.

TANZANIA. Mpwapwa, 3 Mar. 1933, *Hornby* 642!
DISTR. T 5; known only from this collection
HAB. Deciduous bushland; 790 m.

NOTE. This species is related to *E. agowensis* Boiss., but is a larger shrubby plant with entire leaves, minutely but distinctly hispid cyme-branches, glabrous instead of pubescent capsules, and wider ovoid seeds which, from the immature ones seen, are not so obviously flattened. The collectors' notes state that the plant is common.

47. E. polyantha *Pax* in E.J. 43: 87 (1909); N.E.Br. in F.T.A. 6(1): 543 (1911); T.T.C.L.: 211 (1949); K.T.S.: 200 (1961); U.K.W.F.: 223 (1974); S. Carter in K.B. 40: 809 (1985). Type: Tanzania, edge of Rift [probably Masai District, W. of Lake Natron], *Merker* 578 (B, holo.†)

Perennial shrubby herb to 1.5(–3) m. high, with woody stem and branches and shiny red-brown bark; young branches usually hispid. Leaves obovate, to 35 × 23 mm., base cuneate, apex rounded, margin entire, lower suface sometimes hispid especially along the midrib; petiole to 5 mm. long, hispid; glandular stipules vestigial. Cymes in terminal 3(–4)-branched umbels with primary rays to 3(–5) cm. long, 1–2-forked; bracts sessile, deltoid, ± 8 × 8 mm., apex acute. Cyathia sessile, 2.5 × 3.5 mm., with broadly funnel-shaped involucres; glands 4, transversely elliptic, 1.5 mm. broad, yellow becoming dark red; lobes broadly rounded, 0.8 mm. long, margin denticulate, outer surface minutely hispid. Male flowers: bracteoles ligulate, deeply 2–3-lobed, margin ciliate; stamens 3.5 mm. long. Female flower: styles 2 mm. long, joined to halfway, erect with spreading shortly bifid apices. Capsule exserted on a pedicel 5 mm. long, oblong-trigonous, 5.5 × 4.5 mm., becoming reddish. Seeds oblong, flattened, 3.5 × 2.3 mm., apex acute, pale brownish grey, lightly wrinkled, with a pale yellow caruncle 1.7 mm. wide. Fig. 83/5–7.

KENYA. Northern Frontier Province: Mathews Range, Ol Doinyo Lengio, 19 Dec. 1958, *Newbould* 3535!; Machakos District: near Kanga, mile 139 from Mombasa on Nairobi road, 10 Jan. 1964, *Verdcourt* 3873!; Masai District: 40 km. from Magadi on Nairobi road, 31 May 1958, *Verdcourt, Baring & Williams* 2196!
TANZANIA. Lushoto District: 8 km. SE. of Mkomazi, 30 Mar. 1953, *Drummond & Hemsley* 2300!
DISTR. K 1, 2, 4, 6, 7; T 2, 3; central Somalia and SE. Ethiopia
HAB. On sandy stony soils in dry deciduous bushland; 200–1600 m.

5. Subgen. **Trichadenia**

Trees or shrubs, or perennial herbs with a fleshy rootstock and often fleshy stems. Leaves entire, sessile or petiolate; stipules apparently absent, or filamentous and soon deciduous, or modified as glands. Cyathia in terminal umbellate cymes; bracts sessile, ± deltoid, or leaf-like below the umbel. Involucres bisexual; glands 4 or 5, entire, or more usually crenulate, 2-horned or with finger-like processes on the outer margin; lobes 5. Stamens clearly exserted from the involucre; bracteoles included, apices feathery. Perianth of ♀ flower reduced to a rim below the ovary, or obvious and 3-lobed in those species with involucral glands producing marginal processes; styles with bifid apices. Capsule exserted, relatively large, deeply 3-lobed to subglobose. Seeds ovoid to subglobose, without a caruncle.

1. Involucral glands entire, crenulate or with 2 horns 2
   Involucral glands with 6–12 finger-like processes on the
   outer margin 8
2. Leaves obovate, to 20 × 6 cm. 3
   Leaves linear to oblanceolate, less than 1.5 cm. wide 5
3. Stems usually unbranched, to 30 cm. high . . . . . *50. E. platycephala*
   Shrubby herbs to ± 3m. high 4
4. Leaves glabrous . . . . *48. E. engleri*
   Leaves pilose on the lower surface . . . . *49. E. goetzei*
5. Plants glabrous; involucral glands with 2 long
   horns . . . . *51. E. dolichoceras*
   Plants puberulous; involucral glands entire, or bilobed but
   without long horns 6
6. Involucral glands distinctly bilobed . . . . . . *52. E. dilobadena*
   Involucral glands entire 7
7. Leaves sparsely pubescent on the lower surface; capsules
   densely papillose-pilose . . . . *53. E. ruficeps*
   Leaves ± densely pubescent; capsule sparsely pilose *54. E. arrecta*
8. Herb to 30 cm. high . . . . *55. E. taboraensis*
   Shrubs or trees 1–9 m. high 9
9. Cyme-branches and cyathia minutely puberulous *57. E. grantii*
   Cyme-branches and cyathia glabrous 10
10. Shrub 1–2 m. high; mature leaves obovate-
    lanceolate to 7 cm. wide; inner surface of involucral
    lobes glabrous . . . . *56. E. pseudograntii*
    Straight-stemmed shrub or tree 2–7 m. high; inner
    surface of involucral lobes densely puberulous *58. E. friesiorum*

48. E. engleri *Pax* in P.O.A. C: 24 (1895); N.E. Br. in F.T.A. 6(1): 531 (1911); T.T.C.L.: 210 (1949); E.P.A.: 446 (1958); K.T.S.: 197 (1961); U.K.W.F.: 222 (1974). Type: Tanzania, Moshi District, Kilimanjaro, Marangu, *Volkens* 847 (B, holo.†)

Glabrous shrubby perennial herb to 3 m. high, woody at base. Leaves obovate, to 20 × 5.5 cm., base cuneate tapering gradually towards the base of the petiole, apex acute and markedly apiculate, margin entire, lamina slightly fleshy, dark green and shiny above with the midrib prominent on the lower surface; petiole to 1 cm. long. Cymes in 3–6-branched umbels, with primary rays to 8(–13) cm. long, forking 3–5 times; bracts subcircular, ± 1 cm. long, apex markedly apiculate. Cyathia on peduncles to 4(–6) mm. long, 3 × 5–7 mm., with cup-shaped involucres; glands 4, transversely rectangular, 1.5 × 2–2.5 mm., outer margin shallowly 2-lobed to crescent-shaped with lobes to 1.2 mm. long; involucral lobes deltoid, 1 mm. long, margin ciliate, apex shortly 2-lobed. Male flowers: bracteoles fan-shaped, divided to the base many times into feathery filaments; stamens 4 mm. long. Female flowers: styles 2.5 mm. long, joined for ⅓, with spreading apices bifid for ⅕. Capsule exserted on a pedicel to 1.2 mm. long, deeply 3-lobed, 5–5.5 × 7–8 mm. Seeds ovoid, 3.5 × 3 mm., surface dark grey, lightly ridged and wrinkled. Fig. 84/1–4.

UGANDA. Mbale District: Elgon, Suam Ridge, 23 Dec. 1954, *Dale* U. 849!
KENYA. Nakuru District: W. Mau, Ndoinet R. valley, 28–30 Mar. 1970, *Gillett* 19053!; N. Nyeri District: Mount Kenya National Park, Naro Moru track, 10 Apr. 1977, *Hooper & Townsend* 1686!; Teita Hills, Ngangao, 8 km. NNE. of Ngerenyi, 15 Sept. 1953, *Drummond & Hemsley* 4359!
TANZANIA. Masai District: Oldeani Crater, 20 June 1969 *Carmichael* 1694!; Arusha National Park, Mt. Meru, 6 Nov. 1969, *Richards* 24628!; Moshi District: W. Kilimanjaro, 1 Jan. 1967, *Richards* 21868!
DISTR. U 3; K 1, 3–7; T 2; not known elsewhere
HAB. Forest undergrowth and dense bushland; 1600–2700 m.
SYN. *E. pseudoengleri* Pax in E.J. 43: 87 (1909). Type: Kenya, Nakuru District, Mau Escarpment, *Thomas* III. 78 (whereabouts unknown)

FIG. 84. *EUPHORBIA ENGLERI* — **1**, flowering branch, × ⅔; **2**, cyathium, × 4; **3**, capsule, × 2; **4**, seeds, × 4. *E. GOETZEI* — **5**, flowering branch, × ⅔; **6**, cyathium, × 4; **7**, capsule, × 2; **8**, seeds × 2. 1, from *Richards* 1868; 2, from *Hooper & Townsend* 1686; 3, 4, from *Rayner* 19486; 5, from *Richards* 7401; 6, from *Greenway & Kanuri* 12678; 7, 8, from *Richards* 5618. Drawn by Christine Grey-Wilson.

VAR. Collections from the southern limits of distribution in Tanzania (particularly from Kilimanjaro and Mt. Meru), possess glands which are more noticeably crescent-shaped than those from the Teita Hills and Aberdares–Mt. Kenya regions of Kenya. Otherwise features vary little, except for very lush specimens, usually from riverine or mist forest, which can produce slightly larger leaves and inflorescences with longer-branched cymes and longer cyathial peduncles than normal.

49. E. goetzei *Pax* in E.J. 28: 420 (1900). Type: Tanzania, Iringa District, Ruaha R., *Goetze* 450 (B, holo. †, K, iso.!)

Shrubby perennial herb to 3(–4) m. high with grey succulent glabrous stems and branches. Leaves obovate, to 15 × 6 cm., base cuneate, apex rounded, margin entire, glaucous especially beneath, lower surface pilose with long (to 1.5 mm.) spreading hairs especially along the midrib, mature leaves rarely completely glabrous; petiole to 4 cm. long, pilose; stipules glandular. Cymes in 3–5-branched umbels on peduncles to 12 cm. long produced from whorls of 5–10 leaves at the swollen apices of branches, with primary rays to 8 cm. long, forking twice; bracts subcircular, ± 1.5 cm. in diameter, the lower ones larger, subsessile to shortly petiolate, glabrous except for a few hairs along the lower margins. Cyathia on peduncles to 1.5 mm. long, 3 × 7 mm., with funnel-shaped involucres; glands 4, or 5 on the central cyathium of the umbel, spreading, transversely elliptic, ± 3 mm. broad, outer margin 2–4-lobed, with each lobe sometimes shortly divided; involucral lobes broadly rounded, ± 1 × 1.5 mm. Male flowers: bracteoles few, strap-shaped, upper margin ciliate and sometimes deeply divided; stamens 4.5–5 mm. long. Female flower: styles 2.5 mm. long, joined for 0.5 mm., suberect with spreading shortly bifid apices. Capsule exserted on a pedicel to 8 mm. long, deeply 3-lobed with a slightly sunken apex, 8 × 10 mm. Seeds ovoid with acute apex, ± 5 × 3.5 mm., brownish grey with longitudinal rows of large flattened tubercles. Fig. 84/5–8, p. 453.

KENYA. Northern Frontier Province: Moyale, 10 Oct. 1952, *Gillett* 14025!; Machakos District: Kibwezi, 25 Mar. 1960, *Bally* 12206!; Teita District: Tsavo National Park, Voi Gate Camp Site, 7 Dec. 1966, *Greenway & Kanuri* 12678!
TANZANIA. Ufipa District: Kawa R. Falls, 30 Dec. 1956, *Richards* 7401!; Iringa District: Ruaha National Park, Great Ruaha R. upstream from ferry, 19 Jan. 1966, *Richards* 21002!; Tunduru District: 11 km. E. of Songea District boundary, 7 June 1956, *Milne-Redhead & Taylor* 10675!
DISTR. **K** 1, 4, 7; **T** 1, 3–8; southern Ethiopia, northern Zambia and Malawi
HAB. On rocky slopes in dense bushland or light woodland, usually near stream beds; 500–1800 m.

SYN. [*E. transvaalensis* sensu N.E. Br. in F.T.A. 6(1): 530 (1911), pro parte quoad specim. Afr. or.; T.T.C.L.: 211 (1949); F.F.N.R.: 199 (1962), pro parte (*White* 3673); Cribb & Leedal, Mount. Fl. S. Tanz.: 77 (1982), *non* Schlechter]

VAR. In drier areas, especially in Kenya, plants are usually shorter, up to 1.5 m., with the apices of the branches distinctly clavate.

NOTE. Since publication of the F.T.A. *E. goetzei* has been considered as synonymous with *E. transvaalensis* Schlechter. However, this is a smaller plant with glabrous leaves except sometimes for a few hairs on the petiole. The styles are longer (4 mm.), joined for nearly half their length and with scarcely bifid tips. Distribution is to the south and southwest of *E. goetzei* in south tropical and southern Africa.

50. E. platycephala *Pax* in E.J. 19: 122 (1894); N.E. Br. in F.T.A. 6(1): 525 (1911). Type: Tanzania, Mwanza District, Kayenzi [Kagehi], *Fischer* 516 (B, holo. †, K, drawing!)

Glabrous perennial herb with a fleshy cylindrical rootstock 20–30 × 2–4 cm. and ± fleshy annual stems to 30 cm. high. Leaves ± fleshy, sessile, obovate to 12 × 4 cm., base cuneate, apex rounded to obtuse and minutely apiculate, margin entire; stipules filamentous ± 1 mm. long, deciduous. Cymes axillary and in terminal 3–4-branched umbels with primary rays to 10 cm. long, forking 2–4 times; bracts deltoid to 4.5 × 4 cm., base subcordate, apex obtuse apiculate, those at the base of the umbel similar to the leaves. Cyathia on peduncles to 1.5 mm. long, 5.5 × 10 mm., with cup-shaped involucres; glands 5, spreading, transversely elliptic ± 1.5 × 2–3.5 mm., outer margin obscurely crenulate; lobes rounded, 1.5 × 3 mm., margin entire and minutely ciliate. Male flowers many: bracteoles laciniate, with feathery apices; stamens 8 mm. long, with anther-thecae minutely hispid. Female flower: styles 2 mm. long, joined at the base for 0.5 mm. then widely spreading, with shortly bifid apices. Capsule on a pedicel 2.5 mm. long, deeply 3-lobed, 1 × 1.2 cm. Seeds globose, 4.5 mm. in diameter, brownish grey, surface reticulately wrinkled.

TANZANIA. Shinyanga District: Huruhuru Plains, 26 Jan. 1936, *B.D. Burtt* 5594!; Ufipa District: Maji-ya-Moto on Mpanda road, 9 Feb. 1962, *Richards* 16031!; Mbeya District: Usangu Plain, 8 Dec. 1963, *Richards* 18606!
DISTR.   T 1, 3–7; Malawi, Zambia, N. Zimbabwe
HAB.   Amongst grass of seasonally wet areas in dry sandy soils and hard sunbaked black cotton pans; 600–1300 m.

SYN.   *E. uhehensis* Pax in E.J. 28: 420 (1900). Type: Tanzania, Iringa District, Uhehe, Muhinde Steppe, *Goetze* 525 (B, holo.†, K, drawing!)

51. E. dolichoceras *S. Carter* in K.B. 35: 418 (1980). Type: Tanzania, Songea District, N. of Songea, *Milne-Redhead & Taylor* 10108 (K, holo.!, B, BR, EA, LISC, iso.!)

Glabrous perennial herb with a thick fleshy rhizomatous rootstock and annual stems to 60 cm. high. Leaves sessile, ± glaucous, obovate-lanceolate, to 9 × 1.5 cm., base cuneate, apex rounded apiculate, margin entire. Cymes axillary and in terminal 3–5-branched umbels, with primary rays to 12 cm. long and forking 3–4 times; bracts sessile, deltoid, ± 2 × 2 cm., the lower ones larger and longer, base truncate to subcordate, apex apiculate. Cyathia on peduncles 2–3 mm. long or the central one to 1.5 cm. long, ± 3 × 5 mm., with cup-shaped involucres; glands 4, or 5 on the central cyathium, transversely elliptic, to 1.5 mm. broad, with 2 horns 1.5–2 mm. long; lobes broadly deltoid, ± 1 mm. long, margin ciliate with long hairs. Male flowers: bracteoles deeply laciniate with feathery apices; stamens 4 mm. long. Female flower: ovary pedicellate; styles 2 mm. long, joined for 0.5 mm., with roughened shortly bifid apices. Capsule exserted on a pedicel to 1 cm. long, shallowly 3-lobed, 4 × 6 mm. Seeds subglobose, 3 × 2.5 mm., grey, with a longitudinal dorsal ridge and obscure rows of small sharply pointed tubercles.

TANZANIA.   Songea District: near Luhimba R., N. of Songea, 6 May 1956, *Milne-Redhead & Taylor* 10108!
DISTR.   T 8; N. Zambia and Malawi
HAB.   *Brachystegia* woodland on sandy loam; 970(–1500 elsewhere) m.

52. E. dilobadena *S. Carter* in K.B. 40: 818 (1985). Type: Tanzania, Mpanda District, Mpanda–Ikola road, *Richards* 11670 (K, holo.!)

Perennial herb with a vertical woody rootstock 12 × 1.3 cm., producing annual sparsely branching, very sparsely pilose stems, erect to 20 cm. high. Leaves sessile, fleshy, linear-lanceolate to 5.5 × 7 cm., base cuneate, apex acute, margin entire, pilose along the margins and the midrib on the lower surface; stipules glandular, reddish. Cymes axillary and terminal, solitary on very sparsely pilose peduncles to 10 cm. long, forking 2–3 times; bracts similar to the leaves. Cyathia on pilose peduncles to 8 cm. long, ± 5 × 5 mm., with cup-shaped involucres; glands 5, ± 2 × 2.8 mm., deeply bilobed, with the apices of the lobes entire to distinctly crenate; involucral lobes transversely elliptic, ± 1 × 2 mm. margin distinctly toothed, ciliate. Male flowers: bracteoles few, strap-shaped, feathery; stamens 7 mm. long. Female flower: ovary glabrous, exserted on a reflexed pedicel 1 cm. long; styles 2.5 mm. long, joined halfway, with spreading minutely bifid thickened apices. Capsule and seeds not seen.

TANZANIA.   Mpanda District: 32 km. along Mpanda–Ikola road, 31 Oct. 1959, *Richards* 11670!
DISTR.   T 4; known only from this collection
HAB.   In sandy soil between flat laterite rocks; 900 m.

53. E. ruficeps *S. Carter* in K.B. 35: 416 (1980). Type: Zambia, Kitwe Province, *Fanshawe* 2639 (K, holo.!)

Perennial herb, with a woody tuberous rootstock, producing a few shortly puberulous stems annually, to 45(–60) cm. high. Leaves deflexed, narrowly lanceolate, to 12(–17) × 1(–1.3) cm., base rounded, apex obtuse, apiculate, margin entire, revolute, midrib prominent on the lower surface which is sparsely pubescent; petiole puberulous, to 1 cm. long. Cymes in 3-branched umbels with primary rays to 12 cm. long, forking twice; bracts similar to the leaves. Cyathia sessile, 4 × 5.5 mm., with shortly puberulous purplish-red broadly cup-shaped involucres; glands 4, transversely elliptic, ± 4 mm. wide, spreading, dark red; lobes subquadrate, 2 mm. long, laciniate with ciliate margin. Male flowers: bracteoles laciniate with feathery apices; stamens 5.5 mm. long, puberulous. Female flower: ovary densely pilose; styles 3 mm. long, joined halfway with bifid apices, sparsely puberulous below. Capsule exserted on a curved densely pilose pedicel 6 mm. long, subglobose, 5 mm. in diameter, very densely papillose-pilose, purplish red. Seeds ovoid, 4 × 2.5 mm., closely tuberculate, greyish brown.

TANZANIA. Mpanda District: Mpanda, 9 Dec. 1956, *Richards* 7194! & Kapapa Camp, 28 Oct. 1959, *Richards* 11625!
DISTR. T 4; Malawi and northern Zambia
HAB. Dry sandy soil in *Brachystegia* woodland; 1050 m.

54. E. arrecta *N.E. Br.* in Wiss. Ergebn. Schwed. Rhod.-Kongo-Exped. 1: 116 (1914); N.E. Br. in Fl. Cap. 5(2): 283 (1915). Type: Zambia, between Lakes Bangweulu and Tanganyika, *Fries* 1078 (UPS, holo., K, drawing!)

Perennial herb, with a woody rootstock, producing annual puberulous sparsely branching stems to 45 cm. high. Leaves usually reflexed, pilose with long spreading hairs, linear-lanceolate, to 5 × 0.8 cm., base rounded, apex obtuse, margin minutely serrate with glandular teeth, midrib prominent on the lower surface; petiole to 8 mm. long; stipules glandular, minute. Cymes in axillary and terminal 3-branched umbels with primary rays to 8 cm. long, forking once; bracts similar to the leaves. Cyathia sessile, 3.5 × 5 mm., with puberulous funnel-shaped involucres; glands 4(-5), transversely elliptic, 2.5 mm. wide, spreading, reddish; lobes subquadrate, 1 mm. long, with fringed, ciliate margin. Male flowers: bracteoles fan-shaped, deeply laciniate, feathery; stamens 4 mm. long. Female flower: ovary densely pilose; styles 2.5 mm. long, joined halfway, with spreading thickened apices. Capsule exserted on a pilose pedicel 3.5 mm. long, obtusely 3-lobed with truncate base, 6.5 × 7.5 mm., sparsely pilose. Seeds ovoid, obscurely 4-angled and apex acute, 4.8 × 3.5 mm., purplish black, surface obscurely tuberculate.

TANZANIA. Chunya District: base of Igila Hill, 22 Mar. 1965, *Richards* 19807!
DISTR. T 7; northern Zambia
HAB. Sandy soil amongst grass in *Brachystegia* woodland; 1500 m.

NOTE. In comparison with the limited material seen from Zambia, the Tanzanian specimen possesses almost glabrous stems and leaves, matching well, however, with the description and drawing of the type. Of the Zambian specimens, those collected shortly after burning are more densely pilose, especially on the stems.

55. E. taboraensis *Hässler* in Bot. Notis. 1931: 331 (1931); S. Carter in K.B. 29: 442 (1974). Type: Tanzania, Tabora District, Nyembe, *Braun* in Herb. *Amani* 5461 (B, holo. †, EA, iso.!)

Glabrous perennial herb, with a fleshy cylindrical rootstock to 20 × 1-2 cm. and annual stems to 30 cm. high. Leaves sessile, linear-lanceolate to 12 × 1.8 cm., apex apiculate, margin entire; stipules filamentous, to 0.5 mm. long, soon deciduous. Cymes axillary and in terminal 3-branched umbels, with rays to 5(-8) cm. long, occasionally forking 1-2 times; bracts ovate-lanceolate to 5 × 2.5 cm., base subcordate, apex acute, those at the base of the umbel larger. Cyathia on peduncles to 3(-6) cm. long, with cup-shaped involucres, ± 7 × 14 mm.; glands 5, transversely rectangular, ± 2 × 4 mm., spreading, the outer margin with up to 10 finger-like processes to 2.5 mm. long which are 2-4-lobed at the tips and minutely hispid; involucral lobes rounded ± 2.5 × 3.5 mm., margin sharply toothed, outer surface finely ciliate and minutely hispid. Male flowers: bracteoles few, divided to the base into 2-3 fine threads with ciliate apices; stamens many, 9.5 mm. long, minutely hispid including the anther-thecae. Female flower: perianth obvious as a rim below the ovary; styles ± 5 mm. long, joined for ⅓, spreading with bifid thickened apices, minutely hispid. Capsule exserted on an erect minutely hispid pedicel to 12 mm. long, subglobose, ± 14 × 16 mm. Seeds ovoid, obtusely 3-angled, ± 5.5 × 4 mm., smooth, pale grey.

TANZANIA. Shinyanga District: Huruhuru, 16 Jan. 1932, *B.D. Burtt* 3513!; Mpanda District: Mpanda–Ikuu road, 1 Dec. 1956, *Richards* 7094!; Chunya District: 40 km. from Chunya towards Rungwa, 4 Dec. 1966, *Richards* 20739!
DISTR. T 1, 4, 5, 7; not known elsewhere
HAB. Amongst grass, usually in marshy ground; 1000-1500 m.

56. E. pseudograntii *Pax* in E.J. 30: 342 (1901); N.E. Br. in F.T.A. 6(1): 528 (1911); T.T.C.L.: 210 (1949). Type: Tanzania, Njombe District, Ukanga Mt., *Goetze* 888 (B, holo. †, K, drawing!)

Glabrous shrub 1-2 m. high with ± woody branches. Leaves subsessile to petiolate, obovate to lanceolate, to 30 × 5.5 cm., base cuneate, apex minutely to shortly apiculate, margin entire; lamina tapering into a petiole to 2.5 cm. long on older larger leaves; stipules apparently absent. Cymes in terminal 2–3-branched umbels on leafless peduncles to 10 cm. long, with primary rays to 6 cm. long, forking 2–4 times; bracts deltoid, ± 2.5 × 2.5 cm. when mature, base subcordate, apex acuminate to ± 2 cm. long; bracts below the umbel a little larger. Cyathia on peduncles to 3 mm. long, ± 10 × 30 mm., with barrel-shaped involucres; glands 4, spreading, transversely oblong, ± 2 mm. × 8 mm., outer margin lobed with 8–12 finger-like processes ± 7 mm. long branching 1–2 times at the tips and terminating in minute knobs; involucral lobes rounded, ± 3 × 5 mm., the margin deeply toothed and minutely ciliate. Male flowers: bracteoles few, ligulate, apex ciliate; stamens many, 14 mm. long, minutely puberulous. Female flower: perianth obvious below the ovary, 1.5–2 mm. wide, with sharply toothed margin; styles 5 mm. long, joined at the base with thickened shortly bifid apices. Capsule exserted on a curved pedicel to 1 cm. long, subglobose, ± 9 mm. in diameter. Seeds ovoid-trigonous, 6 × 4.5 mm., smooth, greyish brown.

TANZANIA. Njombe District: Lupembe, by Ruhudji R., 1931, *Schlieben* 130! & Lumbila, 16 Aug. 1958, *Gilli* 279!; Songea District: escarpment near Lukumburu on Njombe–Songea road, 1 Mar. 1963, *Richards* 17697!
DISTR. **T** 7, 8; not know elsewhere
HAB. Rocky slopes with open woodland; 500–1800 m.

NOTE. Known only from a very restricted area of southern Tanzania. The Kenya plant, which has been called *E. pseudograntii*, is a distinct species, *E. friesiorum* (Species 58).

57. **E. grantii** *Oliv.* in Trans. Linn. Soc., Bot. 29: 144 (1875); N.E. Br. in F.T.A. 6(1): 527 (1911), excl. *Hildebrandt* 2632; T.T.C.L.: 210 (1949); I.T.U., ed. 2: 127 (1952). Type: Tanzania, Tabora District, near Tabora [Kazeh], *Speke & Grant* (K, holo.!)

A few-stemmed bush or sparingly branched tree 1.5–9 m. high, with a smooth but horizontally grooved grey bark; branches semi-succulent, with large prominent closely-set leaf-scars. Leaves sessile, linear to linear-lanceolate, to 30 × 3 cm., base rounded, apex acuminate, margin entire, midrib prominent on the lower surface, lamina glabrous, pale green and slightly glaucous; stipules glandular, minute, deciduous. Cymes in terminal 3-branched umbels usually on a leafless peduncle to 10(–15) cm. long, with primary rays to 5(–7) cm. long, forking up to 8 times, peduncle, rays and branches minutely puberulous; bracts sessile, deltoid, ± 4 × 4 cm. when mature, base subcordate, apex acuminate to ± 2 cm. long, lower surface minutely puberulous at least towards the base; bracts below the umbels larger and longer. Cyathia on peduncles to 1 cm. long, ± 1 × 3 cm., with barrel-shaped involucres, all parts except the upper surface of the glands minutely puberulous sometimes densely so; glands 4, spreading, transversely elliptic, ± 4 × 8 mm., outer margin lobed, with 6–10 finger-like processes ± 8 mm. long branching several times at the tips and terminating in minute knobs, the glands yellowish green becoming brown or sometimes purple, with red to purple processes having yellow to red tips; involucral lobes rounded, ± 3 × 5 mm., margin sharply toothed often with a longer central tooth to 1.5 mm. long, densely puberulous on both surfaces. Male flowers: bracteoles many, fan-shaped, deeply divided and densely ciliate at the apices; stamens 14.5 mm. long. Female flower: perianth obvious below the ovary, irregularly and acutely lobed with lobes to 1.5 mm. long; ovary minutely puberulous; styles 1 cm. long, joined for 4 mm., with shortly bifid thickened rugulose apices. Capsule exserted on a curved pedicel to 1.5 cm. long, subglobose, ± 13 × 17 mm., glabrous when mature, tinged purplish. Seed subglobose, slightly compressed laterally and obscurely 3-angled, ± 6 × 5 mm., greyish brown, with minutely roughened surface. Fig. 85/1, 2, p. 458.

UGANDA. Teso District: Serere, Aug.–Sept. 1932, *Chandler* 946!; Masaka District: Buddu, 1905, *Dawe* 274!; Mengo District: Entebbe, Nakiogo, Jan. 1932, *Eggeling* 153 in *Brasnett* 365!
TANZANIA. Ngara District: Bushubi, Muganza, 6 Sept. 1960, *Tanner* 5138!; Kondoa District: Kolo, 10 Jan. 1962, *Polhill & Paulo* 1123!; Iringa District: Ruaha National Park, top of Mpululu Mt., 21 May 1968, *Renvoize & Abdallah* 2314!
DISTR. **U** 2–4; **T** 1, 2, 4–7; Burundi, Zaire and northeast Zambia
HAB. On red sandy gritty soil often among rocks on hillsides, in open woodland; 800–1600 m.

SYN. *E. mulemae* Rendle in J.L.S. 37: 209 (1905). Types: Uganda, Ankole District, Mulema, *Bagshawe* 321 & 340 (BM, syn.!)

NOTE. The latex appears to be particularly copious in this species, and is often used in local medicines for stomach disorders, sometimes with fatal results. The plant is used for hedging purposes and sometimes cultivated as an ornamental.

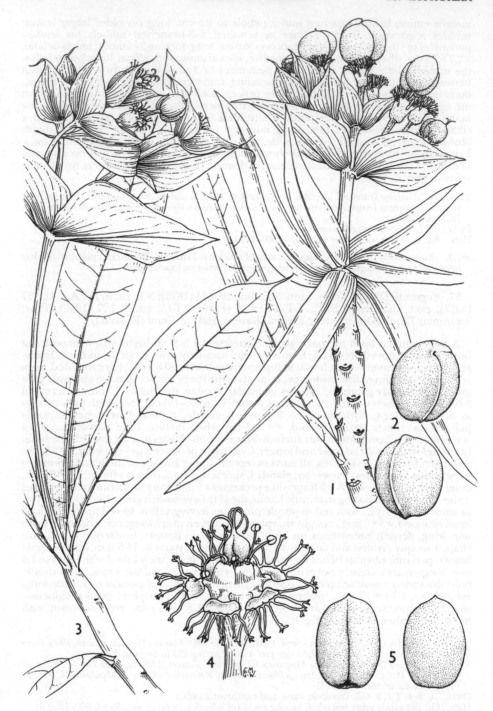

FIG. 85. *EUPHORBIA GRANTII* — 1, flowering branch, × ⅔; 2, seeds, × 4. *E. FRIESIORUM* — 3, flowering branch, × ⅔; 4, cyathium, × 2; 5, seeds, × 4. 1, 2, from *Renvoize & Abdallah* 2314; 3, from *Faden* 74/725; 4, from *MacDonald* in *C.M.* 9204; 5, from *Verdcourt* 3049A. Drawn by Christine Grey-Wilson.

58. E. friesiorum *(Hässler) S. Carter* in K.B. 40: 818 (1985). Type: Kenya, Embu District, Mt. Kenya, Nithi R., *Fries* 1947 (UPS, holo.).

Glabrous straight-stemmed shrub or small tree 2–7 m. high, with a smooth pale grey bark; branches semi-succulent with large leaf-scars. Leaves sessile, oblanceolate to lanceolate, to 20 × 6 cm., base cuneate, apex distinctly acuminate, margin entire, midrib prominent on the lower surface; stipules apparently absent. Cymes in terminal 3-branched umbels on leafless peduncles to 12 cm. long, with primary rays to 6 cm. long forking to to 5 times, occasionally more; bracts sessile, deltoid, ± 4 × 4 cm. when mature, base subcordate, apex acuminate to ± 1.5 cm. long; bracts below the umbels larger and longer. Cyathia on peduncles to 7 mm. long, or the central one of the umbel to 1 cm. long, ± 1 × 2.5 cm., with barrel-shaped involucres; glands 4, noticeably stipitate, widely spreading, transversely rectangular, ± 2.5 × 7–10 mm., green becoming yellow then red, outer margin crenulate with 6–10 red finger-like processes ± 7 mm. long branching 2–3 times at the tips and terminating in yellow lobed knobs; involucral lobes rounded, ± 3 × 5 mm., margin acutely toothed, the central teeth longer to 1.5 mm., inner surface shortly and densely puberulous. Male flowers: bracteoles very many, fan-shaped, deeply laciniate, apices ciliate; stamens 16 mm. long. Female flower: perianth an irregularly lobed rim below the ovary; styles 9.5 mm. long, joined halfway, spreading, with thickened rugulose minutely bifid apices. Capsule exserted on a curved pedicel to 12 mm. long, subglobose, ± 12 × 14 mm. Seeds subglobose, slightly compressed laterally and obtusely trigonous, ± 6.5 × 5.5 mm., greyish brown, with minutely roughened surface. Fig. 85/3–5.

KENYA. Fort Hall District: Maragua, Apr. 1939, *MacDonald* in *C.M.* 9204!; Embu District: Embu–Kangondi road, 3.5 km. towards Embu from Siakago turn-off, 6 June 1974, *R.B. & A.J. Faden* 74/725!; Kitui District: Mutomo Hill, 24 Apr. 1969, *Napper & Kanuri* 2072!
DISTR. **K** 4; not known from elsewhere
HAB. Sandy soils, usually on rocky slopes, in *Acacia-Commiphora* bushland and deciduous woodland; 700–2000 m.

SYN. [*E. grantii* sensu N.E. Br. in F.T.A. 6(1): 527 (1911), pro parte (*Hildebrandt* 2632); W.K.F.: 37 (1948), *non* Oliv.]
    E. grantii Oliv. var. *friesiorum* Hässler in N.B.G.B. 11: 42 (1930)
    [*E. pseudograntii* sensu Dale & Greenway, K.T.S.: 200 (1961); U.K.W.F.: 222 (1974), *non* Pax]

NOTE. This Kenyan species has usually been identified as *E. pseudograntii*, but differs from that species in numerous ways, most noticeably in its habit, involucral glands and lobes, styles and seeds. Both species are very restricted in distribution.

6. Subgen. **Lyciopsis**

Erect perennial herbs, shrubs or trees, with woody or sometimes ± fleshy branches. Leaves shortly petiolate; stipules modified as conspicuous glands, often sharply pointed. Cymes in terminal or axillary 1–7-branched umbels, with rays simple or rarely branching dichotomously, or the cyathia solitary; bracts scarious or leafy, occasionally coloured white or cream and conspicuous. Cyathia bisexual, with 5(–8) subcircular saucer-shaped, funnel-shaped or 2-lipped involucral glands, sometimes with finger-like processes on the outer margin. Stamens and sometimes the bracteoles well-exserted from the involucre. Perianth of ♀ flower occasionally obvious and 3-lobed. Capsule subsessile, shallowly 3-lobed and pubescent, or rarely (in East Africa) exserted on a recurved pedicel, deeply lobed and glabrous. Seeds globose or ovoid, occasionally angular or laterally compressed, smooth, with or without a caruncle.

6a. Sect. **Somalica**

Shrubs with semi-succulent or softly woody stems and branches; inflorescences in 1–5-branched terminal umbels with simple rays; bracts scarious or leafy; involucral glands 5, pectinate or rarely crenulate; capsules large, subsessile, often ridged; seeds usually laterally compressed, without a caruncle.

One species in East Africa.

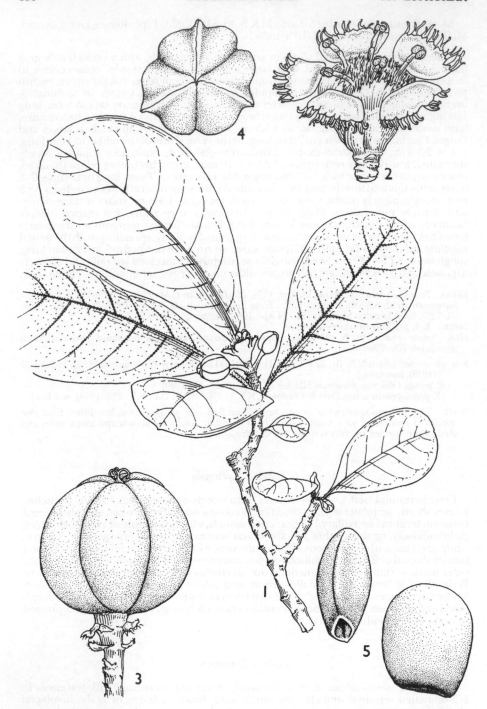

FIG. 86. *EUPHORBIA SCHEFFLERI* — 1, flowering branch, × ⅔; 2, cyathium, × 3; 3, capsule, × 2; 4, capsule from above, × 2; 5, seeds, × 4. 1, 4, 5, from *Polhill & Paulo* 933; 2, 3, from *Drummond & Hemsley* 4415. Drawn by Christine Grey-Wilson.

59. E. scheffleri *Pax* in E.J. 43: 88 (1909); N.E. Br. in F.T.A. 6(1): 549 (1911); T.T.C.L.: 210 (1949); E.P.A.: 457 (1958); K.T.S.: 201 (1961). Type: Kenya, Machakos District, Kibwezi, *Scheffler* 104 (BM, K, iso.!)

Spreading densely branching shrub or small tree 1–6 m. high, with smooth grey bark and semi-succulent glabrous reddish brown branches. Leaves ± fleshy, petiolate, clustered at the ends of the branchlets, obovate to 10 × 7 cm., base cuneate, apex broadly rounded, occasionally minutely apiculate, margin entire, both surfaces pilose especially the lower, the upper surface of mature leaves occasionally almost glabrous; petiole to 1 cm. long, puberulous; glandular stipules dark red. Cymes terminal, reduced to a central cyathium with (1–)2(–3) lateral cyathia on rays 1–1.5 cm. long; bracts scarious, 4–8 enveloping the cyme, soon deciduous, sessile with a broad base, subquadrate ± 1 × 1 cm., margin densely pilose, outer (lower) surface with some hairs along the midrib. Cyathia subsessile, ± 7 × 18 mm., with funnel-shaped involucres, the central cyathium usually larger; glands 5, transversely rectangular, ± 5 × 8 mm., outer margin crenulate with 10–15 finger-like processes 1–2 mm. long, simple or with the tips slightly thickened or minutely bifid, green; lobes transversely elliptic, ± 2.5 × 3.5 mm., margin fimbriate, reddish. Male flowers: bracteoles very broadly fan-shaped, much divided, feathery; stamens 9 mm. long, with pilose anther-thecae. Female flower: ovary subsessile, glabrous, enclosed in an irregularly toothed perianth 3 mm. long; styles 6.5 mm. long, spreading, joined at the base, with thickened bifid apices. Capsule subsessile on a thickened pedicel 7 × 7 mm., subglobose, 1.8–2 × 2–2.5 cm., eventually with 3 ± prominent longitudinal ribs at each valve. Seeds ovoid and laterally compressed, ± 7 × 6.5 mm. and 4.5 mm. thick, pale brownish grey, smooth. Fig. 86.

KENYA. Northern Frontier Province: Kichich, 23 Dec. 1958, *Newbould* 3569!; Kitui District: Kindaruma dam, 26 Dec. 1970, *Gillett* 19258!; Teita District: Voi, 9 Dec. 1961, *Polhill & Paulo* 933!
TANZANIA. Masai District: Mt Longido, 3 Jan. 1969, *Richards* 23622!; Mbulu District: Mwembe, 5 Jan. 1962, *Polhill & Paulo* 1058!; Pare District: S. Pare Mts., Kisiwani, 23 June 1942, *Greenway* 6487!
DISTR. K 1, 4, 6, 7; T 1–3, 5; southern Ethiopia (but see Note)
HAB. On sandy stony soils in open *Acacia-Commiphora* bushland; 300–1600 m.

SYN. *E. monocephala* Pax in E.J. 43: 223 (1909). Type: Tanzania, Masai District, between Kiparbara and Kitwei, *Jaeger* 64 (B, holo.†, K, drawing!)

VAR. As usual with a fairly widespread species, *E. scheffleri* shows some variation. Plants from the Ndoto Mts. and Matthews Range generally have leaves which are less hairy, occasionally almost completely glabrous at maturity, whilst those from Marsabit northwards have usually smaller cyathia and capsules. However, such extremes can also be found further south amongst more typical plants, with all gradations between, precluding the recognition of distinct varieties.

NOTE. *Rawlins* 159 from Lamu District, Kui I., Kiunga, just above sea-level is a sterile specimen with completely glabrous leaves. It may prove to represent an undescribed species, or may be identified with similar specimens of *E. giumboensis* Hässler (in Bot. Notis. 1931: 328 (1931)) from the adjoining southern tip of Somalia. Also, specimens from the Ethiopian border may be distinct: *Gillett* 13243, from between Bururi and El Mole, has hairy capsules, and *Gilbert & Thulin* 1575, from 30 km. NW. of Ramu, has immature cyathia with shorter gland appendages. *Ruspoli & Riva* 440[843]455 from Beila, S. bank of Daua R., a leafless specimen, is reported (fide Gilbert) to be *E. grosseri* Pax (in E.J. 33: 288 (1904)), with gland-margins crenulate and with hairy capsules.

6b. Sect. Lyciopsis

Woody herbs, shrubs or small trees; cymes terminal or axillary, in 2–7-branched umbels with rays forking up to 3 times, or the cyathia solitary; bracts scarious or leafy and occasionally conspicuous; involucral glands 5(–8), entire; feathery tips of the bracteoles exserted from the involucre; capsules subsessile, shallowly 3-lobed; seeds without a caruncle.

1. Bracts conspicuous, white or creamy yellow 2
   Bracts small and scarious or green and leaf-like 3
2. Leaves subsessile; cyathia glabrous . . . . 60. *E. jatrophoides*
   Leaves petiolate; cyathia puberulous . . . . . . 61. *E. joyae*
3. Branching regularly trichotomous . . . . 68. *E. matabelensis*
   Branching not trichotomous 4
4. Branchlets spine-tipped, and/or of equal length at right-
   angles to the branch . . . . 67. *E. cuneata*
   Branchlets not as above 5

60. E. jatrophoides *Pax* in E.J. 33: 287 (1903); N.E. Br. in F.T.A. 6(1): 544 (1911); E.P.A.: 450 (1958). Types: Ethiopia, Harar Province, *Ellenbeck* 1081 (B, syn. †, K, drawing!) & Bale Province, *Ellenbeck* 1164 (B, syn. †, K, fragment & drawing!)

Erect shrub or small tree to 5 m. high, with dark brown bark; branches puberulous when young. Leaves alternate or occasionally fasciculate, subsessile, panduriform, to 6 × 3 cm., base cuneate, apex apiculate to rounded, margin entire, lamina puberulous when young, upper surface ± glabrous at maturity, midrib prominent beneath; petiole to 2 mm. long; stipules glandular. Cymes terminal, consisting of a central cyathium with 2(–3) lateral cyathia on glabrous rays to 1 cm. long; bracts sessile, subcircular, to 1.7 mm. wide, pubescent at first becoming glabrous, white to yellow, conspicuous. Cyathia subsessile, or the central one on a peduncle to 5 mm. long, ± 6.5 × 12 mm.,with cup-shaped involucres, glabrous; glands 5, transversely elliptic, ± 2 × 3.5 mm., shallowly saucer-shaped, yellow; lobes transversely elliptic, ± 2 × 3 mm., margin deeply fringed, ciliate. Male flowers many: bracteoles fan-shaped, deeply divided, apices feathery; stamens ± 8 mm. long, anther-thecae with a tuft of fine hairs 1 mm. long. Female flower: ovary sessile, densely pubescent; styles 5 mm. long, joined for 2.5 mm., pubescent, with spreading apices bifid for 2 mm., glabrous. Capsule and seeds not seen.

KENYA. Northern Frontier Province: 5–10 km. SSE. of Ramu, 2 May 1978, *Gilbert & Thulin* 1338!
DISTR. K 1; limited distribution in S. and SE. Ethiopia
HAB. Limestone hills with open *Acacia-Commiphora* bushland; 400–500 m.

61. E. joyae *Bally & S. Carter* in K.B. 40: 820 (1985). Type: Kenya, Tana River District, Garissa, near Tana R., *Joy Bally* in *Bally* 1988 (K, holo.!)

Much-branched, spreading, woody shrub to 2.5 m. high, with greyish brown bark; young branches puberulous. Leaves alternate, occasionally fasciculate, petiolate, obovate to panduriform, to 5 × 3 cm., base rounded, apex truncate to shallowly emarginate, lamina pubescent, sparsely so at maturity, midrib prominent beneath; petiole to 8 mm. long; stipules filamentous to 1 mm. long. Cymes terminal, consisting of a central cyathium with (2–)3 lateral cyathia on pubescent rays to 3 cm. long, forking once; bracts sessile, obovate, ± 1.5 × 1 cm., pubescent beneath, greenish yellow. Cyathia sessile, or the central one on a peduncle ± 2 (–10) mm. long, ± 5.5 × 10 mm., with minutely pubescent cup-shaped involucres; glands 5, spreading, transversely elliptic, ± 2.5 × 3.5 mm., flat to very shallowly saucer-shaped, yellow; lobes transversely elliptic, ± 1.5 × 3 mm., margin fringed. Male flowers many: bracteoles fan-shaped, deeply divided, apices feathery; stamens ± 6.5 mm. long, anther-thecae shortly pilose. Female flower: ovary densely pubescent, sessile; styles ± 6 mm. long, pubescent, joined up to halfway, with spreading, deeply bifid, thickened apices. Capsule subglobose, pubescent, ± 8 mm. in diameter. Seeds ovoid, angular, with truncate base, ± 3 × 4 mm., smooth, pale grey. Fig. 87/1–3.

KENYA. Northern Frontier Province: 20 km. N. of Isiolo–Wajir road on Garba Tula–Merti turn-off, 19 Dec. 1971, *Bally & Smith* 14694!; Kitui District: 111 km. from Garissa on Thika road, 16 Dec. 1977, *Stannard & Gilbert* 1111!; Teita District: Tsavo National Park East, 33 km. from Voi Gate on Sobo road, 20 Dec. 1966, *Greenway & Kanuri* 12805!
DISTR. K 1, 4, 7; not known elsewhere
HAB. On sandy gravelly soils and lava, with open *Acacia-Commiphora* bushland; 180–750 m.

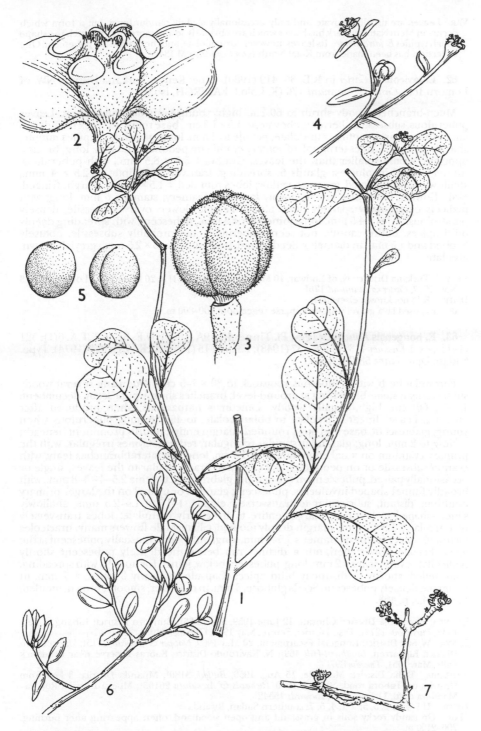

FIG. 87. *EUPHORBIA JOYAE* — 1, branch detail, × ⅔; 2, cyathium, × 3; 3, capsule, × 3. *E. BONGENSIS* — 4, inflorescence, × ⅔; 5, seeds, × 4. *E. INTRICATA* — 6, branch in leaf, × ⅔; 7, flowering branch, × 2. 1, 2, from *Stannard & Gilbert* 1111; 3, from *Adamson* 594; 4, from *Bullock* 3188; 5, from *Richards* 26128; 6, from *Gilbert et al.* 7999; 7, from *Bally & Smith* 14613. Drawn by Christine Grey-Wilson.

VAR. Leaves are usually obovate and only occasionally slightly panduriform, but a form which occurs in Meru National Park has leaves which are obviously so and more closely match the shape which typifies *E. jatrophoides*. Its leaves, however, are petiolate, and the cyathia are pubescent. One collection has been made from Kulal North (*van Swinderen* M.134!).

62. E. lavicola S. *Carter* in K.B. 35: 419 (1980). Type: Kenya, Turkana District, NW. of Lomoru Itae, *Carter & Stannard* 176 (K, holo.!, EA, SRGH, iso.!)

Much-branched woody shrub to 60 cm. high; young branches puberulous. Leaves puberulous, subsessile, alternate, obovate, to 1.5 × 1.3 cm., base rounded, apex rounded to emarginate, margin entire, ± undulate; petiole to 2 mm. long; glandular stipules minute, sharply pointed. Cymes terminal, of solitary cyathia on peduncles 1.5 mm. long; bracts ± opposite, a little smaller than the leaves. Cyathia ± 5.5 × 8.5 mm., with puberulous, funnel-shaped involucres; glands 5, spreading, transversely elliptic to 2.5 × 4 mm., shallowly saucer-shaped, greenish yellow; lobes rounded, ± 1.5 × 2 mm., margin fringed. Male flowers: bracteoles linear, deeply laciniate, feathery; stamens 6 mm. long, with pedicels occasionally sparsely hairy above. Female flower: ovary subsessile, densely puberulous; styles ± 4.5 mm. long, joined for ± 3 mm., pubescent, with spreading deeply bifid apices. Entire capsule not seen, puberulous and probably subsessile, obtusely 3-lobed and ± 5 mm. in diameter. Seeds ovoid, base truncate, 3 × 2.5 mm., greyish brown, areolate.

KENYA. Turkana District: N. of Lodwar, 10 May 1954, *Popov* 1551! & 20 km. NW. of Lomoru Itae, 3 Nov. 1977, *Carter & Stannard* 176!
DISTR. K 2; not known elsewhere
HAB. Exposed lava gravel with very sparse vegetation; 600–800 m.

63. E. bongensis *Kotschy & Peyr.*, Pl. Tinné: 40, 19A (1867); N.E. Br. in F.T.A. 6(1): 521 (1911), excl. *Kassner* 413; W.F.K.: 37 (1948); F.P.S. 2: 15 (1952); U.K.W.F.: 223 (1974). Type: Sudan, Djur, *Tinñe* 58 (W, holo.)

Perennial herb with a tuberous rootstock to 30 × 1–5 cm., producing several woody subterranean stems branching at ground-level; branches shortly pubescent, decumbent, 15–40(–60) cm. high, usually woody, sometimes herbaceous when produced after burning. Leaves linear-lanceolate to oblanceolate, to 45 × 6 mm., puberulous when young, glabrescent, base and apex rounded, margin entire, midrib prominent beneath; petiole to 2 mm. long; glandular stipules triangular, reddish. Cymes irregular, with the primary cyathium on a naked peduncle to 4.5 cm. long, the lateral branches leafy, with cyathia subsessile or on peduncles to 2 cm. long; bracts similar to the leaves, single or occasionally paired, pubescent on the margin, glabrescent. Cyathia 2.5–4 × 5–8 mm., with broadly funnel-shaped involucres, pubescent; glands 5 or up to 8 on the larger primary cyathium, distant, subcircular to transversely elliptic, 1–1.5 × 1.5–2.5 mm., shallowly saucer-shaped to 2-lipped, margin entire to irregularly crenulate; lobes transversely rectangular, ± 1.5 × 2 mm., margin deeply toothed, ciliate. Male flowers many: bracteoles laciniate, apices feathery; stamens 3.5–4 mm. long, with pedicels usually pubescent at the apex. Female flower: perianth a distinct rim below the densely pubescent shortly pedicellate ovary; styles ± 2 mm. long, pubescent below, joined for up to ⅓, with spreading, channelled, shortly to distinctly bifid apices. Capsule shallowly 3-lobed, ± 7 mm. in diameter, densely pubescent. Seeds globose, 3 mm. in diameter, smooth, brown, mottled. Fig. 87/4,5, p. 463.

UGANDA. Karamoja District: Chosan, 12 June 1959, *Symes* 563!; Bunyoro District: Kibangya, Feb. 1943, *Purseglove* 1274!; Teso District: Serere, May 1932, *Chandler* 540!
KENYA. W. Suk District: Kongelai Escarpment, 20 May 1969, *Napper & Tweedie* 2115!; Uasin Gishu District: Kipkarren, *Brodhurst-Hill* 469!; N. Kavirondo District: Kabras Reserve near Broderick Falls, Mar. 1951, *Tweedie* 897!
TANZANIA. Buha District: Matende, 15 Aug. 1950, *Bullock* 3180!; Mpanda District: 3 km. from Uruwira on Tabora road, 30 Sept. 1970, *Richards & Arasululu* 26193!; Mbeya District: Rujewa–Madibira road, 11 Dec. 1961, *Richards* 15542!
DISTR. U 1–4; K 2, 3, 5; T 1, 4, 5, 7; southern Sudan, Rwanda
HAB. On sandy rocky soils in grassland and open woodland, often appearing after burning; 700–2100 m.

SYN. *E. djurensis* Pax in E.J. 43. 86 (1909). Type: Sudan, Djur, *Schweinfurth* 1591 (B, holo.†, K, iso.!)

NOTE. *Eggeling* 786 in *Brasnett* 1171, a leafless gathering from Uganda, Acholi District, Agoro, seems closer to *E. intricata* S. Carter, and may represent that species.

64. E. intricata S. *Carter* in K.B. 40: 821 (1985). Type: Ethiopia, Bale Province, *Gilbert, Vollesen & Ensermu* 7999 (K, holo.!, C, U, iso.!)

Intricately branched woody puberulous shrublet to 30(–80) cm. high, from a subfleshy rootstock 1–2 cm. thick; terminal branchlets very numerous, thin and usually contracted, with leaves subfasciculate. Leaves oblanceolate, to 18(–35) × 8 mm., base cuneate, apex obtuse, margin entire, lamina puberulous, glabrescent beneath; petiole 1 mm. long; glandular stipules minute, pointed. Cymes terminal, reduced to a solitary sessile cyathium, surrounded by 2–5 obovate puberulous bracts ± 2.5 mm. long. Cyathia 2.5 × 4 mm., with very finely puberulous broadly cup-shaped involucres; glands 5, transversely elliptic, ± 1 × 2.5 mm., ± 2-lipped with a slightly wavy margin; lobes rounded, ± 1.2 mm. long, margin fringed. Male flowers: bracteoles strap-shaped, apices feathery; stamens 2.5 mm. long, with pedicels puberulous above. Female flower: ovary sessile, puberulous; styles 1.8 mm. long, puberulous, joined to halfway, with spreading, flattened, rugose, shortly bifid apices. Capsule obtusely 3-lobed, ± 3.5 × 3.5 mm., minutely pilose. Mature seeds not seen. Fig. 87/6, 7, p. 463.

KENYA. Northern Frontier Province: Danissa Hills, 15 Dec. 1971, *Bally & Smith* 14613!
DISTR. **K** 1; southern Ethiopia
HAB. Dense *Acacia-Commiphora* bushland; (500–)800 m.

SYN. *E. sp. nov. aff. E. cuneata* Vahl sensu S. Carter in K.B. 35: 430 (1980)

NOTE. *Gilbert & Thulin* 1684, from between Wajir and El Wak, may represent this species. It differs slightly from other specimens seen in that the terminal branchlets are less contracted, and the few leaves present are linear-lanceolate, to 3.5 × 5 mm. It more closely matches some specimens of *E. kassneri*, but its locality is against definite identification with that species.

65. E. kassneri *Pax* in E.J. 43: 86 (1909). Type: Kenya, Kwale District, Gadu, *Kassner* 413 (B, holo.†)

Perennial shrub, with a fleshy tuberous root to 30 cm. long, producing numerous erect woody stems and branches to 1 m. high, puberulous when young. Leaves sessile, oblanceolate, to 25 × 6 mm., base cuneate, apex rounded to minutely apiculate, lamina glabrous except for a few hairs at the base; glandular stipules minute, pointed. Cymes axillary, reduced to a solitary subsessile cyathium surrounded by several pubescent obovate bracts ± 3 × 2 mm. Cyathia ± 3 × 5 mm., with pubescent cup-shaped involucres; glands 5, the sides inrolled to form an erect tube open on the inner side, transversely elliptic, ± 1.5 × 2 mm. when opened out; lobes transversely rectangular, ± 1.3 × 2 mm., margin fringed. Male flowers many: bracteoles deeply laciniate and feathery; stamens 4 mm. long, with pedicels pubescent at the apex. Female flower: ovary subsessile, densely pubescent; styles 1.5 mm. long, joined to ± halfway, with spreading thickened rugulose apices. Capsule obtusely 3-lobed, pubescent, ± 4 × 5.5 mm. Seeds globose, 2.5 mm. in diameter, smooth, grey.

KENYA. Kwale District: Mrima road, 8 Sept. 1957, *Verdcourt* 1933!; Tana River District: 20 km. S. of Garsen on road to Ngao, 20 Sept. 1974, *B. Adams* 90!
DISTR. **K** 7; not recorded from elsewhere but possibly occurring in **T** 3, Tanga District
HAB. In grassland and bushland near the coast; 20–100 m.

SYN. *E. cuneata* Vahl form B sensu S. Carter in K.B. 35: 430 (1980)

NOTE. N.E. Brown reduced *E. kassneri* to synonymy under *E. bongensis*, but from Pax's description and the locality of the type it is clearly not this species. It is more closely related to *E. cuneata* subsp. *spinescens*, from which it is distinguished by its non-spiny branches and large tuberous root. However, since seedling plants of *E. cuneata* also possess a somewhat tuberous root, Pax's rather ill-defined species may prove to be more suitably placed as another form of *E. cuneata*, possibly identifiable with some specimens now included under subsp. *spinescens* var. *pumilans*.

66. E. handeniensis S. *Carter* in K.B. 40: 822 (1985). Type: Tanzania, Handeni District, Kwamkono, *Archbold* 2747 (K, holo.!)

Woody shrub 30–40 cm. high, with spreading branches puberulous when young. Leaves alternate or fasciculate on much abbreviated branchlets, sessile, obovate to oblanceolate, to 4.5 × 1 cm., base cuneate, apex obtuse, margin entire, lamina glabrous except for occasional hairs at the base; glandular stipules pointed. Cymes reduced to solitary, terminal cyathia; bracts leaf-like, ± 2 × 4 mm. Cyathia subsessile, ± 2.5 × 5.5 mm., with cup-shaped involucres; glands 5, circular, ± 2 mm. in diameter, funnel-shaped to deeply saucer-shaped; lobes transversely oblong, ± 1.5 × 2 mm., margin fringed. Male flowers: bracteoles deeply laciniate, with feathery apices; stamens ± 3.5 mm. long. Female flower: ovary subsessile, densely pubescent; styles ± 2 mm. long, joined halfway, with spreading thickened bifid apices. Capsule obtusely 3-lobed, ± 6 × 5 mm., pubescent. Seeds subglobose, 2.5 × 3 mm., smooth, grey, mottled.

TANZANIA.   Handeni District: Kideliko, 7 Apr. 1954, *Faulkner* 1388! & Nyongolo Hills, 20 Aug. 1961, *Semsei* 3272!
DISTR.   **T** 3; not known from elsewhere
HAB.   In soil pockets on granite rocks; 600–700 m.

SYN.   *E. cuneata* Vahl form A sensu S. Carter in K.B. 35: 430 (1980)

NOTE.   A difficult species to define, apparently limited in distribution to a small area of Handeni District. It differs from *E. cuneata*, especially subsp. *spinescens* var. *pumilans* to which it is similar in size, by its complete lack of spine-tipped branchlets, and from *E. kassneri* by its thick, woody but non-tuberous root, as well as its spreading habit.

67.   E. cuneata *Vahl*, Symb. Bot. 2: 53 (1791); N.E. Br. in F.T.A. 6(1): 545 (1911); T.T.C.L.: 209 (1949), pro parte (excl. *Greenway*, Longido); F.P.S. 2: 68 (1952); E.P.A.: 445 (1958); K.T.S.: 197 (1961), pro parte (*Graham* 1674); S. Carter in K.B. 35: 423 (1980). Type: N. Yemen, *Forsskål* (C, holo., K, microfiche!)

Woody shrub to 4 m. high, with greyish to purple-brown, dull or shiny peeling bark; branches with alternating spine-tipped branchlets to 12 cm. long, spreading at right-angles, minutely puberulous when young. Leaves alternate on young growth, fasciculate on older branches, cuneate-spathulate, to 4.5 × 1.5 cm., base cuneate, apex rounded to emarginate, margin entire, glabrous or pubescent; petiole to 5 mm. long; stipules glandular, minute and sharply pointed. Cymes reduced to subsessile solitary cyathia, or clustered in terminal 2–5-branched umbels with primary rays to 7 mm. long, 2–3-forked; bracts minute, leaf-like, 1–3 mm. long, minutely puberulous. Cyathia 2–5 × 4–7 mm., with broadly funnel to cup-shaped involucres, minutely puberulous; glands 5, subcircular, funnel-shaped and distant, to flat and almost touching, to ± 2.5 mm. in diameter, yellow; lobes rounded, ± 1.5 × 2.5 mm., with deeply fringed margin. Male flowers: bracteoles ligulate, with feathery apices; stamens ± 4–4.5 mm. long. Female flower: ovary subsessile, puberulous; styles ± 2 mm. long, joined for up to ⅔ of their length, with recurved, thickened, channelled apices, bifid for ⅓. Capsule obtusely 3-lobed, to 7 × 8 mm., pubescent. Seeds subglobose, ± 2.5 mm. in diameter, smooth, reddish brown, faintly speckled.

KEY TO INFRASPECIFIC VARIANTS

1. Cymes of 4 or more cyathia   2
   Cymes of 1–3 cyathia   3
2. Shrubs to 4 m. high; cymes of 4–5 cyathia, seldom more;
      cyathia 5 mm. or more in diameter   .   .   .   .   .   .   a. subsp. **cuneata**
   Shrubs to 2 m. high; cymes of up to 30 cyathia in a cluster;
      cyathia 3 mm. in diameter   .   .   .   .   b. subsp. **wajirensis**
3. Shrubs to 3 m. high; trunk bark yellow, branch bark dark
      purplish brown, shiny, conspicuously peeling; cyathial
      glands tubular, occasionally flattening out at
      maturity   .   .   .   .   c. subsp. **lampro-**
                                                                              derma

   Shrubs to 2 m. high; bark pale grey to greyish brown, not
      shiny   4
4. Shrubs to 2 m. high, branches erect; mature leaves to 2.5
      cm. long; cyathia solitary or occasionally with 2
      lateral cyathia   .   .   .   .   d. subsp. **spinescens**
                                                                              var. **spinescens**

   Shrubs to 1 m. high, branches horizontally spreading;
      mature leaves 1.5–2 cm. long; cyathia always solitary   d. subsp. **spinescens**
                                                                              var. **pumilans**

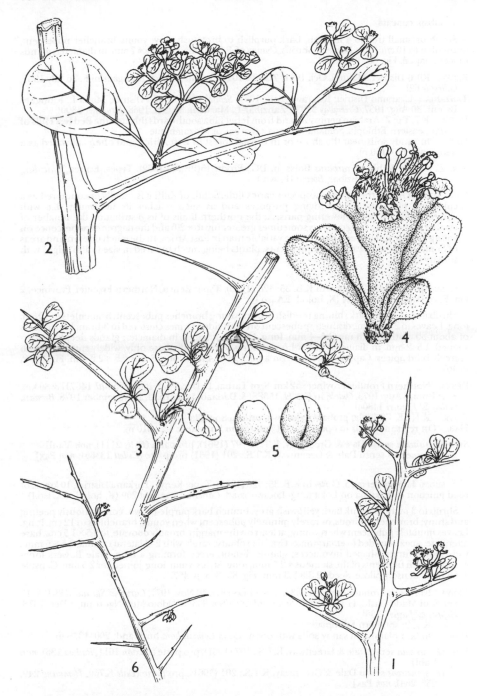

FIG. 88. *EUPHORBIA CUNEATA* subsp. *CUNEATA* — **1**, branch detail, × ⅔. Subsp. *WAJIRENSIS* — **2**, branch detail, × 2. Subsp. *LAMPRODERMA* — **3**, branch detail, × ⅔; **4**, cyathium, × 4; **5**, seeds, × 4. Subsp. *SPINESCENS* var. *PUMILANS* — **6**, branch detail, × ⅔. 1, from *Tweedie* 1024; 2, from *Brenan & Gillett* 14804; 3, 4, from *Carter & Stannard* 209; 5, from *Gilbert et al.* 5230; 6, from *Bally* 13412. Drawn by Christine Grey-Wilson.

a. subsp. **cuneata**

Shrub or small tree to 4 m. high; bark purplish to brownish grey; young branches pubescent, branchlets to 10 cm. long. Leaves glabrous. Cymes of 4–5 cyathia. Cyathia ± 7 mm. in diameter; glands saucer-shaped. Fig. 88/1, p. 467.

KENYA. Kilifi District: Malindi, Oct. 1951, *Tweedie* 1024!; Lamu District: Kiungamini I., 26 July 1961, *Gillespie* 62!

TANZANIA. Uzaramo District: Kisarawe, Aug. 1953, *Semsei* 1340!; Rufiji District: Mafia I., Kifingi to Baleni, 30 Aug. 1937, *Greenway* 5193!; Zanzibar I., Mazizini, 30 Oct. 1968, *Faulkner* 3300!

DISTR. **K** 7; **T** 6; **Z**; Arabian peninsula and from Jebel Elba southwards through the Red Sea Hills of Sudan, eastern Ethiopia, Somalia and coastal areas of Mozambique

HAB. On sandy soils near the shore or in coastal forest in East Africa, and often cultivated as a hedge plant; 0–160 m.

SYN. *E. cuneata* Vahl var. *carpasus* Boiss. in DC., Prodr. 15(2): 97 (1862). Types: Eritrea, *Ehrenberg* (K, isosyn.!) & Zanzibar, *Boivin* (P, syn.)

VAR. In East Africa the typical subspecies varies little. South of Kilifi it is usually encountered as a cultivated plant used for hedging purposes and as such reaches its maximum size with correspondingly larger flowering parts. At the northern limits of its distribution the number of cyathia in the branching cymes is sometimes greater (up to ± 20) and the degree of pubescence on young growth and leaves is much more variable than in East Africa. In the much drier inland areas of Somalia a probably distinct form occurs, plants being much reduced in size (± 1 m. high) with solitary cyathia.

b. subsp. **wajirensis** *S. Carter* in K.B. 35: 428 (1980). Type: Kenya, Northern Frontier Province, 2 km. E. of Wajir, *Gillett* 21274 (K, holo.!, EA iso.!)

Shrub to 3 m. high; bark shining reddish brown; young branches pubescent; branchlets to 10 cm. long. Leaves to 2 × 1.3 cm., distinctly pubescent on both sides. Cymes clustered in 3-branched umbels of about 30 cyathia, with rays 1.5–4 mm. long. Cyathia ± 5 mm. in diameter; glands deeply saucer-shaped, 1.5–2 mm. in diameter. Styles 1.5 mm. long, joined at the base only, widely spreading with scarcely bifid apices. Capsule 5 × 6 mm. on a pedicel to 4 mm. long. Seeds 2.5 × 2.2 mm. Fig. 88/2, p. 467.

KENYA. Northern Frontier Province: 32 km. S. of Tarbaj, 17 Dec. 1971, *Bally & Smith* 14671! & 80 km. E. of Bura, 6 Aug. 1973, *Oxtoby* in *E.A.H.* 15388! & Dadaab–Wajir road, 28 November 1978, *Brenan, Gillett & Kanuri* 14804!

DISTR. **K** 1; SE. Ethiopia; probably also in southern Somalia

HAB. On red sandy soils in open *Acacia-Commiphora* bushland; 135–310 m.

SYN. [*E. cuneata* sensu Dale & Greenway, K.T.S.: 197 (1961), pro parte (*Bally* 2111), *non* Vahl]
     [*E. spinescens* sensu Dale & Greenway, K.T.S.: 201 (1961) pro parte (*Gillett* 13365), *non* Pax]

c. subsp. **lamproderma** *S. Carter* in K.B. 35: 428 (1980). Type: Kenya, Turkana District, 10 km. S. of road junction to Kakuma on Lokitaung–Lodwar road, *Carter & Stannard* 209 (K, holo.!, EA, iso.!)

Shrub to 3 m. high; trunk bark yellowish grey, branch bark purple-brown, conspicuously peeling and shiny; branches glabrous, or rarely minutely pubescent when young; branchlets to 12 cm. long. Leaves minutely pubescent when young, at least on the margin, broadly obovate, to 2.5 × 1.5 cm., base attenuate, apex rounded to emarginate. Cyathia solitary, rarely with 2–3 lateral cyathia, 3.5 × 5 mm., with widely funnel-shaped involucres; glands distant, erect, forming a tube. Male flowers long-exserted, with filaments of the stamens ± 2.5 mm. long. Styles 3 mm. long, joined for 2.5 mm. Capsule 7 × 8 mm., densely pilose. Seeds 2.8 × 2.3 mm. Fig. 88/3–5, p. 467.

KENYA. Northern Frontier Province: 33 km. S. of Lodwar, 10 Nov. 1977, *Carter & Stannard* 294! & 13 km. S. of Mado Gashi, 11 Dec. 1977, *Stannard & Gilbert* 933! & Ndoto Mts., Ngoronit, 3 Dec. 1978, *Hepper & Jaeger* 7238!

DISTR. **K** 1, 2; southern Ethiopia

HAB. On lava plains and sandy soils with open *Acacia-Commiphora* bushland; 200–1375 m.

SYN. [*E. cuneata* sensu Dale & Greenway, K.T.S.: 197 (1961), pro parte (*Buxton* 1011, *Padwa* 136), *non* Vahl]
     [*E. spinescens* sensu Dale & Greenway, K.T.S.: 201 (1961), pro parte (*Dale* K.760, *Hemming* 249, 255, 264), *non* Pax]

VAR. The tubular formation of the involucral glands is particularly noticeable with plants occurring on the arid lava deserts of Turkana. Further east, where the soil is more sandy and the vegetation less sparse, the glands are sometimes flattened out to a deep saucer-shape on cyathia of shrubs which are also somewhat larger.

d. subsp. **spinescens** *(Pax) S. Carter* in K.B. 35: 429 (1980). Type: Tanzania, Massaini [NW. of Kilimanjaro], *Fischer* 524 (B, holo.†, K, drawing!)

Shrub from 15 cm. to a small tree 4 m. high, with reddish grey bark; young branches and leaves pubescent. Leaves obovate. Cyathia solitary, 4–6 mm. in diameter; glands deeply funnel-shaped when young, expanding to become suacer-shaped when mature. Styles ± 1.5 mm. long, spreading, joined at the base only. Capsule subsessile, pubescent.

var. **spinescens**

Shrub or small tree to 4 m. high; branchlets to 10 cm. long. Leaves to 3 × 1 cm. Cyathia ± 6 mm. in diameter; glands expanding to 3.5 mm. wide. Capsule to 7 × 8 mm. Seeds to 2.8 × 2.5 mm.

KENYA. Machakos District: Mtito Andei, 29 Aug. 1959, *Verdcourt* 2381!; Masai District: between Naibor and Kawadie, 23 Feb. 1963, *Glover & Cooper* 3444!; Teita District: Voi, 27 Aug. 1969, *Bally* 13444!

TANZANIA. Masai District: Longido, 18 Dec. 1969, *Richards* 24952!; Mbulu District: Tarangire National Park, Mtete, 13 Mar. 1969, *Richards* 24353!; Lushoto District: Mbalu Mt., near Mkundi Mtae, 29 Jan. 1974, *Bally & Carter* 16398!

DISTR. K 4, 6, 7; T 2, 3, 5; not known elsewhere

HAB. Dry sandy soil in open *Acacia-Commiphora* bushland; 400–1750 m.

SYN. *E. spinescens* Pax in E.J. 19: 120 (1894); N.E. Br. in F.T.A. 6(1): 546 (1911); T.T.C.L.: 210 (1949)
    E. lyciopsis Pax in P.O.A. C: 242 (1895); N.E. Br. in F.T.A. 6(1): 544 (1911); T.T.C.L.: 210 (1949); E.P.A.: 451 (1958). Type: Tanzania, Moshi District, Lake Chala, *Volkens* 1781 (B, holo.†, K, drawing!)

VAR. Height varies considerably, from occasionally less than 1 m. for low, usually weak growth, regenerating after cutting for fodder or during road maintenance, to mature shrubs or small trees of up to 4 m. Pubescence is also extremely variable, being most noticeable on young leaves and weak branches.

var. **pumilans** S. *Carter* in K.B. 35: 430 (1980). Type: Kenya, Teita District, Maktau Hill, *Bally* 13412 (K, holo.!, EA, iso.!)

Small semi-prostrate shrub to less than 1 m. high, with branchlets usually less than 5 cm. long. Leaves to 10 × 5 mm. Cyathia 4–5 mm. in diameter; glands 1–1.5 mm. wide. Capsule to 4 × 4.5 mm. Seeds 2.5 × 2.1 mm. Fig. 88/6, p. 467.

KENYA. Teita District: Maktau Hill, 16 Feb. 1980, *Gilbert* 5818! & Buchuma, 16 Sept. 1961, *Polhill & Paulo* 478!; Kwale District: between Samburu and Mackinnon Road, 30 Aug. 1953, *Drummond & Hemsley* 4048!

TANZANIA. Masai District: 55 km. N. of Arusha on Nairobi road, 10 Dec. 1964, *Leippert* 5327!; Arusha District: foot of Nagasseni Mt., 13 Dec. 1969, *Richards* 24930!; Moshi District: Masailand, track to Moshi, *Richards* 23670!

DISTR. K 7; T 2, 3; not known elsewhere

HAB. In grassland and open *Acacia-Commiphora* bushland; 300–1500 m.

SYN. [*E. spinescens* sensu Dale & Greenway, K.T.S.: 201 (1961), pro parte (*Drummond & Hemsley* 4048), non Pax]

NOTE. Although distinct at its type locality, herbarium specimens seen suggest that distribution of this form overlaps that of what seem to be depauperate plants of var. *spinescens*. Since it is often extremely difficult to distinguish between different specimens of the two forms, they are accorded varietal status only.

68. **E. matabelensis** *Pax* in Ann. Naturhist. Mus. Wien 15: 51 (1900); N.E. Br. in F.T.A. 6(1): 546 (1911); T.T.C.L.: 210 (1949); E.P.A.: 452 (1958); K.T.S.: 198 (1961); F.F.N.R.: 198 (1962); S. Carter in K.B. 40: 823 (1985). Type: Zimbabwe, *Penther* 944 (BM, iso.!)

Woody shrub or small tree to 3(–5) m. high, with peeling greyish brown bark; young branches densely puberulous, glabrescent; branching trichotomous, with branchlets spine-tipped. Leaves alternate or more usually fasciculate, oblanceolate to 6 × 1.7 cm., base cuneate, apex obtuse to rounded, margin entire, lamina densely pubescent when young especially beneath, glabrescent; petiole 1–3 mm. long; stipules glandular, minute, linear. Cymes terminal in densely pubescent 3–7-branched umbels, with rays to ± 1 cm. long, forking usually twice; bracts leaf-like, ± 5 mm. long, often yellowish green. Cyathia densely pubescent, subsessile, or the central one of the umbel on a peduncle to 5(–10) mm. long, ± 3.5 × 6 mm., with cup-shaped involucres; glands 5, spreading, shallowly saucer-shaped, rounded, 2–2.5 mm. wide, yellow; lobes rounded, ± 1.5 × 2 mm., deeply fringed. Male flowers many: bracteoles fan-shaped, deeply divided with feathery apices; stamens ± 4.5 mm. long. Female flower: ovary subsessile, densely pubescent; styles 2 mm. long, joined at the base, widely spreading, with flattened, rugose, bifid apices. Capsule slightly exserted on an erect pedicel to 5 mm. long, obtusely 3-lobed, ± 7 × 8 mm., densely pubescent. Seeds globose, 3.5 mm. in diameter, smooth, brown and obscurely speckled.

KENYA. Machakos District: near Sultan Hamud, Kenia Estate, 28 Nov. 1965, *Archer* in *E.A.H.* 13408!; Kitui District: Thika–Garissa road, 26 km. E. of road to Kabaa, 15 Dec. 1977, *Stannard & Gilbert* 1148!; Masai District: Nguruman Hills, Lenyora, 27 Sept. 1944, *Bally* 3887!
TANZANIA.   Shinyanga District: between Kisapu and Mango, 23 Jan. 1936, *B.D. Burtt* 5527!; Arusha District: Anata Sika, 2 Jan. 1971, *Richards & Arasululu* 26447!; Kondoa, 16 Jan 1962, *Polhill & Paulo* 1190!
DISTR. **K** 1, 4, 6; **T** 1–5, 7; Zambia, Malawi and Zimbabwe extending into Botswana, Angola, the Caprivi Strip and Mozambique
HAB.   Dry sandy soils and rocky slopes with fairly open mixed deciduous bushland; 700–1900 m.

SYN. *E. jaegeriana* Pax in E.J. 43: 87 (1909). Type: Tanzania, Mbulu District, Lake Eyasi, *Jaegar* 339 (B, syn. †) [*Jaeger* 99 (K, syn. fragment!) from Tanzania, Lushoto District, Pangani swamp, is *E. cuneata* var. *spinescens*]
      E. currori N.E. Br. in F.T.A. 6(1): 545 (1911). Type: Angola, *Curror* 29 (K, holo.!)
      E. inelegans N.E. Br. in F.T.A. 6(1): 547 (1911). Type: Tanzania, Chunya District, Usafwa, Songwe valley, *Goetze* 1052 (K, holo.!, BM, iso.!)

VAR. In the drier regions of the species distribution, in northern Tanzania and Kenya, mature leaves retain their pubescence and branching is compact and usually strictly trichotomous. In the moister regions of central and southern Tanzania, mature leaves are usually quite glabrous and branching can sometimes be so lax as to almost lose the characteristic trichotomy.

NOTE.   The latex is used in bird-lime, and occasionally as chewing-gum.

6c. Sect. **Espinosae**

Woody shrubs, sometimes scandent. Cyathia solitary, axillary, surrounded at the base by a cluster of small leafy or scarious bracts; involucral glands entire, spreading. Capsule far-exserted on a reflexed pedicel, deeply 3-lobed. Seeds ovoid, with a cap-like caruncle.

One species in East Africa.

69. E. espinosa *Pax* in E.J. 19: 120 (1894); N.E. Br. in F.T.A. 6(1): 547 (1911); T.T.C.L.: 211 (1949); K.T.S.: 198 (1961); F.F.N.R.: 198 (1962). Type: Tanzania without precise locality, *Fischer* 285 (B, holo. †, K, drawing and fragment of holo.!)

Erect or sometimes scandent woody shrub to 2(–4) m. high, with smooth shiny bark; young branches pubescent. Leaves petiolate, elliptic, to 4.5 × 2.5 cm., base cuneate, apex rounded and minutely apiculate, margin entire, midrib prominent beneath, young leaves pubescent towards the base; petiole to 1 cm. long, usually pubescent on young leaves; glandular stipules large, reddish. Cymes of solitary, axillary cyathia, surrounded at the base by 4–8 scarious oblong bracts, ± 2 × 1.5 mm., with finely ciliate margins. Cyathia subsessile, ± 3 × 5 mm., with funnel-shaped involucres; glands 5, transversely elliptic, ± 1.5 × 2.5 mm., yellowish green, touching; lobes subquadrate, ± 1 mm. long, margin denticulate. Male flowers: bracteoles deeply laciniate, margins ciliate; stamens 3.5 mm. long. Female flower: ovary glabrous, subtended by an obvious 3-lobed perianth; styles 2 mm. long, joined for ⅓, with spreading, bifid apices. Capsule exserted on a reflexed usually puberulous pedicel to 1 cm. long, deeply 3-lobed with slightly depressed apex, to 8.5 × 10 mm., green often tinged with purple. Seeds ovoid, ± 5.5 × 4.5 mm., slightly dorsi-ventrally compressed, smooth, pale grey with a yellow cap-like caruncle 2.5 mm. wide.

KENYA. Machakos District: Kanga, 29 Aug. 1959, *Verdcourt* 2387!; Teita District: Maungu–Ndara road, 12 Sept. 1953, *Drummond & Hemsley* 4274! & Ndi Hill, 13 June 1972, *Cheseny* 14/72!
TANZANIA.   Mbulu District: Tarengire, 13 Sept. 1958, *Mahinde* HSM/250!; Pare District: Makanya, 1934, *Bally* in *C.M.* 11954!; Mbeya District: 64 km. from Mbeya on Iringa road, Sept. 1959, *Procter* 1378!
DISTR. **K** 4, 7; **T** 2, 3, 5, 7; Malawi, Zambia, Zimbabwe
HAB.   In *Acacia-Commiphora* bushland; 500–1500 m.

SYN. *E. gynophora* Pax in E.J. 34: 374 (1904). Types: Tanzania, Pare District, between Kisiwani and Maji ya Juu, *Engler* 1579 & 1586 (B, syn.†, K, drawings!)

VAR. Specimens from southern tropical Africa appear to be from somewhat larger shrubs which have a conspicuously peeling bark, and larger leaves with longer petioles, scarcely pubescent when young.

7. Subgen. **Tirucalli**

Trees or shrubs, with fleshy green cylindrical branches. Leaves sessile, entire, lanceolate, small and quickly deciduous; stipules modified as glands, or absent but with the leaf-scar becoming conspicuously calloused and apparently glandular. Cyathia in terminal umbellate cymes, with rays simple or branching dichotomously; bracts sessile, scarious or leaf-like, usually quickly deciduous. Involucres bisexual or sometimes unisexual; glands 4–5, or 6–8 on central cyathia which are larger and produce only ♂ flowers; lobes 5. Male flowers clearly exserted from the involucre, the bracteoles often with conspicuously feathery apices. Perianth of ♀ flower sometimes obvious and 3-lobed. Capsule exserted on a reflexed pedicel. Seeds ovoid, with a caruncle.

1. Branches with longitudinal striations; cyathia in dense
   clusters . . . . *70. E. tirucalli*
   Branches without striations; cyathia in umbels with 3–8
   simple rays 2
2. Central cyathium of the umbel similar in size to the
   laterals, with 5 involucral glands and usually bisexual *72. E. calamiformis*
   Central cyathium larger than the laterals, with 6–8
   involucral glands and producing only ♂ flowers 3
3. Branchlets ± parallel to the main branches; cyathia 10 mm.
   or more in diameter; capsule ± 10 × 15 mm. . . . *71. E. nubica*
   Branchlets spreading; cyathia to 8 mm. in diameter;
   capsule ± 5.5 × 7 mm. . . . . *73. E. gossypina*

70. E. tirucalli L., Sp. Pl.: 452 (1753); Boiss. in DC., Prodr. 15(2): 96 (1862); N.E. Br. in Fl. Cap. 5(2): 293 (1915); F.P.N.A. 1: 478 (1948); T.T.C.L.: 211 (1949); U.O.P.Z.: 255 (1949); I.T.U., ed. 2: 128 (1952); E.P.A.: 460 (1958); K.T.S.: 201 (1961); F.F.N.R.: 199 (1962); F.P.U.: 113 (1962); Leach in Kirkia 9: 69 (1973). Lectotype: Ceylon, J. Commelin, Hort. Med. Amstel. Rar. Pl. 1: 27, t. 14 (1697)

Densely branched often apparently dioecious shrubs to 4 m. or trees to 12 (–15) m. high, with brittle succulent branchlets ± 7 mm. thick often produced in whorls, green and longitudinally finely striated, with white to yellowish latex. Leaves few, fleshy, linear-lanceolate, to 15 × 2 mm., present only at the tips of young branchlets and very quickly deciduous; extreme tips of young leafy branchlets sparsely tomentose, with curled brown hairs, soon glabrescent; glandular stipules minute, dark brown. Cymes 2–6 congested at the apices of the branchlets, forking 2–4 times, with rays less than 1 mm. long producing a dense cluster of cyathia developing only ♂ flowers, or occasionally a few ♀ flowers also present, or cyathia fewer and only ♀ flowers developing, the whole cyme ± glabrous or tomentose, with curled brown hairs, especially the involucres and lobes; bracts rounded, ± 2 × 1.5 mm., ± sharply keeled, usually glabrous except on the margin. Cyathia subsessile, ± 3 × 4 mm., with cup-shaped involucres; glands 5, subglobose to transversely elliptic, 0.5 mm. in diameter to 1.5 × 2 mm., bright yellow; lobes triangular, ± 0.5 mm long. Male involucres: bracteoles linear with plumose apices; stamens 4.5 mm. long; an aborted ♀ flower is occasionally present. Female involucres: bracteoles present and occasionally a few ♂ flowers; ♀ perianth distinctly 3-lobed below the tomentose ovary, with lobes 0.5 mm. long; styles 2 mm. long, joined at the base, with thickened deeply bifid recurved apices. Capsule glabrescent, exserted on a tomentose pedicel to 1 cm. long, subglobose, ± 8 × 8.5 mm. Seeds ovoid, 3.5 × 2.8 mm., smooth, buff speckled with brown and with a dark brown ventral line; caruncle 1 mm. across. Fig. 89/1, 2, p. 472.

UGANDA. Karamoja District: Koputh, *Brasnett* 164!; Teso District: Serere, Dec. 1932, *Chandler* 1047!; Mengo District: Entebbe, Oct. 1931, *Eggeling* 38 in *Brasnett* 204!
KENYA. W. Suk District: Ortum, Dec. 1960, *Tweedie* 2101!; Fort Hall District: Thika, 1 Apr. 1971, *Bally & Carter* 14130!; Teita District: Mwatate, 5 Apr. 1971, *Bally & Carter* 14136!
TANZANIA. Masai District: Olduvai, 26 Sept. 1977, *Raynal* 19302!; Lushoto District: E. Usambara Mts., near Amani, 27 Oct. 1930, *Greenway* 2568!; Tanga District: Sawa, 1 Feb. 1955, *Faulkner* 1575!; Zanzibar I., Massazine, 8 Mar. 1962, *Faulkner* 3014!
DISTR. U 1–4; K 1–7; T 1–8; Z; widespread throughout tropical Africa, also in the Arabian peninsula, Madagascar, and India to the Far East
HAB. In grassland and thin woodland, and often naturalised around habitation; 0–2000 m.

SYN. *E. rhipsalioides* Lem., Ill. Hort. 4, Misc.: 72 (1857). Type: Madagascar or Zanzibar, cult. Paris (no specimen)

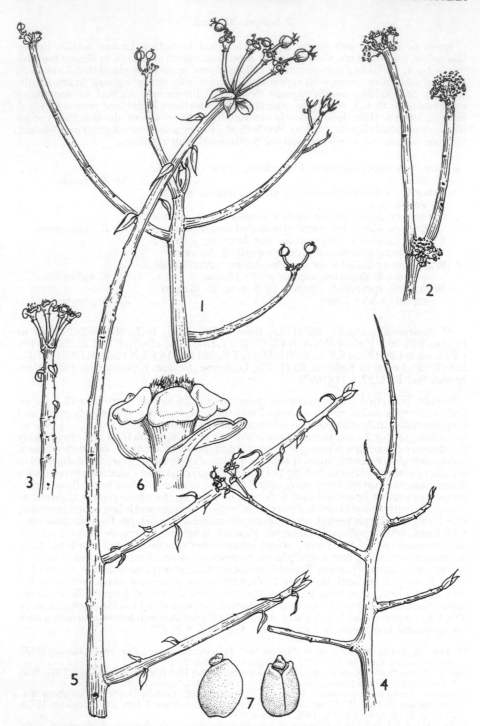

FIG. 89. *EUPHORBIA TIRUCALLI* — **1**, branch detail, with ♀ cyathia, × ⅔; **2**, flowering branch, with ♂ cyathia, × ⅔. *E. NUBICA* — **3**, flowering branch, × ¾. *E. CALAMIFORMIS* — **4**, branch detail, × ⅔. *E. GOSSYPINA* var. *GOSSYPINA* — **5**, branch detail, × ⅔; **6**, cyathium, × 4; **7**, seeds, × 4. 1, from *Batty & Carter* 14136; 2, from *Brasnett* B.164; 3, from *Tweedie* 4238; 4, from *Carter & Stannard* 558; 5, from *Polhill & Paulo* 630; 6, from *Drummond & Hemsley* 4044; 7, from *Faden* 74/1054. Drawn by Christine Grey-Wilson.

E. rhipsaloides N.E. Br. in F.T.A. 6(1): 556 (1911). Type: Angola, *Welwitsch* 630 (BM, lecto., K, isolecto.!)

E. media N.E. Br. in F.T.A. 6(1): 556 (1911). Types: Tanzania, Tabora, *Speke & Grant* (K, syn.!) & Dodoma District, Mukondokwera, *Busse* 150 (K, syn.!) & Lindi District, Mtange, *Busse* 2463 (EA, syn.!) & Malawi, Karonga *Scott* (K, syn.!); also Tanzania, *Fischer* 525, *Volkens* 191, *Merker* 576, *Mildbraed* 1082, *Kassner* 20 (not seen)

E. media N.E. Br. var. *bagshawei* N.E. Br. in F.T.A. 6(1): 556 (1911). Types: Mozambique, Tete, *Kirk* (K, syn.!) & Uganda, Bunyoro, *Bagshawe* 898 & Toro, *Bagshawe* 1196 & Zanzibar, *Stuhlmann* 642 (not seen)

E. scoparia N.E. Br. in F.T.A. 6(1): 557 (1911); E.P.A.: 458 (1958). Types: Sudan, Roseires, *Muriel* 67 (K, syn.!) & Ethiopia, Eritrea, *Schweinfurth* 345 (K, drawing of syn.!) & Schahagenne, *Schimper* 896 (not seen)

NOTE. Because of the ease with which branch cuttings take root and quickly form dense bushes, this species has been used extensively as a hedging plant. If left, it soon becomes naturalised and eventually forms a small tree. Plants usually produce only ♂ flowers, with ♀ flowers (or plants) much less common. Some with bisexual cyathia also occur, although the ♀ flower apparently often aborts. The latex can cause severe eye injury, hence the use of hedges as an impenetrable barrier. It is also used as a fish poison and in local medicines as a purgative. Its use as a source of hydro-carbon oils is being investigated. During the Second World War the latex was used in South Africa in the development of a rubber substitute, but this proved to be unstable.

71. E. nubica N.E. Br. in F.T.A. 6(1): 554 (1911); F.P.S. 2: 68 (1952); E.P.A.: 453 (1958). Types: Sudan, Nubia coast, *Bent* & Ethiopia, Eritrea, near Accrur, *Schweinfurth & Riva* 1083 (both K, syn.!)

Sprawling or scrambling glabrous perennial shrub forming tangled masses to 2 m. high, densely branched with branchlets eventually ± parallel to the main branches; branches and branchlets ± 1 cm. thick, green, with leaf-scars made prominent by the formation of a dark brown callus developed as the leaf falls. Leaves sessile, lanceolate, to ± 18 × 5 mm., becoming reflexed, caducous. Cymes in 3–7-branched umbels surrounding a central cyathium, with simple rays to 2 cm. long; bracts broadly deltoid, to 5 mm. long, caducous. Cyathia on peduncles to 2 mm. long, ± 3.5 × 10 mm., with broadly funnel-shaped involucres, but twice as large in the deciduous central cyathium which develops only ♂ flowers; glands 4, or 7–8 in the central cyathium, circular to transversely elliptic, to 3.5 × 5 mm., golden yellow; lobes broadly triangular, 1 × 2 mm., with apex deeply notched, margin finely ciliate. Male flowers: bracteoles very numerous, spathulate, apices plumose; stamens 4.5 mm. long. Female flower: ovary pedicellate; styles ± 3 mm. long, joined at the base with spreading thickened apices bifid for 1 mm. Capsule obtusely 3-lobed, to ± 1 × 1.5 cm., exserted on a reflexed pedicel ± 1 cm. long. Seeds ovoid, ± 4 × 3.3 mm., grey, faintly rugulose, with a caruncle 1.5 mm. across. Fig. 89/3.

KENYA. Northern Frontier Province: Mt. Kulal on road to Mission, Mar. 1972, *Tweedie* 4238! & Ol Lolokwe [Lolokwi], 15 Apr. 1979, *M. Gilbert* 5354!
DISTR. U 2; K 1, 2; Sudan, Ethiopia, Somalia
HAB. In dry open deciduous woodland and bushland; 1350–1525 m.

NOTE. There appear to be several forms of this species around the area of the type locality and extending into northern Somalia, all of which are reasonably distinct from the obviously related *E. schimperi* Presl and its allies occurring in the Arabian peninsula. Without thorough investigation in the field it is difficult to establish which form represents the true *E. nubica*, since differences appear to lie in mode of branching, in colour and thickness of stems and branches, and especially in habit, all characters which are not easily apparent from herbarium specimens and which are seldom recorded adequately by collectors. From the material available measurements of inflorescence details appear to be variable, although, when present, some reliance may be placed upon capsule shape and size and especially in seed indumentum. Until more is known of this common and widespread complex the form occurring in Kenya must remain as *E. nubica* sensu lato.

72. E. calamiformis *Bally & S. Carter* in K.B. 40: 824 (1985). Type: Ethiopia, Sidamo Region, Moyale–Mega road, *Gillett* 14174 (K, holo.!, EA, iso.!)

Glabrous perennial shrub scrambling to 3 m. high, forming tangled masses of semi-woody branches, with very slender brittle branchlets to ± 15 cm. long arising at right-angles to the main branches and possibly deciduous; branches less than 1 cm. thick, glaucous, with leaf-scars made ± prominent by the formation of a dark brown callus produced as the leaf falls. Leaves sessile, linear-lanceolate, to 30 × 3 mm., reflexed, caducous. Cymes terminal on lateral branchlets, in 3–5-branched umbels surrounding a central cyathium, with simple rays to 3 cm. long, sometimes with 1–2 rays replaced by sterile leafy branchlets; bracts deltoid, ± 4.5 × 3.5 mm., but resembling the leaves below the umbel, caducous. Cyathia on peduncles to 2 mm. long, ± 3 × 5 mm., with funnel-shaped involucres; glands 4, or 5 on the deciduous central cyathium which often develops only ♂ flowers, subcircular to transversely elliptic, ± 2 × 3.5 mm., yellow; lobes transversely rectangular, ± 0.7 × 1.5 mm., apex deeply notched, margin finely ciliate. Male flowers: bracteoles numerous, spathulate, plumose; stamens 4.3 mm. long. Female flower: ovary pedicellate; styles 2.5 mm. long, joined at the base, with spreading, thickened, distinctly bifid apices. Capsule obtusely 3-lobed, ± 5.5 × 6.5 mm., exserted on a pedicel ± 6 mm. long. Seeds ovoid, ± 3 × 2.5 mm., brownish grey, rugose, with a caruncle 1 mm. across. Fig. 89/4, p. 472.

UGANDA. Karamoja District: Katikelele, May 1956, *Wilson* 246!
KENYA. Northern Frontier Province: El Barta Plains, 1.5 km. S. of Bartagwet, 19 Nov. 1977, *Carter & Stannard* 558! & Mt. Kulal, Mar. 1972, *Tweedie* 4304! & 48 km. from Ramu on El Wak road, 9–10 May 1978, *Gilbert & Thulin* 1609!
DISTR. U 1; K 1; southern Ethiopia and southern Somalia
HAB. Dry rocky slopes and sandy soils with open *Acacia* woodland; 400–1875 m.

73. E. gossypina Pax in E.J. 19: 119 (1894); N.E. Br. in F.T.A. 6(1): 553 (1911); W.K.F.: 37 (1948); T.T.C.L.: 211 (1949); E.P.A. 447 (1958); K.T.S.: 198 (1961); U.K.W.F.: 223 (1974); S. Carter in K.B. 40: 823 (1985). Type: Tanzania, Mwanza District, Kagehi [Kayenzi], *Fischer* 514 (B, holo.†)

Glabrous much-branched perennial shrub, sprawling to 1.5 m. high or scrambling in bushes and small trees to 4 m. high; branches spreading, ± succulent, green, ± 1 cm. thick, with dark brown leaf-scars becoming ± prominent on older branches. Leaves sessile, lanceolate to linear-lanceolate to 4 × 1 cm., reflexed, caducous. Cymes in 4–8-branched umbels surrounding a central sessile cyathium, with simple rays to 3 cm. long, rarely more; bracts deciduous or persistent, green or red, broadly deltoid to 5 × 6 mm., with apex occasionally ciliate, the umbellate bracts ovate to 15 × 10 mm. Cyathia on peduncles to 4 mm. long, ± 3 × 7.5 mm., with funnel-shaped involucres, the central cyathium a little larger and usually deciduous, developing only ♂ flowers; glands 4(–5), or 6–8 on the central cyathium, subcircular to transversely elliptic, ± 3 × 4.5 mm., ± reflexed; lobes subquadrate, ± 1 × 1.5 mm., margin ciliate, with deeply notched apex. Male flowers: bracteoles very numerous, spathulate, deeply laciniate, plumose; stamens 5 mm. long. Female flower: ovary pedicellate; styles ± 3 mm. long, joined at the base, with spreading, deeply bifid apices. Capsule obtusely 3-lobed, ± 5.5 × 7 mm. on a pedicel to ± 8 mm. long. Seeds ovoid, 3 × 2.5 mm., slightly laterally compressed, pale greyish brown, faintly mottled, almost smooth; caruncle 1.3 mm. across.

var. gossypina

Bracts green, usually deciduous, the umbellate bracts longer than broad, to 15 × 8 mm.; involucral glands yellow with rounded margins. Fig. 89/5–7, p. 472.

UGANDA. Busoga District: Kaliro, 4 Aug. 1971, *Kakasko & Ferneira* 136!; Mengo District: Mawokota, cult. Entebbe, Dec. 1907, *E. Brown* 414!
KENYA. Teita District: Tsavo National Park East, Mazinga Hill, 15 Aug. 1969, *Bally* 13398!; Kwale District: between Samburu & Mackinnon Road, 30 Aug. 1953, *Drummond & Hemsley* 4044!; Tana R. District: 48 km. S. of Garsen, 9 Oct. 1961, *Polhill & Paulo* 630!
TANZANIA. Masai District: E. Serengeti, Oldiang'arangar, Nov. 1962, *Oteke* 245!; Arusha District: Mkuru road on N. side of Mt. Meru, 29 Dec. 1968, *Richards* 23502!; Pare District: near Kihurio, 8 July 1960, *Leach & Brunton* 10210!
DISTR. U 3, 4; K 1, 3, 4, 6, 7; T 1–3; disjunct population in Zimbabwe
HAB. In *Acacia-Commiphora* bushland; 15–1900 m.

SYN. *E. implexa* Stapf in K.B. 1908: 408 (1908). Type: Uganda, Mengo District, Mawokota, *E. Brown* 414, cult. Entebbe (K, holo.!)
E. merkeri N.E. Br. in F.T.A. 6(1): 555 (1911); T.T.C.L.: 211 (1949). Types: Kenya, Masai District, Morijo, *Uhlig* 292 (B, syn. †) & Tanzania, Masai District, between Engaruka and Lake Natron, *Merker* 580 (B, syn.†, K, fragment and drawing!) & Sonjo to Sale, *Merker* 581 (B, syn. †, K, fragment!)

var. coccinea Pax in E.J. 19: 120 (1894). Type: Kenya, Massalni, *Fischer* 515 (B, holo.!)

Bracts crimson, always persistent, the umbellate bracts ± as long as broad, to 12 × 10 mm.; involucral glands greenish yellow with the outer margin often emarginate to ± crenulate.

KENYA. Nairobi District: Dagoretti, Oct. 1956, *Tweedie* 1417!; Machakos District: Uuni Hill, 23 Nov. 1960, *Wilkinson* in *E.A.H.* H.313/60!; Masai District: Ngong Hills, Apr. 1960, *Bally* 12202!
TANZANIA. Arusha District: Ngare Nanyuki, 31 Dec. 1970, *Richards & Arasululu* 26420!
DISTR. **K** 3–6; **T** 2; not known elsewhere
HAB. In lightly wooded grassland; 1375–2300 m.

VAR. The most intense colouring of var. *coccinea* occurs at the higher altitudes of the species distribution, especially around Nairobi, where the persistent bracts are bright crimson and the glands green. Further south towards Mt. Meru and south-east in Machakos District, the variety occurs with red-tinged bracts, together with the entirely green-bracted var. *gossypina*, colouration depending apparently upon shade conditions, being most intense in full sun. Towards the coast reddening disappears and bracts are invariably deciduous. A form occurs around Voi and the Pare Hills with cyathial bracts distinctly ciliate on the apical margins, but differing in no other respect.

8. Subgen. **Euphorbia**

Glabrous perennial herbs, shrubs or trees, with spiny succulent cylindrical or longitudinally ridged (angled) and often toothed (tuberculate) stems and branches (except for *E. monadenioides*, a hysteranthous geophyte). Leaves opposite (apparently) or spiral, usually minute, sessile and soon deciduous, occasionally large, petiolate and more persistent; leaf-scars at the apex of horny pads (spine-shields) which are situated on the angles and crown the teeth, bearing a pair of spines below the leaf, sometimes united to form a single spine, and stipules usually modified as prickles or often vestigial. Cyathia in axillary cymes, usually 1-forked, rarely more, subsessile or pedunculate, solitary, or in horizontal or vertical rows of 2–5, or in groups of 5–8, usually developing successively, the point of emergence (flowering eye) situated shortly above the spine-shield, or sometimes enclosed by the continuous horny margin of the stem-angles formed by the elongation of the spine-shields and then sometimes flanked by small subsidiary spines; bracts shorter than the cyathia, with denticulate margins, soon deciduous. Involucres bisexual, with 5 entire glands and 5 fringed lobes. Stamens clearly exserted; bracteoles enclosed. Perianth of the ♀ flower reduced to a rim below the ovary, or sometimes of 3 distinct toothed lobes. Capsule subsessile or exserted and then often on a reflexed pedicel, distinctly 3-lobed or occasionally subglobose and ± fleshy (before dehiscence in some tree species). Seeds subglobose or ovoid, usually minutely tuberculate, without a caruncle.

1. Hysteranthous geophyte, spineless . . . . *159. E. monadenioides*
   Herbs, shrubs or trees, spiny 2
2. Spines paired, divergent, occasionally joined at the base,
   rarely absent 3
   Spines single, with or without forked tips 78
3. Trees, or shrubs with a definite central stem at least 1 m.
   high and more than 5 cm. thick 4
   Herbs or shrubs, sometimes scandent 24
4. Trees 5–30 m. high; spine-shields separate at least on
   young growth; spines 0–10 mm. long 5
   Large shrubs, or trees to 15 m. high; spine-shields
   contiguous, usually forming a continuous horny
   margin; spines 0.5–8 cm. long 16
5. Terminal branches not more than 3 cm. wide 6
   Terminal branches at least 3.5 cm. wide 11
6. Slender laxly-branched trees to 8 m. high; leaves 5–20 cm.
   long, persistent at the branch apices 7
   Trees to 20 m. high; leaves less than 1.5 cm. long, soon
   deciduous 8
7. Branches obtusely 4-angled; leaves to 25 × 10 cm.   *74. E. teke*
   Branches sharply 3-angled or narrowly winged; leaves to
   9 × 5 cm. . . . .   *75. E. bwambensis*
8. Trees to 15 m. high, with branches ± horizontally
   spreading to form a large crown; terminal branches
   obtusely 3–4-angled to cylindrical; spines stout or
   spinescence absent on mature trees 9
   Trees to 20 m. high, with small crowns on slender naked
   trunks; terminal branches sharply 3–4-angled or
   winged; spines needle-like 10

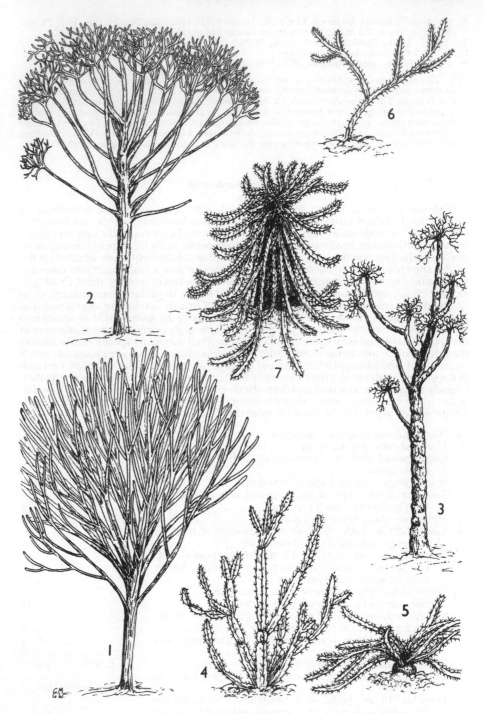

FIG. 90. Habit diversity in *EUPHORBIA* subgen. *EUPHORBIA* — **1**, *E. candelabrum*, × ¹/₁₅₀; **2**, *E. robecchii*, × ¹/₁₅₀;
3, *E. quadrialata*, × ¹/₁₅₀; **4**, *E. heterochroma*, × ¹/₁₅; **5**, *E. uhligiana*, × ¹/₁₀; **6**, *E. glochidiata*, × ¹/₁₅; **7**, *E. schizacantha*,
× ¹/₁₀. 1, from *Greenway* 6481; 2, from *Bally* 1949; 3, from 'Trop. African Plants'; 4, from *Greenway* 6475; 5,
from *Bally* E52; 6, from *Bally* cc.xxx.v.6, 7, from *Bally* 9321. Drawn by Christine Grey-Wilson.

9. Branches of mature trees cylindrical and spinescence
   absent; involucral glands yellow; capsules purplish grey   92. *E. robecchii*
   Branches of mature trees 3–4-angled, with spines to 5 mm.
   long, rarely absent; involucral glands and capsules
   reddish purple . . . .   93. *E. lividiflora*
10. Branches 3-winged, conspicuously sinuate-toothed; cyathia
   reddish yellow . . . .   95. *E. wakefieldii*
   Branches 4-angled, ± straight edged; cyathia ivory-white   96. *E. quadrialata*
11. Terminal branches 5–6-angled; cymes 2-forked, with rays
   to 2.5 cm. long; cyathia and capsules crimson   94. *E. tanaensis*
   Terminal branches 2–5-angled; cymes 1-forked, with rays
   to 1.5 cm. long; cyathia yellow 12
12. Leaves persistent on young growth, to 20 cm. long   77. *E. obovalifolia*
   Leaves semi-persistent on young growth, less than 2.5 cm.
   long 13
13. Branches stoutly (3–)4(–5)-angled . . . .   79. *E. candelabrum*
   Branches deeply 2–4-winged 14
14. Trees to 25 m. high; terminal branches 3–4-winged; spines
   to 2.5 mm. long . . . .   78. *E. cussonioides*
   Trees to 8 m. high; terminal branches 2–3-winged; spines
   to 10 mm. long 15
15. Spine-shields triangular to 5 mm. long; capsule subglobose
   and fleshy before dehiscence, exserted on a reflexed
   pedicel ± 10 mm. long . . . .   76. *E. dawei*
   Spine-shields decurrent for 10 mm. or more; capsule
   deeply 3-lobed, erect on a pedicel ± 5 mm. long   80. *E. nyikae*
16. Branches deeply constricted into ± rounded or obovate-
   oblong segments, 5–20 cm. wide 17
   Branches only slightly constricted into oblong segments
   with parallel sides, 2–4 cm. wide 23
17. Branches all 3-winged, with segments often wider than
   long; longest spines to 8 cm. long . . . . . .   86. *E. breviarticulata*
   Branches 2–8-winged, with segments rounded or longer
   than wide; longest spines 1–4 cm. long 18
18. Branches thinly 2–4-winged; horny margin 1 mm. or less in
   width . . . .   80. *E. nyikae*
   Branches 3–8-winged; horny margin at least 1.5 mm. wide 19
19. Branch-segments at least 8 cm. wide, deeply winged 20
   Branch-segments to 7 cm. wide, shallowly winged 22
20. Branches rarely rebranching, 4–6(–8)-winged; capsules ± 6
   × 10 mm. . . . .   83. *E. cooperi* var.
                                              ussanguensis

   Branches rebranching, 3–5-winged; capsules at least
   9 × 19 mm. 21
21. Branches predominantly 3–4-winged; capsules deeply 3-
   lobed, ± 9 × 19 mm. . . . .   81. *E. bussei*
   Branches predominantly 4–5-winged; capsules shallowly
   3-lobed, 12 × 22–25 mm. . . . .   82. *E. magnicapsula*
22. Terminal branches 3–7 cm. wide; cyathia subsessile;
   capsules ± 6 × 9 mm. . . . .   84. *E. hubertii*
   Terminal branches 2–5 cm. wide; cyathia on peduncles
   2–7.5 mm. long; capsules ± 7.5 × 12 mm. . . . .   85. *E. adjurana*
23. Terminal branches 4-angled; capsules subsessile . . .   97. *E. dumeticola*
   Terminal branches (4–)5(–6)-angled; capsules exserted on
   a reflexed pedicel ± 7 mm. long . . . .   98. *E. quinquecostata*
24. Capsule subsessile, or only shortly exserted on an erect
   pedicel 25
   Capsule exserted on a reflexed pedicel 30
25. Stems and branches 2–5-winged or deeply angled, 2.5–12
   cm. wide, usually constricted into segments 26
   Stems and branches ± cylindrical or 4–8-angled, rarely
   more than 2 cm. wide, not constricted into segments 43
26. Spine-shields forming a continuous horny margin 27

Spine-shields separate or rarely contiguous at the widest
part of the branch-segments 29
27. Shrubs to 4.5 m. high; branches deeply 3-winged; longest
spines to 8 cm. long . . . . *86. E. breviarticulata*
Shrubs to 50 cm. high; branches shallowly (3–)4–5-winged;
longest spines 1.5–2.5 cm. long 28
28. Cyathia and capsules crimson . . . . *87. E. biharamulensis*
Cyathia yellow; capsules green flushed with red . . . . *88. E. pseudoburuana*
29. Stems 3-winged, to 50 cm. long, constricted into segments;
spine-shields decurrent, occasionally contiguous *89. E. buruana*
Stems 2–3-winged, to 15 cm. long, not constricted into
segments; spine-shields triangular, never contiguous *90. E. brevitorta*
30. Shrubs, with branches decumbent, erect or subscandent, at
least 50 cm. high 31
Low herbs, sometimes decumbent, to 30 cm. high 40
31. Branches 6–8-angled; spine-shields quite separate *99. E. classenii*
Branches 4–6(–8)-angled; spine-shields forming a
continuous horny margin, or almost touching the
flowering eye below 32
32. Branches 4–5-angled, 1–2 cm. thick; longest spines 6–8
(–10) mm. long 33
Branches 4–6(–8)-angled, 1.5–4 cm. thick; longest spines
10–15 mm. long 38
33. Spines reduced or often obsolete on the upper flower-
bearing branches 34
Spines of even length, not reduced on flowering branches 36
34. Shrubs to 2 m. high, with decumbent branches, distinctly
variegated with yellow-green . . . . . . *100. E. heterochroma*
Shrubs erect or subscandent to 3.5 m. high, uniformly
green or only indistinctly variegated with blue-green 35
35. Branches 4–5-angled; basal spines to 10 mm. long;
capsules red or red-flushed . . . . *103. E. heterospina*
Branches 4-angled; basal spines to 8 mm. long; capsules
buff-coloured . . . . *104. E. borenensis*
36. Shrubs to 4 m. high; spines to 6 mm. long *101. E. stapfii*
Shrubs to 1 m. high; spines to 8 mm. long 37
37. Spreading shrub, with irregular decumbent branching;
cymes 2-forked; pedicel of capsule 10 mm. long *102. E. petraea*
Compact shrub, with regular erect branching; cymes
1-forked; pedicel of capsule 5 mm. long . . . *105. E. vulcanorum*
38. Shrubs to 1.5(–2) m. high, with branches spreading from
the base, sparsely rebranched; spines to 1.5 cm. long,
rarely shorter on upper flowering branches . . . *108. E. tescorum*
Much-branched erect shrubs to 2.5–3 m. high; spines to 1
cm. long, reduced on upper flowering branches 39
39. Branches (4–)5–6-angled; mature cyathia and glands
bright red; capsules dark red *106. E. scarlatina*
Branches 4–5(–6)-angled; mature cyathia and glands dark
red; capsules reddish black . . . . *107. E. atroflora*
40. Stems with 8–10 longitudinal ribs (angles) . . . . *109. E. baioensis*
Stems 4-angled, or cylindrical with spine-shields in 5–7
spiral series 41
41. Stems 4-angled . . . . *110. E. colubrina*
Stems cylindrical 42
42. Stems smooth, branched, forming tangled mats; prickles
half as long as the spines . . . . *111. E. quadrispina*
Stems tessellated, unbranched, in tufts; prickles much
shorter than the spines . . . . *112. E. ellenbeckii*
43. Stems and branches cylindrical, with (3–)5–10 longitudinal
ribs formed by the contiguous spine-shields 44
Stems and branches 4-angled, or subcylindrical with spine-
shields in 4(–5) longitudinal series 45

44. Stems (3–)5–6-ribbed; spines obvious, 1–4 mm. long,
      recurved; involucral glands brownish pink   . . .   *150. E. erlangeri*
    Stems 5–10-ribbed; spines very weak, often apparently
      absent to the naked eye, 1–3 mm. long; cyathia deep
      crimson . . . .   *151. E. cryptospinosa*
45. Erect or subscandent shrubs to at least 1 m. high, with
      branches at least 1 cm. thick 46
    Herbs or shrubs, occasionally subscandent but with
      branches no more than 1 cm. thick, or if more then
      plant procumbent 49
46. Lateral cyathia of each cyme usually 3, with red glands   *113. E. proballyana*
    Lateral cyathia of each cyme 2, with crimson or yellow
      glands 47
47. Plants in branching clumps; cyathia and capsules deep
      crimson   . . . .   *117. E. elegantissima*
    Plants single-stemmed and sparsely branched; cyathia and
      capsules not crimson 48
48. Involucral glands greenish yellow . . . .   *115. E. quadrilatera*
    Involucral glands yellowish green, with a distinct red
      border 1 mm. wide   . . . .   *116. E. quadrangularis*
49. Spine-shields separate longitudinally or contiguous, often
      expanded at the apex but not divergent into 2 arms and
      stem-teeth not prominent for more than 5 mm.; cyathia
      always glabrous 50
    Spine-shields separate longitudinally, the upper part
      dividing each side of the flowering eye into 2 distinct
      arms, or stem-teeth very prominent to 1 cm. long;
      cyathia usually minutely papillose 73
50. Spine-shields contiguous, or their bases decurrent to just
      above the flowering eye below 51
    Spine-shields quite separate, usually with their bases ±
      halfway to the flowering eye below 61
51. Cyathial peduncles at least 3.5 mm. long; involucral glands
      quite separate 52
    Cyathia subsessile, or on peduncles never more than 2.5
      mm. long; involucral glands contiguous laterally to
      form a flat ring 55
52. Stems decumbent, branching, ± 10 mm. thick, forming
      clumps 30–120 cm. high 53
    Stems erect, branched only at the base, to 6 mm. thick 54
53. Stems bright green, sinuately toothed; spinescence bright
      red-brown . . . .   *118. E. rubrispinosa*
    Stems glaucous, usually distinctly variegated, sharply
      toothed; spinescence purplish black . . . . . .   *121. E. greenwayi*
54. Spine-shields extremely slender; spines minute, apparently
      absent to the naked eye . . . .   *119. E. asthenacantha*
    Spine-shields slender, but widening at the apex; spines
      and prickles obvious   . . . .   *120. E. torta*
55. All parts of the cyathia, including capsules, deep crimson   *142. E. saxorum*
    Cyathial parts and capsules not deep crimson 56
56. Spines 0–1 mm. long but rarely developed; prickles 1–3
      mm. long, extremely fine . . . .   *143. E. taruensis*
    Spines obvious, 2–20 mm. long; prickles 1–5 mm. long 57
57. Spine-shields forming a continuous dark-brown horny
      margin, even on young growth . . . .   *145. E. odontophora*
    Spine-shields almost touching but not forming a
      continuous horny margin 58
58. Spines to 12 mm. long or more; prickles at least 2 mm.
      long 59
    Spines no more than 6 mm. long; prickles no more than
      1.5 mm. long 60
59. Spine-shields extended 1 mm. above the spines   . . .   *114. E. reclinata*
    Spine-shields elongated 2–6 mm. above the spines   *144. E. tenuispinosa*

60. Stems densely branching, greyish green, with dark grey
    spinescence; involucral glands yellow . . . . .   *124. E. discrepans*
    Stems sparsely branched, bright green with whitish
    spinescence; involucral glands yellow becoming red   *128. E. cataractarum*
61. Involucral glands transversely rectangular, contiguous
    laterally to form a flat ring 62
    Involucral glands transversely elliptic or subcircular, quite
    separate or only just touching laterally 68
62. Branches 10–15 mm. thick, with prominent rounded teeth 63
    Branches 5–10 mm. thick, with obscure or pointed teeth 64
63. Stems semi-prostrate, densely branching to form tangled
    mats 20–30 cm. high   *123. E. tetracanthoides*
    Stems erect to 1 m. high, sparsely branched, or semi-
    prostrate with trailing stems . . . . . .   *125. E. nyassae*
64. Prickles ± equal in length to the spines, arranged in the
    form of an X . . . .   *127. E. isacantha*
    Prickles much shorter than the spines 65
65. Spine-pairs 5–8 mm. apart; involucral glands yellow   *126. E. exilispina*
    Spine-pairs 10–20 mm. apart; involucral glands brownish
    or red 66
66. Stems greyish green tinged with purple; involucral glands
    yellowish brown . . . .   *129. E. eyassiana*
    Stems blotched pale and darker green; involucral glands
    bright brick-red 67
67. Stems 5–7 mm. thick, with shallowly pointed teeth;
    spinescence reddish brown to grey . . . .   *141. E. gemmea*
    Stems 8–10 mm. thick, with prominently pointed teeth;
    spinescence dark brown to blackish . . . .   *146. E. dauana*
68. Spinescence purplish black; involucral glands quite
    separate, pinkish yellow; styles 4–7 mm. long,
    conspicuously exserted from the involucre 69
    Spinescence pale grey or copper-coloured; involucral
    glands just touching, bright yellow; styles ± 2 mm. long,
    not much exserted from the involucre 70
69. Stems 4-angled, with prominent rounded teeth; involucres
    barrel-shaped . . . .   *122. E. angustiflora*
    Stems subcylindrical, with obscure teeth; involucres
    funnel-shaped . . . .   *140. E. subscandens*
70. Stems few, becoming prostrate, to 90 cm. long and 10–20
    cm. thick . . . .   *136. E. samburuensis*
    Stems densely tufted, to 50 cm. long and 5–8 mm. thick 71
71. Stems distinctly toothed, with teeth 5–10 mm. apart; spine-
    shields 2–2.5 mm. wide above the spines; spines sturdy
    to 10 mm. long . . . .   *138. E. dichroa*
    Stems obscurely toothed, with teeth 10–20 mm. apart;
    spine-shields 1–1.5 mm. wide above the spines; spines
    slender, 5–17 mm. long 72
72. Spinescence pale grey . . . . . .   *137. E. septentrionalis*
    Spinescence copper-coloured . . . . .   *139. E. cuprispina*
73. Stem-teeth extremely pointed, to 1 cm. long; diverging
    arms of the spine-shields short . . . .   *130. E. furcata*
    Stem-teeth shallow, or prominent but not extremely
    so; diverging arms of the spine-shields obvious 74
74. Stems to 8 mm. thick, subcylindrical, with shallow teeth 75
    Stems 10–20 mm. thick, 4–5-angled, with shallow or
    prominent teeth 76
75. Stems uniformly greyish green; spines needle-like to 12
    mm. long, prickles very fine to 5 mm. long . .   *132. E. petricola*
    Stems variegated with yellowish green; spines slender to 18
    mm. long, prickles to 2 mm. long . . . .   *133. E. marsabitensis*
76. Stems uniformly dark green or only obscurely mottled;
    cyathia glabrous . . . .   *131. E. uhligiana*
    Stems distinctly streaked and mottled; cyathia papillose 77

77. Stems tufted and semi-prostrate to 30 cm. long, 4–5-angled,
    12–20 mm. thick; spines to 20 mm. long, prickles to 2.5 mm.
    long . . . .   *134. E. vittata*
    Stems crowded, erect to 45 cm. high, 4-angled, to 15 mm.
    thick; spines very sturdy to 30 mm. long, prickles to 8
    mm. long . . . .   *135. E. turkanensis*
78. Spines with forked tips 79
    Spines without forked tips 82
79. Plants with a central stem to 6 cm. thick and numerous
    trailing cylindrical branches to 12 mm. thick . . .   *156. schizacantha*
    Plants subscandent with stems and branches 4-angled, not
    differentiated 80
80. Spinescence sturdy, with spine-shields contiguous or
    almost touching the flowering eye below . . . .   *147. E. glochidiata*
    Spinescence often weak, with the spine-shield bases 5–10
    mm. above the flowering eye below 81
81. Stems and branches variously mottled, variegated or
    striped; cyathia brownish yellow to brick-red . . .   *148. E. fluminis*
    Stems and branches bluish green, with a darker stripe
    down the angles; cyathia vivid carmine . . . . .   *149. E. meridionalis*
82. Stout shrub or tree 2–5 m. high, with branches to 3.5 cm.
    thick . . . .   *91. E. venenifica*
    Low spreading herbs, or if erect, never more than 50 cm.
    high 83
83. Stems erect or tufted 84
    Branches radiating from the apex of a short thick stem ± at
    ground-level 87
84. Stems few, erect to 30(–50) cm. high . . . . . .   *152. E. ballyana*
    Stems many, tufted from a thick fleshy rootstock 85
85. Stems 10–15 mm. thick, 4–5-angled; spine-shields decurrent
    to 8 mm. or ± halfway to the flowering eye below   *153. E. similiramea*
    Stems 5–10 mm. thick, 4-angled or subcylindrical; spine-
    shields shortly decurrent to 4 mm. 86
86. Branches to 25 cm. long, with ± prominent teeth; spines to
    20 mm. long, prickles to 1 mm. long . . . . . .   *154. E. graciliramea*
    Branches to 10 cm. long, obscurely toothed; spines to 12
    mm. long, prickles absent . . . .   *155. E. laikipiensis*
87. Branches to 1 m. long and 2 cm. thick; spines to 7 cm. long;
    involucral glands yellow . . . .   *157. E. kalisana*
    Branches to 15 cm. long and 1 cm. thick; spines to 2 cm.
    long; involucral glands red . . . .   *158. E. actinoclada*

74. E. teke *Pax* in E.J. 19: 118 (1894); N.E. Br. in F.T.A. 6(1): 563 (1911); F.P.S. 2: 67 (1952); I.T.U., ed. 2: 127 (1952). Type: Zaire, Niamniam, Juru, *Schweinfurth* III. 143 (K, iso.!)

Laxly branched shrubby tree to 6(–10) m. high; trunk to 10 cm. in diameter, smooth, with flaking grey to pale brown bark; branches fleshy, obtusely 4-angled, 1–1.5 cm. thick; angles with small teeth 2–5 cm. apart. Spine-shields obtusely triangular or subcircular, to 5 mm. in diameter, sometimes not quite enclosing the large circular leaf-scar (3–4 mm. in diameter) below on young growth, eventually ± enclosing the flowering eye immediately above; spines stout, 1–4 mm. long; prickles vestigial, just evident on young growth. Leaves persistent at the branch apices, fleshy, obovate, to 25 × 10 cm., margin entire, midrib prominent beneath, base tapering to a petiole 2–4 cm. long. Cymes 1–5 in a horizontal line, 1-forked with peduncles to 6 cm. long and cyme-branches to 2 cm. long; bracts deltoid, ± 3.5 × 4.5 mm. Cyathia ± 3.5 × 10 mm., with widely funnel-shaped involucres; glands transversely elliptic, ± 2.5 × 5 mm., almost touching, greenish yellow; lobes transversely elliptic, ± 1.5 × 2.5 mm. Male flowers; bracteoles spathulate, plumose; stamens 5.8 mm. long. Female flower: perianth distinct, with 3 or more irregular lobes to ± 1 mm. long; styles 2.5 mm. long, joined at the base, apices rugose. Capsule exserted on a pedicel 5–10 mm. long, acutely 3-lobed, with truncate base, ± 15 × 25 mm., green becoming brown. Seeds subglobose, slightly compressed laterally, 5 × 4.5 mm., pale brown, smooth.

UGANDA. Toro District: Bwamba, Nyabarango Forest, Apr. 1940, *Sangster* 640!; Masaka District: Nkose I., 21 Jan. 1956, *Dawkins* 871!; Mengo District: Semunya Forest, 16 June 1950, *Dawkins* 601!

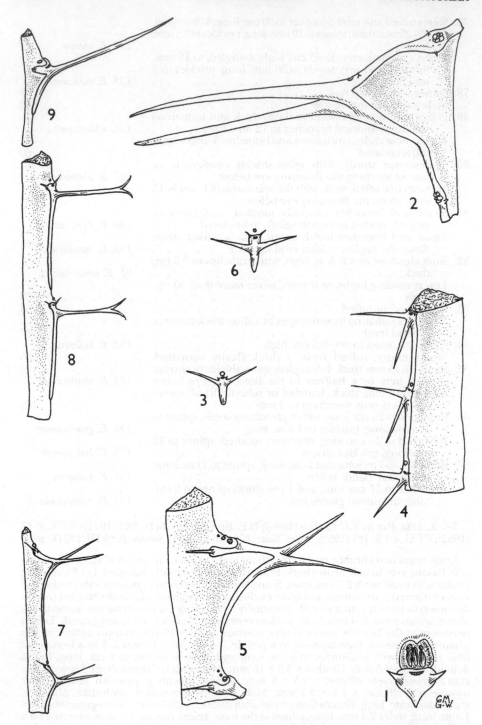

FIG. 91. Spine-shield formation in *EUPHORBIA* subgen. *EUPHORBIA* — 1, *E. candelabrum*; 2, *E. bussei* var. *kibwezensis*; 3, *E. wakefieldii*; 4, *E. heterospina* subsp. *baringensis* 5, *E. uhligiana*; 6, *E. septentrionalis*; 7, *E. tenuispinosa* var. *tenuispinosa*; 8, *E. glochidiata*; 9, *E. graciliramea*; all × 2. 1, from *Greenway & Kanuri* 11191; 2, from *Bally* 14211; 3, from *Mabberley* 716; 4, from *Gilbert* 5061A; 5 from *Bally* 10666; 6 from *Bally & Carter* 16533; 7, from *Bally* 13415; 8, from *Bally & Smith* 14598. Drawn by Christine Grey-Wilson.

TANZANIA. Bukoba District: Munene Forest Reserve, 6 Jan. 1975, *Balslev* 527!
DISTR. U 2, 4; T 1; NE. Zaire, southern Sudan, Central African Republic
HAB. Clay soil with swamp forest; 900–1200 m.

75. E. bwambensis *S. Carter* in K.B. 42: 681 (1987). Type: Uganda, Toro District,
Bwamba Forest, *Carcasson* 31 in *Bally* 10588 (E. 443) (K, holo.!, EA, G, iso.!)

Tree to ± 6.5 m. high, with ± drooping branches sparsely rebranched to form a loose
untidy crown; branches fleshy, shortly 3(–5)-angled, or narrowly winged, 1–2.5 cm. wide;
angles shallowly toothed, with teeth 1–4 cm. apart. Spine-shields obtusely triangular, to ± 4
× 3 mm., eventually extended to enclose the flowering eye just above; spines to 3 mm.
long; prickles vestigial. Leaves persistent at the branch apices, obovate, to 9 × 5 cm.,
tapering to a petiole to 7 mm. long, margin entire, midrib prominent beneath. Cymes
solitary, 2-forked, with peduncle ± 2 mm. long and cyme-branches 5–7 mm. long; bracts
deltoid, ± 3 × 3 mm. Cyathia ± 2.5 × 6.5 mm., with cup-shaped involucres; glands
transversely elliptic, ± 1.5 × 3 mm., almost touching, yellow; lobes subcircular, ± 1.5 × 1.5
mm. Male flowers: bracteoles spathulate, laciniate; stamens 4 mm. long. Female flower:
perianth distinct, with 3 or more irregular lobes 0.5–1.5 mm. long; styles 2.5 mm. long,
joined at the base, apices rugose, bifid. Capsule exserted on a stout pedicel to 8 mm. long,
acutely 3-lobed, ± 9 × 14 mm., reddening. Seeds subglobose, 3.8 mm. in diameter, reddish
brown with pale speckles, smooth.

UGANDA. Toro District: Bwamba, Jan. 1939, *Sangster* 508!; Kidongo, Aug. 1937, *Eggeling* 3368!
DISTR. U 2; known only from Bwamba Forest
HAB. In *Cynometra* forest; ± 800 m.

SYN. [*E. neglecta* sensu Dale & Greenway, I.T.U., ed. 2: 127 (1952), *non* N.E. Br.]

76. E. dawei *N.E. Br.* in F.T.A. 6(1): 583 (1912); F.P.N.A. 1: 480 (1948); I.T.U., ed. 2: 126
(1952). Types: Uganda, Toro District, Burule, Lake Ruisamba [George], *Dawe* 677 (K,
syn.!) & Wimi R., *Bagshawe* 1019 (K, syn.!)

Tree to 15(–25) m. high, with a simple trunk rarely branching, to 60 cm. in diameter and
with fissured greyish brown bark marked with 4(–5) vertical rows of persistent spines and
pit-scars resulting from fallen branches; branches spreading horizontally, to ± 4 m. long,
densely and irregularly rebranching to form a rounded crown; terminal branchlets
fleshy, 2–3(–4)-angled, 4–10 cm. wide, deeply and thinly winged, ± constricted at irregular
intervals of 5–40 cm. into oblong segments; angles sinuately toothed with teeth 1–2.5 cm.
apart. Spine-shields triangular to 5 × 3 mm., eventually extended to enclose the flowering
eye just above; spines 2–6 mm. long; prickles rudimentary, triangular. Leaves persistent
on young growth, obovate, to 6 × 4 mm. Cymes 3–7, crowded together, 1(–2)-forked, with
peduncles and cyme-branches 3–4 mm. long; bracts deltoid ± 2.5 × 2.5 mm. Cyathia ± 3 ×
6.5 mm., with funnel-shaped involucres; glands transversely elliptic, ± 1.5 × 3 mm., almost
touching, yellowish green; lobes transversely elliptic, ± 1.5 × 2 mm., deeply fringed. Male
flowers: bracteoles laciniate, plumose; stamens 4.3 mm. long. Female flower: perianth 1.5
mm. long irregularly divided into 5–9 short lobes; styles 2 mm. long, joined at the base,
apices rugose, bifid. Capsule exserted on a curved pedicel to 1 cm. long, subglobose, ± 7 ×
10 mm., fleshy, green flushed with red, hardening to become obtusely 3-lobed. Seeds
subglobose, 4.5 mm. in diameter, smooth.

UGANDA. Toro District: between Katwe and Fort Portal on equator, 16 Apr. 1948, *Bally* 6192 (E. 318)!;
  Ankole District: Ruizi R., 15 Nov. 1950, *Jarrett* 149!; Mengo District: Bussi I., Mawakota, Feb. 1932,
  Eggeling 200!
TANZANIA. Bukoba District: 110 km. S. of Bukoba, 15 Sept. 1974, *Balslev* 30; Biharamulo District: 1.5
  km. on Nyamirembe road from junction with Biharamulo–Mwanza road; Kigoma District: 104 km.
  S. of Kigoma, 1 Sept. 1959, *Harley* 9498!
DISTR. U 2, 4; T 1, 4; Burundi, Rwanda, E. Zaire
HAB. Deciduous woodland, sometimes open with grassland; 800–1300 m.

SYN. [*E. nyikae* sensu de Witte, Expl. Parc Nat. Albert 1, t. 5/2, 6/1 (1937) & Lebrun, Expl. Parc. Nat.
  Albert 1: 318, t. 50/2, 51/1 (1947), *non* Pax]

NOTE. Hybrids with *E. candelabrum* are known to occur in the area of the Kazinga Channel (U 2,
  Ankole District, see Lock in E.J. 98: 428 (1977))

77. E. obovalifolia *A. Rich.*, Tent. Fl. Abyss. 2: 239 (1851); DC., Prodr. 15(2): 80 (1862);
N.E. Br. in F.T.A. 6(1): 594 (1912); E.P.A.: 454 (1958); K.T.S.: 200 (1961); S. Carter in Cact.
Succ. J. Gt. Brit. 38: 66 (1976). Type: Ethiopia, Shoa & Shire, *Quartin Dillon* (P, holo.)

Tree to 30 m. high, with a simple trunk to ± 90 cm. in diameter and roughened grey bark; primary branches ascending, rebranching irregularly and ± densely at the apices to form a spreading rounded crown; terminal branchlets fleshy, 3(–4)-angled, 5–17 cm. wide, deeply and thinly winged with wings 2–7 cm. wide, ± constricted at irregular intervals into oblong segments 15–40 cm. long; angles straight to sinuately toothed, with teeth 1.5–3 cm. apart. Spine-shields rounded, to 5 mm. in diameter, eventually extended to 5 mm. above the spines to enclose the flowering eye; spines stout, 1–3 mm. long; prickles absent or vestigial; spine-shields and spines becoming corky and eventually disintegrating. Leaves persistent for some time on young growth, obovate, to 15 × 6 cm., fleshy, midrib keeled on the lower surface, base tapering to a petiole ± 1 cm. long. Cymes 1–3 in a horizontal line or occasionally 4 crowded together, 1-forked, with stout peduncles, and cyme-branches 3–4 mm. long arranged in a vertical plane; bracts rounded, ± 4 × 7 mm. Cyathia ± 4 × 7 mm., with cup-shaped involucres; glands transversely elliptic, ± 2 × 4 mm., with undulate margin, not quite touching, yellow; lobes transversely elliptic, ± 2 × 3.5 mm. Male flowers: bracteoles spathulate, plumose; stamens 6 mm. long. Female flower: perianth irregularly divided into 5–12 lanceolate lobes with toothed margins, to 6 mm. long; styles 3.5 mm. long, joined at the base, apices thickened, ± bifid. Capsule on a stout pedicel 3 mm. long, subglobose, to 12 × 16 mm., fleshy, red, hardening before dehiscence to become obtusely 3-lobed with woody walls 2–3 mm. thick. Seeds subglobose, slightly compressed laterally, 4.5 mm. in diameter, yellowish brown speckled with black, smooth.

UGANDA. Karamoja District: Morongole, 11 Nov. 1939, *A.S. Thomas* 3328!; Mbale District: Buginyanya, Oct. 1937, *Hancock* 1991! & Buluganya, 27 Aug. 1932, *A.S. Thomas* 280!
KENYA. W. Suk District: Chepinat Forest, 9 Dec. 1970, *Mabberley* 497!; Elgeyo District: Kipkunurr Forest, 14 Apr. 1975, *Hepper & Field* 4975!; Masai District: Ngong Hills, 3 Nov. 1962, *Verdcourt* 3292!
TANZANIA. Ufipa District: Chala Mt., Oct. 1943, *Bredo* in *Bally* E. 243! & Mbizi Forest, 17 June 1960, *Leach & Brunton* 10081!; Mbeya District: 26 km. N. of Mbeya, 7 Feb. 1974, *Bally & Carter* 16468!
DISTR. U 1, 3; K 2–6; T 3, 4, 7; Ethiopia, northern Malawi-Zambia border
HAB. On well-drained rocky slopes with montane evergreen forest; 1800–2500 m.

SYN. *E. amplophylla* Pax in Ann. Ist. Bot. Roma 6: 186 (1897). Type: Ethiopia, between Alghe & Oi, *Riva* 1636 [1337] (FT, holo.!)
E. winkleri Pax in E.J. 30: 342 (1901); N.E. Br. in F.T.A. 6(1): 593 (1912); T.T.C.L.: 214 (1949). Type: Tanzania, Njombe District, Lipanga [Lipanye] Mts., *Goetze* 1000 (B, holo. †, K, fragment & drawing of holo.!)

NOTE. *E. neglecta* N.E. Br. in F.T.A. 6(1): 593 (1912) should not be included in synonymy (as by S. Carter in Cact. Succ. J. Gt. Brit. 38: 66 (1976)). It was described from a plant of unknown origin cultivated at Kew. The type material has thinly fleshy 5-winged branches bearing a few small leaves 2–3 cm. long. Its original identification, under cultivation at Kew, as *E. abyssinica* Gmel. is more probably correct, the thin wings and semi-persistent leaves being a result of growth under very unnatural conditions.

78. E. cussonioides *Bally* in Bothalia 7: 29 (1958); K.T.S.: 197 (1961). Type: Kenya, Masai District, Ngong, *McDonald* in *Bally* E.42 (EA, holo., K, iso.!)

Tree to 25 m. high, with a simple trunk to ± 80 cm. in diameter and smooth wrinkled greyish brown bark; primary branches spreading-ascending, sparsely rebranched, with clusters of fleshy branchlets at the apices forming a loose crown; branchlets irregularly rebranching, 3–4-angled, 2–6 cm. wide, deeply and thinly winged, ± constricted at irregular intervals into pear-shaped or oblong segments to ± 20 cm. long; angles sinuately toothed with teeth ± 1.5 cm. apart. Spine-shields obtusely triangular, 3–5 mm. long and wide, eventually extended to enclose the flowering eye just above; spines 1–2.5 mm. long; prickles rudimentary. Leaves rounded, to 9.5 × 10 mm. Cymes 1–3 in a horizontal line or occasionally 4 crowded together, 1-forked with stout peduncles, and cyme-branches 2–3 mm. long arranged in a vertical plane; bracts rounded, 3 × 4 mm. Cyathia ± 3.5 × 7 mm., with cup-shaped involucres; glands transversely elliptic, ± 2 × 4 mm., not quite touching, yellowish green; lobes subcircular, ± 2 × 2 mm. Male flowers: bracteoles spathulate, plumose; stamens 4.5 mm. long. Female flower: perianth irregularly divided into 3–5 lobes, to 1.5 mm. long; styles 1.2 mm. long, joined at the base, apices rugose, bifid. Capsule subsessile, obtusely 3-lobed, ± 12 × 15 mm. Seeds subglobose, 4 × 3.5 mm., brownish grey, mottled, smooth.

KENYA. Fort Hall District: Thika, Chania Gorge, 1 Apr. 1971, *Bally & Carter* 14127!; Masai District: Ngong, 8 Jan. 1939, *McDonald* in *Bally* E.42!
DISTR. K 4, 6; not known elsewhere

HAB.   Steep rocky slopes in dry evergreen forest; 1300–1800 m.

NOTE.   The last remaining tree at the type locality of Ngong Boma was cut down in 1980. This species is endangered, and is now restricted to a few small pockets of remnant forest near Embu, Thika and in the Ngong Hills.

79. **E. candelabrum** *Kotschy* in Mitt. Geog. Ges. Wien 1(2): 169 (1857); N.E. Br. in F.T.A. 6(1): 598 (1912); F.P.S. 2: 66 (1952); I.T.U., ed.2: 126 (1952); E.P.A.: 444 (1958); K.T.S.: 196 (1961); S. Carter in K.B. 42: 680 (1987). Type: Sudan, Dar Foq [Fung], Jebel Kaçane [Qeissan], 1847–48, *Trémaux*, Voy. Soudan Orient., Atlas, t. 13 & 14 (1853)

Tree to 12(–20) m. high, with a simple trunk to ± 90 cm. in diameter and rough fissured grey bark; branches persistent from ± 3 m. upwards, suberect, densely rebranching to form eventually a large broadly rounded crown; terminal branchlets fleshy, (3–)4(–5)-angled, 5–10 cm. wide, ± square in cross-section to distinctly winged, with wings to 2.5 cm. wide, usually ± constricted at irregular intervals into oblong segments 15–25 cm. long; angles straight to shallowly sinuately toothed, with teeth 1–1.5 cm. apart. Spine-shields very obtusely triangular, to ± 8 × 7 mm., extending ± 5 mm. above to include the flowering eye; spines stout, to 5 mm. long; prickles flexible, triangular, 1.5 mm. long, soon deciduous; spines and spine-shields soon becoming corky, rusty-brown and disintegrating. Leaves oblanceolate on seedlings and young growth, to 7 × 1.5 cm., deltoid on older growth and soon deciduous, ± 5 × 5 mm. Cymes 1–6 crowded together, 1-forked, with stout peduncles 5–20 mm. long and cyme-branches ± 5 m. long; bracts rounded, ± 5 × 6 mm. Cyathia ± 4 × 9 mm., with broadly cup-shaped involucres; glands transversely elliptic, ± 2 × 4 mm., almost touching, golden yellow; lobes transversely elliptic, ± 2.2 × 3 mm. Male flowers many: bracteoles spathulate, plumose; stamens 5.8 mm. long. Female flower: perianth irregularly divided into 3 or more filiform lobes 2–4 mm. long, sometimes with 1 or 2 teeth; styles 2 or 3, ± 3 mm. long, joined at the base, apices thickened, rugulose, bifid. Capsule shortly exserted on a stout pedicel ± 5 mm. long, 2–3-locular, subglobose, ± 8 × 12 mm., fleshy, green becoming red, hardening immediately before dehiscence to ± 6 × 9 mm. and very obtusely 2–3-lobed. Seeds subglobose, slightly compressed laterally, 3 mm. in diameter, greyish brown speckled with paler brown, smooth.

var. **candelabrum**

Styles 3, with 3-locular fruits.

UGANDA.   Karamoja District: Lodoketemit, Nov. 1962, *Kerfoot* 4500!; Toro District: Mohokya, Sept. 1941, *A.S. Thomas* 3768!; Mengo District: Mawakota, Nov. 1930, *Laboratory Services Dept.* 1825!
KENYA.   Baringo District: 17 km. from Kabarnet on Marigat road, 30 Oct. 1964, *Leippert* 5248!; Central Kavirondo District: Port Victoria, 25 Oct. 1948, *Glasgow* 48/1!; Masai District: Mara Masai Game Reserve, Egerok, 24 Sept. 1947, *Bally* 5448!
TANZANIA.   Maswa District: Moru Kopjes, 30 Aug. 1962, *Greenway* 10785!; Moshi District: Ngare Nairobi R., 25 Aug. 1968, *Bigger* 2157!; Tanga District: Mtimbwani, 6 Dec. 1935, *Greenway* 4233!
DISTR.   U 1–4; K 1–7; T 1–8; Z; Sudan, Ethiopia, Somalia, eastern Zaire, Malawi, Zambia
HAB.   Rocky slopes and well-drained flat-land, sometimes on termite mounds, with open wooded grassland; (15–)900–2000 m.

SYN.   *E. reinhardtii* Volk. in N.B.G.B. 2: 263 (1899); N.E. Br. in F.T.A. 6(1): 590 (1912); T.T.C.L.: 214 (1949). Type: Tanzania, Lushoto District, Usambara Mts., *Holst* 8821, partly (B, holo.†)
   E. confertiflora Volk. in N.B.G.B. 2: 266 (1899); N.E. Br. in F.T.A. 6(1): 592 (1912); T.T.C.L.: 213 (1949); S. Carter & M.G. Gilbert in K.B. 42: 387 (1987). Type: Tanzania, Lushoto District, Usambara Mts., *Holst* 8821, partly (B, holo.†)
   E. murielii N.E. Br. in F.T.A. 6(1): 589 (1912). Type: Sudan, Bahr-el-Jebel, *Muriel* E (K, holo.!)
   E. calycina N.E. Br. in F.T.A. 6(1): 597 (1912); F.P.N.A. 1: 478 (1948). Types: Sudan, Meshra el Rek, *Schweinfurth* 1259 (K, syn.!, BM, isosyn.!) & Legbi, *Schweinfurth* 2824A (K, syn.!, BM, isosyn.!)

NOTE.   Hybrids with *E. dawei*, which have predominantly 3-angled winged branches and are virtually sterile, are known to occur in the region of the Kazinga channel (U 2, Ankole District). *E. confertiflora* should be placed in synonymy here rather than under *E. heterochroma* as its description of a 1 m. high shrub suggests. Volkens' description adopted by subsequent authors, was a composite one, based on vegetative characters of the shrub, for which there was no specimen, and inflorescence characters of juvenile material removed from *Holst* 8821. Cyathial measurements and especially the sessile ovary with lobed perianth indicate that this material should not have been separated by Volkens from the rest of the specimen which he used to describe *E. reinhardtii*.

var. **bilocularis** (*N.E. Br.*) *S. Carter* in K.B. 42: 681 (1987). Type: Kenya, Machakos District, Kibwezi, *Scheffler* 335 (K, holo.!, BM, iso.!)

Styles predominantly 2, with 2-locular fruits.

UGANDA. Karamoja District: Lotome, 6 Oct. 1952, *Verdcourt* 774!
KENYA. Northern Frontier Province/Tana River District: Bura–Ijara, 15 Jan. 1943, *Bally* E.212!; Machakos District: Kiu, Wilson's Farm, 15 Sept. 1961, *Polhill & Paulo* 459!
TANZANIA. Pare District: Gonja, 21 June 1942, *Greenway* 6481!; Ulanga District: Kwiro, May 1960, *Haerdi* 526/0!; Mbeya District: 11 km. N. of Chimala R., 16 June 1938, *Pole Evans & Erens* 771!
DISTR. U 1; K 1, 4, 6, 7; T 1, 3, 5–7; Malawi, northern Zambia
HAB. On rocky slopes and in open wooded grassland; 550–1800 m.

SYN. *E. bilocularis* N.E. Br. in F.T.A. 6(1): 594 (1912); T.T.C.L.: 213 (1949); E.P.A.: 443 (1958)

VAR. A widely distributed species exhibiting considerable variation in overall size, width of the branches, wing formation of the angles and in inflorescence characters especially the length of the cyme-peduncle. The development of 2 styles with 2-locular fruits cannot be linked consistently with any other feature, either of the plant, or of habitat, or in distribution, and as such is treated as a variety only. Some 3-locular fruits, with 3 styles, can usually be found on otherwise 2-locular individuals, often on the same cyme, but, when present, 2 styles and 2-locular fruits appear to predominate.
    Distribution of the very closely related species *E. ingens* E. Mey. from Natal, is taken here to extend northwards into Mozambique, Zimbabwe and possibly southern Zambia and the Shaba Province of Zaire. It may eventually prove to be conspecific at least over part of this area, especially as 2-locular fruits are found on some specimens from as far south as Barberton, Transvaal (*Pole Evans* 2929).

NOTE. There is some controversy over when the name *E. candelabrum* was first validly published. Leach maintains (in Taxon 30: 483 (1981) & 35: 711 (1986)) that Kotschy's publication was predated by Welwitsch (in Proc. Linn. Soc. 2: 329 (1855)) for a different species from Angola, but I consider this to be invalid and a synonym of *E. conspicua* N.E. Br. (in F.T.A. 6(1): 600 (1912)). An explanation of this view is given in Taxon 34: 699 (1985).

80. E. nyikae *Pax* in P.O.A. C: 242 (1895) & E.J. 23: 533 (1897); N.E. Br. in F.T.A. 6(1): 585 (1912), pro parte excl. *E. bussei*; U.O.P.Z.: 255 (1949); L.C. Leach in Journ. S. Afr. Bot. 36: 49 (1970); S. Carter in K.B. 42: 677 (1987). Type: Tanzania, Lushoto District, near Magila, *Volkens* 51 (B, holo.†, K, drawing of holo.!)

Tree to 7 m. high, with branches persistent from ground-level upwards, or to 15(–20) m., with the trunk naked below; trunk to 30(–50) cm. in diameter, ± smooth, marked by 4–6 vertical rows of spines and pit-scars resulting from whorls of fallen branches; seedlings 2-angled; branches to 3.5 m. long, 2–5-angled, rebranching and curving upwards to form a loose ± ovate crown; terminal branchlets fleshy, 2–3(–4)-angled, 3–15 cm. wide, very deeply and thinly winged, constricted at irregular intervals into ± subcircular, obconic or oblong segments 5–20 cm. long; angles ± straight or sinuately toothed with teeth ± regularly or irregularly spaced 0.5–5 cm. apart. Spine-shields separate or forming a continuous horny margin, often extremely narrow; spines 5–40 mm. long; prickles vestigial or absent. Leaves deltoid, ± 1.3 × 2 mm. Cymes 1–7 crowded together, 1-forked, with peduncles and cyme-branches 4–7 mm. long; bracts deltoid, ± 3 × 5 mm. Cyathia ± 3 × 8 mm., with broadly funnel-shaped involucres; glands tranversely elliptic, ± 2.5 × 4 mm., just touching, golden yellow; lobes subquadrate, ± 1.5 × 1.5 mm. Male flowers many: bracteoles spathulate, deeply laciniate, plumose; stamens 4.5 mm. long. Female flowers: perianth distinctly 3-lobed, with lobes 1.5–2 mm. long; styles ± 3 mm. long, joined for ⅓ to halfway, apices thickened, rugulose, distinctly bifid. Capsule exserted on a pedicel ± 5 mm. long, deeply 3-lobed, ± 8 × 17 mm., base and apex truncate, green becoming red. Seeds subglobose, 4 mm. in diameter, buff speckled with brown, smooth.

var. nyikae

Tree to 15(–20) m. high; branchlets ± constricted to form mostly oblong segments to 7 × 20 cm.; angles straight to shallowly toothed with teeth ± regularly spaced 5–15 mm. apart; spine-shields separated at least on young growth, no more than 2 mm. wide at the widest point; spines subequal in length to 1 cm. long; cymes solitary, rarely 2–3 arranged horizontally. Fig. 92/1.

KENYA. Kwale District: Mwachi creek, 5 Feb. 1948, *Bally* 5781!; Kilifi District: Kombeni R. between Rabai and Ribe, 22 Oct. 1974, *Balslev* 47!
TANZANIA. Lushoto District: Soni Falls, 6 July 1960, *Leach & Brunton* 10188!; Handeni District: 8 km. NE. of Sindeni, 1 Feb. 1974, *Bally & Carter* 16411!; Bagamoyo District: 32 km. from Bagamoyo on Handeni road, 5 July 1960, *Leach & Brunton* 10183!
DISTR. K 7; T 3, 6; not known elsewhere
HAB. Sandy soils with fairly dense deciduous woodland; 250–700 m.

NOTE. A 2-winged seedling and material of 3-winged branches have been collected from a small population in the Great Ruaha Gorge, Kilosa District, Tanzania (*Bally & Gilbert* 14248). The

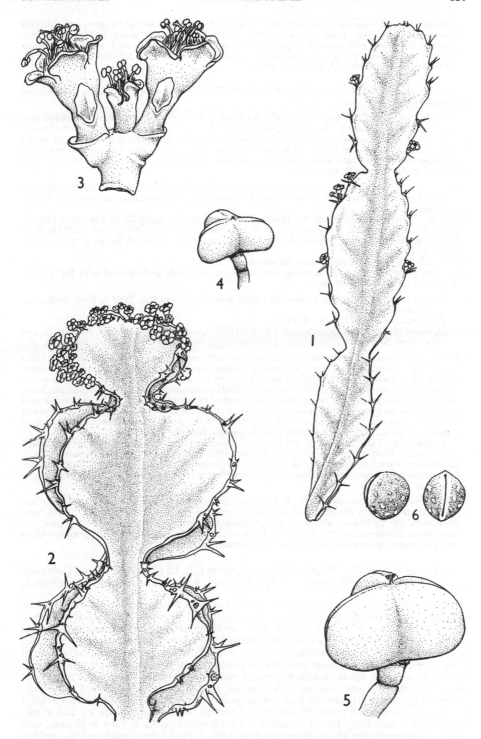

FIG. 92. *EUPHORBIA NYIKAE* var. *NYIKAE* — **1**, flowering branch, × ⅔. *E. BUSSEI* var. *KIBWEZENSIS* — **2**, flowering branch, × ⅔; **3**, cyme, × 3. Var. *BUSSEI* — **4**, capsule, × 2. *E. MAGNICAPSULA* var. *LACERTOSA* — **5**, capsule, × 2; **6**, seeds, × 2. 1, from *Bally* 11467; 2, from *Bally & Smith* 14423; 3, from *van Someren* in *Bally* E. 79; 4, from *Cribb & Grey-Wilson* 10215; 5, 6, from *Carter & Stannard* 562. Drawn by Christine Grey-Wilson.

branches are very wide, to 17 cm., with thin wings to 7.5 cm. wide, ± straight angles, spine-shields contiguous on the older branches but separate on young growth, and spines to 1 cm. long. One segment is fruiting with capsules exserted on slightly recurved pedicels 1 cm. long. Further material and information are needed before this can be established as a distinct taxon or as an abnormally robust form of *E. nyikae* var. *nyikae*.

var. **neovolkensii** (*Pax*) S. *Carter* in K.B. 42: 677 (1987). Type: Zanzibar, *Werth* (B, holo. †, K, drawing of holo.!)

Shrubby tree to 4(–7) m. high, rarely with the trunk naked below; branchlets constricted very irregularly into subcircular to obconic segments to 12 × 15 cm.; angles prominently toothed at the widest part of the segments, with teeth to 5 cm. apart; spine-shields forming a continuous horny margin, extremely narrow to 3.5 mm. wide, rarely separated on young growth, enlarged at the base of the longest spines to ± 5 mm. thick; spines very irregular in length, 5–40 mm. long; cymes (1–)3–7 crowded together at each flowering eye.

KENYA. Kwale District: 8 km. N. of Lunga Lunga, 8 July 1960, *Leach & Brunton* 10220!; Kilifi District: Bamba, 11 Feb. 1953, *Bally* 8536!; Lamu District: Duduntu, 48 km. from Lamu on Ijara road, 30 Jan. 1943, *Bally* E. 213!
TANZANIA. Tanga District: Mtotohova–Makajambi, 8 Dec. 1935, *Greenway* 4254!; Bagamoyo District: 72 km. S. of Mkata on Dar es Salaam road, 25 Nov. 1974, *Balslev* 349!; Uzaramo District: 25 km. N. of Dar es Salaam, 2 July 1960, *Leach & Brunton* 10167!; Zanzibar I., Pwani Mchangani, 26 Jan. 1929, *Greenway* 1188!
DISTR. **K**7; **T** 3, 6, 8; **Z**; **P**; not known elsewhere
HAB. On cliffs and sand-dunes along the sea-shore, or sandy soils inland with fairly open deciduous woodland; 2–250 m.
SYN. *E. volkensii* Werth, Veg. Ins. Sansib.: 50 (1901), *nom. illegit., non* Pax. Type as for *E. nyikae* var. *neovolkensii*
E. neovolkensii Pax in E.J. 34: 74 (1904)
[*E. nyikae* sensu Dale & Greenway, K.T.S.: 200 (1961), pro parte quoad *Bally* E. 288 & 8536, *non* Pax sensu stricto]

VAR. Var. *neovolkensii* occurs normally as a shrub or small tree with persistent branches, only older specimens sometimes shedding their lower branches. Principally 2-winged, branches of mature trees produce 3 wings or occasionally 4. Branch segmentation and sturdiness of the spination vary considerably, especially amongst the more southerly populations around Dar es Salaam. Some older robust specimens can be confused with weaker plants of *E. bussei* and it is possible that hybridisation may occur between the two species. A combination of habit, the number of branch angles, sturdiness of the spination, subtle differences in the shape of the branch segments, and capsule size presents the most reliable way of distinguishing the two species.
A population of tall shrubs with very long wavy-margined branch segments has been recorded from Newala District, Lupaso, near the Mozambique border. It appears to be similar to *E. halipedicola* Leach (in Journ. S. Afr. Bot. 36: 42 (1970)) from further south in central Mozambique, but material and more information are needed before it can be considered as distinct from *E. nyikae* var. *neovolkensii*.

NOTE. *E. lemaireana* Boiss. in DC., Prodr. 15(2): 81 (1862) was described from a living sterile plant cultivated at the Paris Museum and was said to originate from Zanzibar. It could have been collected there by Boivin, and if so was probably the plant described here as *E. nyikae* var. *neovolkensii*. However, since no further details are known, and especially since no specimen exists for verification, application of the name must remain uncertain.

81. E. bussei *Pax* in E.J. 33: 286 (1903); S. Carter in K.B. 42: 677 (1987). Type: Tanzania, Handeni District, near Magofu, *Busse* 317 (B, holo.†, K, fragment and drawing of holo.!)

Tree to 10(–15) m. high, with a trunk to ± 30 cm. thick and grey fissured bark marked with usually 6 vertical rows of persistent spines and pit-scars resulting from fallen branches; seedlings 3–4-angled; branches spreading to 3(–5) m., rebranching several times to form a rounded crown; terminal branchlets fleshy, 3–4-angled, to 15 cm. wide, deeply winged and deeply constricted into ovate to subcircular segments to 15 cm. long or more, elongated with undulate margins on young plants; angles sinuately toothed with teeth 1–3 cm. apart. Spine-shields forming a continuous horny margin 1.5–8 mm. wide; spines robust, 0.5–2 cm. long or to 3.5 cm. on young plants; prickles to 1 mm. long; flowering eye usually flanked by a pair of secondary spines to 2 mm. long. Leaves ovate, ± 2 × 2 mm. Cymes 3–8 crowded together, 1-forked, with stout peduncles and cyme-branches 3–8 mm. long; bracts rounded, ± 3.5 × 6 mm. Cyathia 4–4.5 × 8–10 mm., with cup-shaped involucres; glands transversely oblong, ± 2 × 4.5 mm., just touching, golden yellow; lobes subquadrate, ± 1.5 × 1.5 2 mm. Male flowers: bracteoles spathulate, laciniate; stamens 5.5 mm. long. Female flower: perianth 3-lobed, with lobes 2–3 mm. long; styles 3.5 mm. long, joined for ⅓ at the base, apices thickened, rugose, ± bifid.

Capsule exserted on a pedicel to 6 mm. long, deeply 3-lobed, ± 9 × 19 mm., base and apex truncate. Seeds subglobose 3.5–4.2 mm. in diameter, buff with dark brown speckles, smooth.

var. bussei

Branch-segments mostly ovate; wings 3(–4), thin; horny margin ± 1.5 mm. wide, enlarging to 4–5 mm. at the base of the spines; cymes usually 5–8, with peduncles 5–8 mm. long; cyathia 6–8 mm. in diameter, with glands ± 2 × 4 mm.; capsule exserted on a pedicel 5–6 mm. long. Fig. 92/4, p. 487.

TANZANIA. Lushoto District: Mbalu Mt. near Mkundi Mtae, 29 Jan. 1974, *Bally & Carter* 16394!; Handeni District: 6 km. W. of Handeni, 1 Feb. 1974, *Bally & Carter* 16412!; Morogoro District: 9.5 km. E. of Morogoro roundabout, 25 Nov. 1974, *Balslev* 350!
DISTR. T 3, 5, 6; not known elsewhere
HAB. Rocky slopes and sandy soils with open deciduous woodland; 350–1000 m.

SYN. *E. mbaluensis* Pax in E.J. 34: 85 (1904); N.E. Br. in F.T.A. 6(1): 586 (1912); T.T.C.L.: 213 (1949). Type: Tanzania, Lushoto District: Mbalu Mt., *Engler* 1472c (B, holo.†)
[*E. nyikae* sensu N.E. Br. in F.T.A. 6(1): 585 (1912) pro parte quoad syn. *Busse* 317, *non* Pax]

var. kibwezensis (*N.E. Br.*) *S. Carter* in K.B. 42: 678 (1987). Type: Kenya, Machakos District, Kibwezi, *Scheffler* 223 (K, holo.!, BM, iso.!)

Branch-segments mostly subcircular; wings (3–)4; horny margin ± 2 mm. wide, enlarging to 5–8 mm. wide at the base of the spines; cymes usually 3 in a horizontal line, occasionally up to 7 crowded together, with peduncles 3–5 mm. long; cyathia 8–10 mm. in diameter, with glands ± 2.5 × 5 mm.; capsule just exserted on a pedicel 3–5 mm. long. Fig. 92/2, 3, p. 487.

KENYA. Machakos District: Kibwezi, 14 July 1960, *Leach & Bayliss* 10242!; Kitui District: 2 km. W. of Mwingi, 30 Nov. 1971, *Bally & Smith* 14423!; Teita District: Mawileni, 7 km. W. of Voi, 31 July 1971, *Bally* 14210!
TANZANIA. Moshi District: 7 km. S. of Himo–Moshi road, 23 July 1960, *Leach* 10335!; Pare District: 3 km. from Mwango on Shigatini road, 6 Nov. 1974, *Balslev* 215! & Vudee, 27 Jan. 1930, *Greenway* 2072!
DISTR. K 1 (known only from near Isiolo), 4, 7; T 2, 3; not known elsewhere
HAB. Usually on steep rocky slopes with fairly dense deciduous woodland; 400–2000 m.

SYN. *E. kibwezensis* N.E. Br. in F.T.A. 6(1): 586 (1912); Blundell, Wild Fl. Kenya: 52 (1982)
[*E. nyikae* sensu Dale & Greenway, K.T.S.: 200 (1961) pro parte, quoad *Scheffler* 223, *non* Pax]

VAR. Var. *kibwezensis* is generally a sturdier tree than var. *bussei*, with notably more robust spination and larger almost subsessile cyathia. However, such differences are less distinct in populations around the northern end of the North Pares and towards Voi in Teita District. An exception is a small population near Himo Bridge, Moshi District, which exhibits the robust spination and large cyathia typical of var. *kibwezensis*.

The spination of young plants (to ± 1 m. high) of both varieties is usually very robust, often bearing a similarity to weaker plants of *E. breviarticulata*. Confusion is most likely among populations in the areas around the Usambara and Pare Mountains and in Teita District where both species occur. The possibility of hybridisation should not be ruled out, but *E. bussei* can usually be reliably distinguished by the presence of a central stem and by the shorter length of the longest spines, i.e. no more than 3.5 cm.

There may also be some overlapping of distinguishing features between var. *bussei* and *E. nyikae* var. *neovolkensii*, older mature plants of the latter sometimes shedding their lower branches to adopt a tree-like rather than shrubby habit, and producing 3-angled rather than 2-angled branches — see notes on variation under *E. nyikae* var. *neovolkensii*.

82. E. magnicapsula *S. Carter* in K.B. 42: 678 (1987). Type: Tanzania, Masai District, Ngorongoro, foot of Windy Gap road, *Greenway & Kanuri* 12532 (K, holo.!)

Tree to 12 m. high, with a trunk to 45 cm. in diameter, rarely branched, and grey-brown roughened fissured bark irregularly marked with pit-scars resulting from fallen branches; seedlings 4–5-angled; branches spreading upwards, to 2.5 m. long, rebranching to form a ± compact rounded or flattened crown; terminal branches fleshy, 3–5-angled, to 15 cm. wide, prominently and stoutly winged, deeply constricted to form subcircular or ovate segments to 20 cm. long, or shallowly constricted into oblong segments with undulate margins in young plants; angles sinuately toothed, with teeth 1–3 cm. apart, ± regularly spaced on older plants. Spine-shields forming a continuous horny margin 3–8 mm. wide; spines stout 5–15(–25) mm. long; prickles vestigial; a pair of secondary spines to 2.5 mm. long often flanking the flowering eye. Leaves deltoid, ± 2 × 3 mm. Cymes 1–3 in a horizontal row, or 4–5 crowded together, 1-forked in a vertical plane, peduncles and cyme-branches very stout, 5–8 × 5–8 mm.; bracts rounded, ± 5 × 9 mm. Cyathia 4.5–6 ×

10–12 mm., with cup-shaped involucres; glands transversely oblong, 2.5–3 × 5–6 mm., just touching, yellow; lobes subquadrate, ± 2 × 2.5 mm., conspicuously toothed. Male flowers many: bracteoles deeply laciniate; stamens 6–7 mm. long. Female flower: perianth conspicuous, 3-lobed, with lobes to 3 mm. long; styles 4.5–7 mm. long, joined at the base for ± ⅓ their length, apices thickened, rugulose, ± bifid. Capsule exserted on a stout pedicel 5–8 mm. long, 3-lobed with truncate base and slightly pointed apex, 12 × 22–25 mm., ± fleshy and becoming bright red. Seeds subglobose, slightly compressed laterally, ± 4.8 × 4.5 mm., buff with dark speckles and streaks, smooth.

var. magnicapsula

Trees 6–12 m. high, usually forming dense stands; crown rounded, generally narrower than the height of the tree in mature specimens, above a trunk naked from 3–10 m.; cymes 1–3 in each flowering eye, rarely 4.

KENYA. Naivasha District: Naivasha Farm, 22 Oct. 1979, *Kagari Muhu* 5!; Kiambu District: Kikuyu Escarpment, on Naivasha road below Lari Forest Guard Post, 16 Dec. 1966, *Perdue & Kibuwa* 8264!; Masai District: N. Loita Hills, Ol Doinyo Loloponi, 16 Jan. 1981, *Kuchar* 14355!
TANZANIA. Masai District: Ngaserai, 18 Sept. 1971, *Bally & Gilbert* 14250!; Mbulu District: Lake Manyara National Park Headquarters, 12 June 1965, *Greenway & Kanuri* 11847!; Arusha District: Mt. Meru, Kasekenyi Gorge, 10 Sept. 1971, *Richards & Arasululu* 27065!
DISTR. K 3, 4, 6; T 2, ?5 (see note); not known elsewhere
HAB. Rocky slopes with open deciduous bushland; 1000–2165 m.

NOTE. A tree, which may be this species, occurs to the northeast of Lake Victoria. It is represented by two specimens, one from Uganda, Mbale District, Bukedi (Budama), Tororo Rock. *A.S. Thomas* 2996!, with spines to 1 cm. long, the other from Kenya, Central Kavirondo District, Port Victoria, *Glasgow* 48/5!, with very short (2 mm.) spines. From a photograph accompanying *Glasgow* 48/5, they are probably identical, but more needs to be known about the trees from both these localities before they can be identified positively.
    Another specimen, *Stewart* H14/48, from T 5, Singida District, Iramba Plateau, appears to be a 5-angled branch of a seedling plant.

var. lacertosa *S. Carter* in K.B. 42: 679 (1987). Type: Kenya, Northern Frontier Province, 5 km. N. of Kisima, *Bally & Carter* 16537 (K, holo.!)

Trees 3–6 m. high, rarely more, usually scattered or forming small groups; crown bulky, broad and flattened, usually about the same width as the height of the tree in mature specimens, above a trunk naked for 1–4 m.; cymes usually 4–5 crowded at each flowering eye. Fig. 92/5, 6, p. 487.

UGANDA. Acholi District: Chua, Amiel, *Eggeling* 2338!; Karamoja District: Kadam (Mt. Debasien), 19 Jan. 1937, *A.S Thomas* 2237! & Kidepo National Park, 11 Aug. 1973, *Synott* 1552!
KENYA. Baringo District: near Marigat, 16 Oct. 1971, *Bally* 14300!; N. Nyeri District: Timau–Meru road, 8 Dec. 1971, *Bally & Smith* 14442!
DISTR. U 1; K 1–4; possibly also in southern Sudan
HAB. Rocky slopes with open, usually sparse, bushland; 1200–1900 m.

SYN. *E. sp. aff. thi* Schweinf. sensu Eggeling, I.T.U., ed. 2: 128 (1952)

VAR. Mature trees of var. *lacertosa* generally present a more bulky appearance, with a broader flattened crown on a shorter stouter trunk, than those of var. *magnicapsula*. They also occur scattered in small groups, or, in larger populations, more widely spaced than the dense stands formed by var. *magnicapsula*. Such distinctions are less obvious in populations from the slopes of the Kamasia Hills and from small hills north of Rumuruti and north of Meru, but the characteristics of var. *lacertosa* predominate.

NOTE. When collecting specimens of this, or any of the related species, it is essential to note details of habit and branch characteristics, otherwise locality offers the only means for positive identification of sterile material. Such notes should also reflect the characteristics of the local population as a whole, and care should be taken to select an average specimen, not one which emphasizes extremes.

83. E. cooperi *Berger*, Sukk. Euph.: 83 (1907); N.E. Br. in Fl. Cap. 5(2): 368 (1915). Type: South Africa, Umgeni valley, *Cooper*, cult. in Kew, 1899 (K, holo.!)

Tree to 9(–12) m. high, or occasionally a shrub to ± 2 m. high; trunk simple, to 35 cm. thick, bark greyish brown, rough and fissured, with whorls of 4–9 pit-scars resulting from fallen branches; branches curving upwards, to 2.5 m. long, occasionally rebranching to form a rounded, flat-topped crown, fleshy, (3–)4–6(–8)-angled, 5–20 cm. wide, distinctly winged, deeply constricted at ± regular intervals into pear-shaped to subcircular segments 10–50 cm. long; angles shallowly sinuately toothed, with teeth ± regularly spaced 5–25 mm. apart. Spine-shields contiguous to form a horny margin 3–10 mm. wide, becoming corky on older branches; spines 3–10 mm. long, or occasionally to 30 mm. on the lower branches; prickles vestigial or absent. Leaves deltoid, ± 1.5 × 1.5 mm. Cymes 1–3 in a horizontal row, 1-forked, subsessile; bracts rounded, ± 4.5 × 6.5 mm. Cyathia ± 4.5 × 8 mm., with cup-shaped involucres; glands transversely oblong, ± 1.5 × 4 mm., just touching, golden yellow; lobes subquadrate, ± 1.5 × 2 mm. Male flowers: bracteoles very deeply laciniate, plumose; stamens 6.8 mm. long. Female flower: perianth obvious, shallowly lobed; styles 2–5.5 mm. long, joined for ½–⅔, apices thickened, rugulose, ± bifid. Capsule exserted on a stout pedicel 4–10 mm. long, distinctly 3-lobed, with truncate base, 6–7.5 × 10–13.5 mm. Seeds subglobose, 2.8–3.5 mm. in diameter, pale greyish brown, speckled, smooth.

var. **ussanguensis** (*N.E. Br.*) *Leach* in Journ. S. Afr. Bot. 36: 31 (1970). Type: Tanzania, Njombe District, Mwigi Mt., *Goetze* 1008 (K, holo.!, BM, iso.!)

Seedlings 4–8-angled; branch-segments usually subcircular; ♂ flowers very many; capsule to 6 × 10 mm., just exserted on a pedicel 3–5 mm. long; seeds ± 3 mm. in diameter.

TANZANIA. Ufipa District: Sopa on Sumbawanga–Mbala road, 16 June 1960, *Leach & Brunton* 10054!; Mbeya District: 51 km. W. of Chimala, 6 Feb. 1974, *Bally & Carter* 16467!; Iringa District: Kitonga Escarpment, 25 June 1960, *Leach & Brunton* 10125!
DISTR. T 4–7; northern Zambia, Malawi
HAB. Rocky slopes with deciduous woodland; 500–1730 m.

SYN. *E. ussanguensis* N.E. Br. in F.T.A. 6(1): 587 (1912); T.T.C.L.: 214 (1949)

NOTE. The typical variety extends from Natal northwards through the Transvaal and Zimbabwe. It can be distinguished by its shorter height, to 6 m., branches mostly 4–5-angled, branch-segments pear-shaped, cyathia with fewer male flowers, capsules to 7.5 × 13.5 mm. exserted on a pedicel to 1 cm. long, and seeds to 3.5 mm. in diameter.

84. **E. hubertii** *Pax* in Jahres. Schles. Ges. 89, Abt. 2b: 1 (1912); T.T.C.L.: 212 (1949). Type: Tanzania, Mwanza, islands in Lake Victoria, *Winkler* 4114a (WRSL, holo.)

Tree to 6 m. high; trunk simple or occasionally forked, to 30 cm. in diameter, with brown, cracking bark; branches spreading then ascending and rebranching to form a large rounded crown; terminal branches fleshy, (3–)4–5(–6)-angled, 3–7 cm. wide, distinctly winged, constricted at irregular intervals into pear-shaped segments 5–15 cm. long; angles sinuately toothed, with teeth ± 1 cm. apart. Spine-shields contiguous to form a horny margin 2–5 mm. wide, becoming corky; spines 3–10 mm. long, shorter at the constrictions; prickles absent or vestigial. Leaves deltoid, ± 1.5 × 1.5 mm. Cymes 2–4 in a horizontal line at the flowering eye 3–5 mm. above the spines, 1-forked, subsessile, with cyathia arranged vertically; bracts rounded, ± 2.5 × 3 mm. Cyathia ± 2.5 × 5 mm., with cup-shaped involucres; glands transversely oblong, ± 1 × 2 mm., almost touching, yellow; lobes subcircular, ± 1 × 1 mm. Male flowers: bracteoles deeply laciniate, plumose; stamens 4.3 mm. long. Female flower: perianth 3-lobed, to 2 mm. long; styles 4.5 mm. long, joined for 2.8 mm., apices thickened, rugose, minutely bifid. Capsule subsessile, distinctly 3-lobed, with truncate base, ± 6 × 9 mm., green becoming red. Seeds subglobose, 2.6 mm. in diameter, yellowish brown, smooth.

TANZANIA. Mwanza District: shore of Lake Victoria at Bweru, 5 Jan. 1936, *B.D. Burtt* 5537!; Musoma District: 6.5 km. S. of Musoma on Mwanza road, 12 Dec. 1974, *Balslev* 408!
DISTR. T 1; not known elsewhere
HAB. On rocky outcrops near the lake shore; 1025–1200 m.

85. **E. adjurana** *Bally & S. Carter* in Hook., Ic. Pl. 39, t. 3875 (1982). Type: Kenya, Northern Frontier Province, Dandu, *Gillett* 13445 (K, holo.!)

Tree to 7 m. high; trunk simple, marked with pit-scars resulting from fallen branches; branches spreading, then ascending and rebranching to form a rounded crown; terminal branches fleshy, 4–6-angled, 2–5(–6) cm. wide, shallowly winged, constricted at irregular intervals into pear-shaped segments 5–15 cm. long; greyish green and usually obliquely streaked with yellowish green; angles straight to shallowly and sinuately toothed, with teeth 5–15 mm. apart. Spine-shields contiguous to form a horny margin 2–4 mm. wide, eventually peeling off in strips; spines sturdy, 2–15 mm. long, a little shorter at the constrictions; prickles weak, triangular, 0.5 mm. long, later obsolescent; secondary spines 0–3.5 mm. long flanking the flowering eye 2–5 mm. above the spines. Leaves acutely deltoid, ± 2 × 1.5 mm. Cymes 1–3 in a horizontal line, 1–2-forked, with stout peduncles and cyme-branches vertically arranged, 2–7.5 mm. long; bracts rounded, ± 4 × 6.5 mm. Cyathia ± 4 × 7 mm., with cup-shaped involucres, all parts golden yellow; glands transversely elliptic, ± 1.8 × 3.5 mm., almost touching; lobes circular, 1.75 mm. in diameter. Male flowers: bracteoles deeply laciniate, plumose; stamens 6.5 mm. long. Female flower: perianth with 3 or more irregular lobes, to 1.5 mm. long; styles 4 mm. long, joined for 2 mm., apices thickened, rugose, distinctly bifid. Capsule subsessile, 3-lobed, with truncate base and apex, ± 7.5 × 12 mm., dark red becoming brown. Seeds subglobose, 3.3 × 3 mm., brown minutely speckled with darker brown, smooth.

KENYA. Northern Frontier Province: Beloble Hill, 125 km. E. of Moyale, 1 Mar. 1974, *Bally & Carter* 16575! & between Kufole and Dandu, 103 km. E. of Moyale, 19 Jan. 1972, *Bally & Carter* 14906!
DISTR. **K** 1; southern Ethiopia
HAB. Rocky hillsides with dense *Acacia-Commiphora* bushland; 700–1200 m.

SYN. [*E. nyikae* sensu Cufod., E.P.A.: 453 (1958), *non* Pax]

VAR. Populations east of Dandu generally possess longer spines than those on hills to the west. Secondary spines, flanking the flowering eyes, are also usually obvious instead of mostly obsolete.

86. E. breviarticulata *Pax* in E.J. 34: 84 (1904); N.E. Br. in F.T.A. 6(1): 582 (1912); I.T.U., ed. 2: 126 (1952); Leach in Journ. S. Afr. Bot. 36: 40 (1970); S. Carter in K.B. 42: 676 (1987). Type: Tanzania, Lushoto District, W. Usambara Mts., Kwai, *Engler* 1184b (B, holo.†)

Large straggling shrub to 4.5 m. high branching from the base or occasionally a shrubby tree to 6 m. high; branches fleshy, erect and spreading, the lowest ± prostrate, loosely rebranching, 3(–4)-angled, to 12 cm. wide, deeply winged, deeply and ± regularly constricted into segments to ± 8 cm. long, usually broader than long, bluish green densely streaked with yellow-green; angles undulate, sinuately toothed, with teeth distant, to ± 4 cm. apart at the base of each segment, closer, to ± 1 cm. or less above, and towards the constrictions. Spine-shields forming a continuous rounded horny grey margin 2–6 mm. wide, enclosing the flowering eye 5–20 mm. above the spines and enlarged at the base of the longest spines to 8 mm. thick; spines extremely robust, the longest (3–)4–8 cm. long, much shorter at the constrictions; prickles weak, 0.5–2 mm. long; flowering eye flanked by a pair of secondary spines 0.5–5 mm. long and occasionally another rudimentary pair just above. Leaves deltoid, ± 1.5 × 2.5 mm. Cymes (1–)3–7 crowded together, 1-forked with stout peduncles 8–12 mm. long and cyme-branches 3–7 mm. long; bracts rounded, ± 2.5 × 3.5 mm. Cyathia ± 3.5 × 9 mm., with broadly funnel-shaped involucres; glands transversely oblong, ± 1.8 × 4 mm., just touching, golden yellow; lobes transversely elliptic, ± 1.2 × 2 mm. Male flowers: bracteoles spathulate, laciniate, plumose; stamens 6.3 mm. long. Female flower: perianth 3-lobed, with lobes ± 2 mm. long; styles 3 mm. long, joined for 0.8 mm., apices thickened, rugulose, bifid. Capsule just exserted on a pedicel 3–5 mm. long, ± deeply 3-lobed, with truncate base and apex, ± 9 × 18 mm., green then reddening. Seeds globose, 3.5 mm. in diameter, pale grey, speckled, smooth.

var. **breviarticulata**

Shrubs 1–4.5 m. high, without an obvious central stem.

UGANDA. Karamoja District: Nabalitak, 11 Jan. 1937, *A.S. Thomas* 2250! & Apule R., 29 Oct. 1939, *A.S. Thomas* 3102!
KENYA. Northern Frontier Province: 94 km. NW. of Isiolo near Wamba, 21 Dec. 1971, *Bally & Smith* 14736!; Teita District: Voi, 14 July 1960, *Leach & Bayliss* 10241!; Tana River District: 30 km. S. of Galole, 27 Nov. 1971, *Bally & Smith* 14404!
TANZANIA. Pare District: 5 km. S. of Makanya, 11 Nov. 1974, *Balslev* 235!; Lushoto District: Mkomazi, 30 Nov. 1935, *B.D. Burtt* 5315!
DISTR. **U** 1; **K** 1, 2, 4, 7; **T** 3; southern Ethiopia, southern Somalia
HAB. In dry open *Acacia-Commiphora* bushland, often forming thickets; 60–1200 m.

SYN. [*E. grandicornis* sensu Brenan, T.T.C.L.: 213 (1949); Cufod., E.P.A.: 447 (1958); Dale & Greenway, K.T.S.: 196 (1961); Lind & Tallantire, F.P.U.: 112 (1962), *non* N.E. Br.]

VAR. This variety is sometimes separated with difficulty from young plants of the tree *E. bussei* Pax, in areas of SE. Kenya and NE. Tanzania where both occur. Segmentation is often less uniform and the spinescence less robust than on plants elsewhere. However, the absence of a central stem and the longest spines of representative specimens reaching at least 4 cm., can be used as reliable characters for identification.

NOTE. *E. grandicornis* N.E. Br., from South Africa and Mozambique, is distinguished primarily by the cyme arrangement, with 1–3 in a horizontal row within each flowering eye, by the shorter cyme-peduncles, smaller fruits, and styles joined for half their length.

The locality of Kwai, at 1600–1700 m. altitude, for the type material of *E. breviarticulata*, seems unlikely. It was more probably collected at the foot of the Usambaras just before Engler's arrival at Kwai. The species is fairly common around the NW. end of the range.

var. **trunciformis** *S. Carter* in K.B. 42: 676 (1987). Type: Kenya, Northern Frontier Province, 17 km. SE. of Moyale, *Bally & Carter* 16572 (K, holo.!)

Shrubs with a central stem 1 m. or more high or trees to 6 m. high with an obvious trunk; lowest branches spreading and ± prostrate, later deciduous.

KENYA. Northern Frontier Province: 15 km. SE. of Moyale, 20 Aug. 1952, *Gillett* 13734! & 17 km. SE. of Moyale, 28 Feb. 1974, *Bally & Carter* 16572!
DISTR. K 1, known in East Africa with certainty only from near Moyale, but reported in I.T.U. also from Karamoja; southern Ethiopia and possibly southern Somalia (see illustration of *Paoli* 534 from Juba R. in Chiov., Miss. Stef.-Paoli Som. Ital. t. 23B (1916) sub *E. mbaluensis*)
HAB. Fairly dense *Acacia-Commiphora* bushland; 700–800 m.

87. E. biharamulensis *S. Carter* in K.B. 42: 675 (1987). Type: Tanzania, Biharamulo, *Balslev* 31 (K, holo.!)

Straggling loosely branched succulent shrub to 50 cm. high and 1 m. in diameter; branches numerous from the base, 4-angled, 1.5–3 cm. wide, constricted at irregular intervals on upper parts of the branches into ± obovate segments 1.5 cm. or more long; angles compressed, sinuately toothed with teeth irregularly spaced 0.5–2 cm. apart. Spine-shields oblong-triangular, contiguous to form a horny margin 1–4 mm. wide or separated between the segments; spines robust, 1–15 mm. long, very irregular in length; prickles rudimentary; flowering eye flanked by a pair of rudimentary secondary spines. Leaves deltoid, ± 1.5 × 1.5 mm. Cymes solitary, apparently unbranched, peduncle ± 3 mm. long; bracts rounded, ± 2 × 3.5 mm. Cyathia ± 3 × 8 mm., with broadly funnel-shaped involucres, all parts dark red; glands transversely oblong, ± 1.8 × 4.5 mm., just touching; lobes transversely elliptic, ± 1.5 × 2 mm. Male flowers: bracteoles laciniate; stamens 4.5 mm. long. Female flower: styles 3 mm. long, joined for 1.7 mm., apices thickened, distinctly bifid. Capsule sessile, distinctly 3-lobed, ± 4.5 × 9 mm., dark red. Seeds subglobose, 2.8 × 2.5 mm., brown, speckled, smooth.

TANZANIA. Biharamulo District: 168 km. S. of Bukoba on Biharamulo road, 15 Sept. 1974, *Balslev* 31! & Bukoba–Biharamulo road, Aug. 1952, *Procter* 243!
DISTR. T 1; known only from the one locality
HAB. In crevices of rocky outcrops in *Brachystegia* woodland; 1200–1400 m.

88. E. pseudoburuana *Bally & S. Carter* in Hook., Ic. Pl. 39, t. 3874 (1982). Type: Kenya, Masai District, 27 km. S. of Uaso Nyiro R. on Narok–Loliondo road, *Gillett* 16306 (K, holo.!)

Succulent perennial with a large tuberous root and a reduced underground stem producing numerous branches erect to 50 cm. high, or weaker branches spreading; branches rebranching above, (3–)4–5-angled, 1.5–5 cm. wide, constricted at irregular intervals into ± obovate segments 2 cm. or more long, greyish green with paler oblique streaks; angles compressed, sinuately toothed, with teeth irregularly spaced 0.5–3.5 cm. apart. Spine-shields contiguous forming a horny margin 0.5–3 mm. wide, occasionally separate at the branch bases; spines sturdy, 0.5–2.5 cm. long, very irregular in length with the shortest at the constrictions; prickles rudimentary; flowering eye flanked by a pair of secondary spines 0.5–2.5 mm. long. Leaves deltoid, ± 2 × 2 mm. Cymes 1–3 in a horizontal line, 1-forked, with stout peduncle and cyme-branches to 2.5 mm. long arranged in a vertical plane; bracts rounded, ± 3 × 3 mm. Cyathia ± 4.5 × 7.5 mm., with cup-shaped involucres, all parts golden yellow; glands transversely oblong, ± 1.5 × 4 mm., just touching; lobes transversely elliptic, ± 1.5 × 2.5 mm. Male flowers: bracteoles laciniate, plumose; stamens 6.5 mm. long. Female flower: perianth obvious as several acute lobes 0.5–1.5 mm. long; styles 4 mm. long, joined to halfway, apices thickened, distinctly bifid. Capsule sessile, distinctly 3-lobed, with truncate base, ± 4 × 7.5 mm., green flushed with red. Seeds subglobose, 2.5 mm. in diameter, grey, mottled, smooth.

KENYA. Masai District: Olorgasailie Mt., 10 June 1956, *Bally* 10571! & 20 km. S. of Ngong, 14 Nov. 1971, *Bally & Smith* 14317!
TANZANIA. Masai District: Lisingita, 9 Jan. 1969, *Richards* 23697!
DISTR. K 6; T 2; not known elsewhere
HAB. Amongst grass often on rocky slopes, with open *Acacia-Commiphora* bushland; 1200–1800 m.

SYN. [*E. buruana* sensu Agnew, U.K.W.F.: 223 (1974), *non* Pax]

89. E. buruana *Pax* in E.J. 34: 85 (1904); N.E. Br. in F.T.A. 6(1): 582 (1912). Type: Kenya, Teita District, between Taveta and Bura Mts., *Engler* 1930a (B, holo.†, EA, iso.!, K, drawing of holo.!)

Dwarf succulent perennial with large tuberous root ± 10 cm. in diameter, and a much reduced underground stem; branches numerous, weakly erect to ± 30 cm. high, or often semi-prostrate to 50 cm. long, simple or occasionally rebranched, grey-green with conspicuous oblique paler streaks, 3(–4)-angled, 1–4 cm. wide, deeply winged, constricted at irregular intervals into ± obovate segments 2 cm. or more long; angles toothed with teeth irregularly spaced 1–5 cm. apart. Spine-shields narrowly oblong, 1–2 mm. wide, extended 1–5 mm. above the spines to enclose the flowering eye, decurrent for 5 mm. or more or sometimes contiguous especially on wider portions of the branch; spines 1–20 mm. long, very irregular in length with the shortest at the constrictions; prickles vestigial; flowering eye flanked by a pair of secondary spines 0.5–1.5 mm. long. Leaves deltoid, ± 1.5 × 1.5 mm. Cymes solitary, 1-forked with peduncles and cyme-branches to 4 mm. long arranged in a vertical plane; bracts rounded, ± 2 × 3 mm. Cyathia ± 4 × 6 mm., with cup-shaped involucres, all parts yellow; glands transversely elliptic, ± 1.5 × 3.5 mm., just touching; lobes transversely elliptic, ± 1.5 × 2 mm. Male flowers: bracteoles laciniate; stamens ± 5mm. long. Female flower: styles 3.5 mm. long, joined for 1.5 mm., apices thickened, distinctly bifid. Capsule sessile, sharply 3-lobed, ± 4 × 6 mm., green flushed with red along the sutures. Seeds subglobose, 2.5 mm. in diameter, buff, mottled, smooth.

KENYA. Teita District: Maktau, 5 Apr. 1971, *Bally & Carter* 14138! & Tsavo W. National Park, 1 km. NE. of Maktau Gate, 16 Feb. 1980, *Gilbert* 5817!
TANZANIA. Masai District: near Loskitok Hill, 19 Sept. 1933, *B.D. Burtt* 4911!; Pare District: Mwembe valley, 30 June 1942, *Greenway* 6508!; Lushoto District: 20 km. NE. of Mkomazi on Mnazi road, 12 Nov. 1974, *Balslev* 237!
DISTR. K 7; T 2, 3; not known elsewhere
HAB. In sandy soil amongst grass in open *Acacia-Commiphora* bushland; 600–1100 m.

90. E. brevitorta *Bally* in Fl. Pl. S. Afr. 33, t. 1288 (1959); U.K.W.F.: 223 (1974). Type: Kenya, Nairobi District, Mbagathi, N. of Mokoyeti R., *Bally* 11475 (E. 162) (K, holo.!, EA, iso.!)

Dwarf very densely tufted succulent perennial to 15 cm. high; root large, tuberous; stem much reduced, subterranean, with short underground branches producing secondary plants crowded together to form a domed 'cushion' 15–100 cm. in diameter; branches usually simple, to 12 cm. long, (2–)3-angled, 1–2.5 cm. wide; angles compressed and often spirally twisted, prominently toothed with the teeth irregularly spaced 0.5–2 cm. apart. Spine-shields oblong-triangular, to 2 mm. wide, extended to 4 mm. above the spines to enclose the flowering eye, shortly decurrent, or sometimes forming a continuous horny margin ± 1 mm. wide; spines to 9 mm. long; prickles absent or occasionally vestigial; flowering eye usually flanked by a pair of minute secondary spines. Leaves deltoid, ± 2 × 1.5 mm. Cymes solitary, 1-forked with peduncles and cyme-branches 3–5 mm. long arranged in a vertical plane; bracts rounded, ± 2.5 × 4 mm. Cyathia ± 3 × 6 mm., with cup-shaped involucres, all parts yellow; glands transversely elliptic, ± 1.8 × 2.7 mm., contiguous; lobes transversely elliptic, ± 1 × 1.7 mm. Male flowers: bracteoles laciniate; stamens 5 mm. long. Female flower: styles 3.5 mm. long, joined for 1.5 mm., apices rugulose, deeply bifid. Capsule sessile, obtusely 3-lobed, ± 7 × 8.5 mm., green flushed with purple. Seeds subglobose, 2.7 × 2.5 mm., grey-mottled, smooth.

KENYA. N. Nyeri District: 40 km. NW. of Nyeri on Rumuruti road, 21 Apr. 1975, *Hepper & Field* 5105!; Masai District: NW. of Kajiado, Oldonyo Narok, 2 Nov. 1948, *Vesey-FitzGerald* 207 in *Bally* E. 338!
DISTR. K 3, 4, 6; not known elsewhere
HAB. In rock crevices on exposed sloping usually swampy ground with free drainage; 1500–2000 m.

NOTE. In the original description the date for the type collection is given as October 1939, but this is the date Bally made his first collection of living material (E. 162). The text makes it quite clear that the type was prepared from material collected at the same locality in 1957, i.e. *Bally* 11475, which he cultivated under the same number of E.162.

91. E. venenifica *Kotschy* in Mitt. Geog. Ges. Wien 1, (2): 173 (1857), as "*E. venefica*"; N.E. Br. in F.T.A. 6(1): 562 (1911); I.T.U., ed. 2: 128 (1952); F.P.S. 2: 67 (1952); E.P.A.: 461 (1958). Type: Sudan, Dar Foq [Fung], Jebel Kaçane, *Trèmaux*, Voy. Soudan Orient., Atlas, t. 13, 14 (1853)

Stout shrub 2–5 m. high, with 1–several stems from ground-level, ± sparsely branched and rebranched; branches spreading-ascending, fleshy, cylindrical, to 3.5 cm. thick, pale grey. Spine-shields ± 1 cm. apart in ± 8 spiral series, obtusely triangular to subcircular, 6–8 mm. in diameter, surmounted by a single spine; spines very stout at the base, abruptly tapering, to 8(–15) mm. long on young growth, becoming obsolete; prickles obsolete or occasionally rudimentary. Leaves terminally crowded, sessile, fleshy, variable in size and shape from lanceolate to obovate, 4–22 × 1–4 cm., apex acute, truncate or occasionally emarginate, apiculate, margin entire, often minutely crisped. Cymes solitary immediately above the spine-shields, 1–2-forked, with peduncles and cyme-branches ± 7 mm. long; bracts rounded, ± 2.5 × 4 mm. Cyathia ± 2.5 × 5 mm., with cup-shaped involucres; glands transversely oblong, ± 1 × 2.8 mm., contiguous, yellowish green; lobes subcircular, ± 1.5 mm. diameter. Male flowers: bracteoles spathulate, deeply laciniate; stamens 3.75 mm. long. Female flower: styles 1.5 mm. long, joined at the base, apices much thickened. Capsule exserted on a reflexed pedicel to 5 mm. long, distinctly 3-lobed, ± 4 × 6 mm., buff. Seed subglobose, 2.5 mm. in diameter, buff, mottled, smooth.

UGANDA. W. Nile District: Leya and Ayo R. junction, 25 Mar. 1945, *Greenway & Eggeling* 7245!; Acholi District: Agoro Rest Camp, *Eggeling* 837!
DISTR. U 1; Sudan, SW. Ethiopia
HAB. Stony slopes in dry grassland with scattered trees; ± 1220 m.

NOTE. *E. venenifica* was originally published as *E. venefica*, an incorrect orthographic variant which was followed in F.P.S. and E.P.A.
  This species is difficult to place in a sequence of E. African species. It is most closely related to the very similar W. African single-spined *E. poissonii* Pax, in turn related to the double-spined species *E. sudanica* Chev. and *E. desmondii* Keay & Milne-Redh. It also shows a striking affinity with the Indian species *E. nivulia* Buch.-Ham. and *E. neriifolia* L. which seem to bear some relation to the Arabian *E. qarad* Defl. An undescribed species occurs in Djibouti which is intermediate between this last species and *F. robecchii*.
  E. decidua Bally & Leach in Kirkia 10: 293 (1975), & Candollea 18: 335 (1963) (description without type) should also be mentioned here. It has been recorded from Tanzania, Rungwe District, Ichinga, by Hargreaves in Cact. & Succ. Journ. Am. 56: 109 (1984), but I have not seen the specimen (*Hargreaves & Mkhoma* 2392). It occurs in Malawi, Zimbabwe and Zambia and is most closely related to a group of mainly Angolan dwarf tuberous-rooted species, including *E. imitata* N.E. Br. and *E. brevis* N.E. Br. The group consists of apparently relict species and seems to have some connections with the W. African species mentioned above, with noticeably leafy branches, small rounded spine-shields, pedunculate cymes, capsules exserted on reflexed pedicels and smooth subglobose seeds. *E. decidua* has a large turnip-shaped root with a stem reduced to a growing-point; branches deciduous, 3–5-angled, toothed, ± 12 cm. long and 5 mm. thick; spine-shields ± 2.5 mm. in diameter; paired spines 1.5–4.5 mm. long; leaves semi-persistent, lanceolate, ± 3 × 1 mm.; cymes produced from the stem apex after the branches are shed, 1–2-forked, peduncle and cyme-branches 1–3 cm. long; cyathia cup-shaped, ± 3 × 4 mm.; glands reddish; capsule exserted on a reflexed pedicel to 6 mm. long, deeply 3-lobed, ± 3.2 × 4 mm.; seeds subglobose, 2 mm. in diameter, smooth.

92. E. robecchii Pax in Ann. Ist. Bot. Roma 6: 186 (1897); N.E. Br. in F.T.A. 6(1): 583 (1912); E.P.A.: 456 (1958); K.T.S.: 201 (1961); Bally in Cact. & Succ. Journ. Am. 38: 6 (1966). Types: Ethiopia, Ogaden, *Robecchi* 332[286] & 370[287] & R. Milmil, *Riva* 1169[1061] (all FT, syn.!, K, photos.!)

Seedlings to ± 1 m. high, with unbranched fleshy stem 4(–5)-angled, 3–4 cm. thick, pale green heavily mottled with darker green; angles ± straight, with spine-shields forming a continuous horny margin ± 2 mm. wide; spines 3–7 mm. long, 1.5–2.5 cm. apart. Young plants to ± 3 m., branching and rebranching irregularly; branches at first ascending then drooping, obtusely 3–4-angled, 1–1.5 cm. thick, uniformly greyish green, ± distinctly toothed, with teeth 3–5 cm. apart; spine-shields stout, 2–4 mm. wide, tapering abruptly above and below the spines up to ± 1 cm., becoming progressively shorter as further branches are produced; spines to 12 mm. long; prickles obsolete. Mature tree 3–10(–13) m. high; trunk simple, to 40 cm. in diameter; bark grey, rough and fissured, with a few pit-scars resulting from fallen branches visible at higher levels; branches to 5 m. long, at first ascending, finally the lower ones spreading horizontally to form a rounded flat-based crown; secondary branches crowded at the apices, densely and irregularly rebranching; terminal branchlets obscurely 3-angled to cylindrical, marked with calloused obtusely triangular leaf-scars ± 2.5 × 2.5 mm. and 1–3 cm. apart immediately below the flowering eyes, rarely with a pair of minute spines on lower branches. Leaves deltoid, ± 5 × 3 mm. Cymes solitary, 1-forked with peduncles to 5 mm. long and cyme-branches to 7 mm. long; bracts rounded, ± 4 × 5 mm. Cyathia ± 4 × 9 mm., with

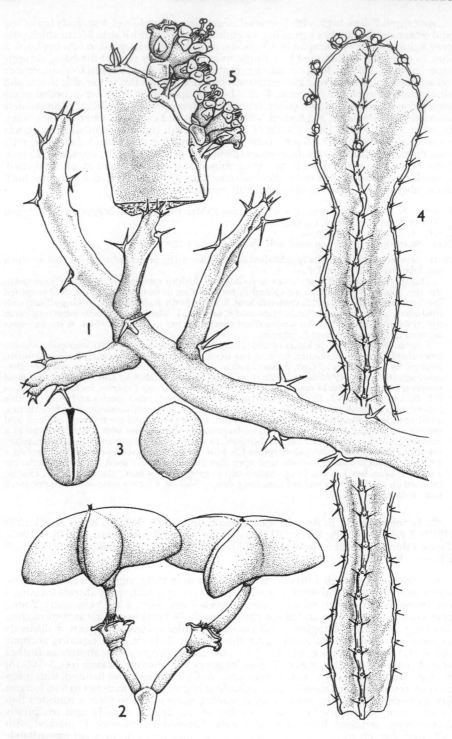

FIG. 93. *EUPHORBIA ROBECCHII* — **1**, branch of young tree, × ⅓; **2**, capsules, × 2; **3**, seeds, × 2. *E. QUINQUECOSTATA* — **4**, flowering branch, × ⅔; **5**, portion of flowering branch with spinescence detail, × 2. 1, from *Greenway* 6596; 2, from *Carter* 805; 3, from *Verdcourt* 3125; 4, from *Verdcourt* 3891; 5, from *Greenway & Kanuri* 12826. Drawn by Christine Grey-Wilson.

funnel-shaped involucres, reddish brown; glands transversely oblong, ± 2 × 4 mm., yellow, just touching; lobes transversely elliptic, ± 1.5 × 2 mm., red. Male flowers: bracteoles spathulate, plumose; stamens 5 mm. long. Female flower: perianth with 3 distinct lobes, 1.5 mm. long; styles 3 mm. long, free to the base, apices slightly thickened, bifid. Capsule exserted on an erect pedicel to 1 cm. long, deeply and acutely 3-lobed with truncate base, ± 8 × 20 mm., dark purplish grey. Seeds subglobose, 3.5–4 mm. in diameter, pale greyish buff, smooth. Fig. 93/1–3.

KENYA.   Turkana District: Oropoi, Feb. 1965, *Newbould* 7021!; Masai District: 14.5 km. from Ol Tukai on Namanga road, 15 May 1961, *Verdcourt* 3123!; Teita District: 8 km. from Voi on Lugards Falls road, 22 Dec. 1966, *Greenway & Kanuri* 12831!
TANZANIA.   Pare District: 8 km. S. of Kifaru, 7 Nov. 1974, *Balslev* 232!; Lushoto District: Mkomazi–Mbalu track, S. of Mt. Lasa, 13 July 1942, *Greenway* 6596!
DISTR.   K 1, 2, 4, 6, 7; T 3; eastern Ethiopia, Somalia
HAB.   On red sandy soils with fairly open *Acacia-Commiphora* bushland; 30–1200 m.

SYN.   *E. pimeleodendron* Pax in Jahres. Schles. Ges. 89, Abh. 2b: 1 (1912). Type: Kenya: Teita District, between Taveta and Taveta R., *Winkler* 4298 (WRSL, holo.)
  E. ruspolii Chiov. in Miss. Stef.-Paoli Somal. Ital.: 159 (1916). Type: Ethiopia, R. Milmil, *Riva* 1169[1061] (syntype of *E. robecchii*) (FT, holo.!, K, photo.!)

93. E. lividiflora *Leach* in Kirkia 4: 20 (1964). Type: Mozambique, Manica e Sofala District, 39 km. S. of Muda, *Leach* 11129A (SRGH, holo., K, iso.!)

Tree to 4(–10) m. high, with a simple trunk 12–25 cm. in diameter; branches at first ascending and 5-angled, finally horizontal and cylindrical to ± 3 m. long, with secondary branches arising loosely in whorls and rebranching to form a ± rounded slightly flattened crown; terminal branchlets fleshy, 3–4-angled and ± winged; angles sinuate-toothed with teeth 1.5–3 cm. apart. Spine-shields oblong-triangular, ± 4 × 2 mm. above the spines and extended to enclose the flowering eye, decurrent 1–3 mm.; spines to 5 mm. long; prickles minute or absent. Leaves deltoid, ± 5(–13) × 3 mm. Cymes solitary 1(–5)-forked, with peduncle 3–6 mm. and cyme-branches 5–9 mm. long; bracts rounded, ± 4 × 5 mm. Cyathia ± 4.5 × 9 mm., with funnel-shaped involucres, reddish purple; glands 5(–6), transversely elliptic, ± 2.5 × 4.5 mm., just touching, dark bluish purple; lobes 5(–6), transversely elliptic, ± 1.5 × 2 mm. Male flowers: bracteoles plumose; stamens 5 mm. long, with blackish anthers. Female flower: perianth 3-lobed, with toothed lobes 1.25–2 mm. long; styles 2 mm. long, free to the base, apices thickened, slightly bifid. Capsule exserted on an erect pedicel to 8 mm. long, deeply and acutely 3-lobed, with truncate base, ± 9 × 25 mm., reddish purple. Seeds subglobose, ± 4.5 mm. long, minutely rugulose.

TANZANIA.   Mikindani District: Mtwara–Mikindani road, 13 Mar. 1963, *Richards* 17878!
DISTR.   T 8, known in East Africa from this collection only; Mozambique, southern Malawi
HAB.   Sandy soils (amongst coral rocks near the coast), with open deciduous woodland; 15 (–1000 in Malawi) m.

NOTE.   The one specimen from Tanzania is from a small plant only 1 m. high; the description has been made from Mozambique and Malawi material. A photograph in Kew, taken by Greenway, shows a shrub, growing on coral cliffs on the east coast of Mafia Island, that may be this species.

94. E. tanaensis *Bally* in Candollea 29: 390 & 19: 160 (1964) [description without type]. Type: Kenya, Lamu District, Mambosasa, *Greenway* 9473 (K, holo.!)

Tree to 30 m. high, trunk simple, ± 40 cm. in diameter, 6-angled above, terminating in a rounded crown of horizontally spreading branches, occasionally with 2 or 3 crowns one above the other; bark yellowish green marked with rough vertical lines; branches rebranching, to 3 m. long; terminal branchlets fleshy, 4–6-angled, 5–10 cm. wide, deeply winged, constricted at intervals of 10–35 cm.; angles straight to shallowly toothed, with teeth 1.5–2.5 cm. apart. Spine-shields obtusely triangular, 3–4 × 3–4 mm., eventually including the flowering eye immediately above, abruptly decurrent below; spines 3–4 mm. long; prickles vestigial. Leaves deltoid, 8–12 × 4–6 mm. Cymes 1–3 in a horizontal line, 2-forked, with peduncles and cyme-branches to 2.5 cm. long; bracts rounded, ± 2.5 × 4.5 mm. Cyathia ± 4 × 12 mm., with very broadly cup-shaped involucres, all parts dark crimson; glands 5–6, transversely elliptic, ± 2.5 × 5 mm., just touching; lobes 5–6, transversely elliptic, ± 2 × 4 mm. Male flowers very many: bracteoles spathulate, deeply laciniate, plumose; stamens 3.75 mm. long. Female flower: perianth 3-lobed with lobes ± 1 mm. long; styles 1 mm. long, thick and fleshy, joined at the base, apices thickened, rugulose, bifid. Immature capsule exserted on a pedicel 14 mm. long, 3-lobed, 6 × 10 mm.; mature capsule and seeds not seen.

KENYA. Lamu District: Mambosasa, ± 10 km. E. of Witu, 7 Nov. 1957, *Greenway* 9473! & 1.5 km. SW. of Mambosasa Forest Post, 17 Oct. 1986, *Robertson* 4327!
DISTR. **K** 7; known only from the type locality
HAB. Semi-deciduous swamp forest; 15 m.

SYN. *E. sp.* sensu Dale & Greenway in K.T.S.: 202 (1961), *Greenway* 9473, Utwani [Mambosasa] Forest Reserve

NOTE. The type locality given by Bally of Mambosasa as S. of Bura, is incorrect.

95. E. wakefieldii *N.E. Br.* in F.T.A. 6(1): 583 (1912); K.T.S.: 202 (1961); Bally in Candollea 21: 368 (1966). Type Kenya, Kilifi District, Ribe, *Wakefield* (K, holo.!)

Seedlings and young plants from ± 30 cm. high producing branches to 1 m. or more long which trail on the ground, take root and produce new plants, eventually forming dense stands of trees. Mature tree to 7 m. high; trunk simple or with a few short side-branches, to 15 cm. in diameter; bark pale grey marked with rings of ± 6 pit-scars resulting from fallen branches at intervals of ± 15 cm.; trunk terminating in a small rounded crown of loosely clustered radiating branches to 1.5 m. long, each with a few whorls of branchlets to ± 15 cm. long, seldom rebranching; terminal branchlets fleshy, 3(–4)-angled, 1–2 cm. wide, deeply winged, not noticeably constricted; angles distinctly sinuately toothed, with teeth ± 1 cm. apart. Spine-shields shortly triangular, 1 × 1.5 mm. above the spines, decurrent for 1–3 mm.; spines slender, 3–8 mm. long, black; prickles minute, triangular, sometimes obsolete. Leaves deltoid, ± 1 × 1 mm. Cymes solitary, 1-forked, subsessile; bracts deltoid, ± 1.5 × 2 mm. Cyathia ± 2 × 4.5 mm., with cup-shaped involucres; glands transversely elliptic, ± 1.5 × 2 mm., greenish yellow, just touching with the inner margin slightly raised; lobes subcircular, ± 1 mm. in diameter. Male flowers: bracteoles spathulate, plumose; stamens 3.8 mm. long. Female flower: styles 1.5 mm. long, joined at the base, apices thickened, rugose, bifid. Capsule exserted on a reflexed pedicel to 6 mm. long, distinctly 3-lobed, ± 4.5 × 6.5 mm. Seeds ovoid, 3 × 2.8 mm., grey and faintly mottled, smooth.

KENYA. Kwale District: Mwachi Creek, 6 Feb. 1948, *Bally* 5780!; Kilifi District: Chasimba, 14 Feb. 1971, *Mabberley* 716!
DISTR. **K** 7; not known elsewhere
HAB. Limestone outcrops with dense xerophytic bushland and forest remnants; 50–275 m.

NOTE. Trees recorded from near the mouth of Kilifi Creek are reported to exhibit a different branching habit and may prove to be distinct.
The specimen *Greenway* 6578, from the S. Pare Mts. at Hedaru cited by Bally, is *E. quadrialata. E. wakefieldii* is an endangered species, at great risk from the felling of coastal forest for charcoal burning. A tree reported to be very similar to *E. wakefieldii* has been photographed by *Powys* at the top of the deep gorge dividing the eastern slopes of Mt. Kulal (**K** 1). More investigation is needed before a positive identification can be made, but it is almost certainly distinct from *E. wakefieldii.*

96. E. quadrialata *Pax* in E.J. 33: 286 (1903); N.E. Br. in F.T.A. 6(1): 188 (1912); T.T.C.L.: 213 (1949); Bally in K.B. 28: 322 (1973). Type: Tanzania, Handeni District, Magofu, *Busse* 319 (B, holo.†, G, iso., K, drawing of holo.! & fragment of iso.!)

Tree to 15(–20) m. high, trunk simple or with a few ascending branches, to 30 cm. in diameter; bark grey and marked with 6–8 vertical lines of pits, being the scars of fallen branches; branches terminating in small fairly dense clusters of secondary branches 1–3 m. long, each with further whorls of branchlets; terminal branchlets fleshy, 3–4-angled, 1.5–3.5 cm. wide, deeply winged, slightly constricted at irregular intervals of 10–30 cm.; angles straight to sinuately toothed, with teeth 1–1.5 cm. apart. Spine-shields shortly triangular, 1 × 2.5 mm. above the spines, decurrent for 1–3 mm.; spines 2–7 mm. long; prickles minute, at first triangular, flexible, attached to the leaf-bases, later hardening and usually vestigial. Leaves deltoid, ± 2 × 1.3 mm. Cymes solitary, 1-forked, subsessile; bracts deltoid, ± 1.5 × 1.5 mm. Cyathia ± 2 × 5 mm., with cup-shaped involucres, all parts ivory-white; glands transversely elliptic, ± 1.5 × 2.2 mm., the inner margin slightly raised, just touching; lobes subcircular, 1 mm. in diameter. Male flowers: bracteoles spathulate, plumose; stamens 3.5 mm. long. Female flower: styles 1.5 mm. long, joined for 0.5 mm., apices thickened, rugose, ± bifid. Capsule exserted on a slightly curved pedicel to 5 mm. long, obtusely 3-lobed, ± 5 × 7 mm. Seeds ovoid, 3.5 × 2.8 mm., grey, smooth.

TANZANIA. Pare District: Kihunda, above Hedaru, 9 July 1942, *Greenway* 6578!; Lushoto District: 13 km. SW. of Soni on Mombo road, 20 Nov. 1974, *Balslev* 325!; Handeni District: Kideleko Mission, 14 Feb. 1948, *Bally* 5806!

DISTR. **T** 3, 6; not known elsewhere

HAB. Rocky slopes and gneiss outrops with open deciduous bushland; 400–1300 m.

NOTE. The specimen from **T** 6, *Wingfield* 3699, is of a young cultivated plant originating from Morogoro District, Kimboza Forest Reserve.

The latex of this species is reputed to be less virulent, and as such is used as a constituent of bird-lime.

97. E. dumeticola *Bally & S. Carter* in Cact. Succ. Journ. Gt. Brit. 38: 66 (1976); S. Carter in K.B. 32: 583 (1978). Type: Tanzania, Kilosa District, Uhehe by Ruaha R., *Goetze* 465 (B, holo.†, K, drawing of holo.!); Great Ruaha Gorge, 45 km. W. of Mikumi, *Leach & Brunton* 10342 (K, neo.!, B, BR, PRE, SRGH, isoneo.)

Tree to 4(–6) m. high, with a simple or rarely branching trunk to 15 cm. in diameter, with smooth grey bark and 4 longitudinal spiny ridges 1 cm. high; branching fairly loose, forming a ± rounded crown; primary branches to 2 m. long, spreading horizontally then upturned at the tips, with whorls of secondary branches which branch again; terminal branchlets fleshy, 4-angled and ± winged, 2–4 cm. wide, slightly constricted at intervals of 10–20 cm., bright green, occasionally variegated; angles sinuately toothed, with teeth ± 1 cm. apart. Spine-shields oblong-triangular to 2 × 1.5 mm. above the spines, narrowly decurrent to just above the flowering eye below and soon contiguous; spines 3–10 mm. long; prickles ± 0.5 mm. long, recurved, at first triangular in shape, flexible and clearly attached to the leaf-base, later hardening and persisting after the leaf has fallen. Leaves deltoid, ± 1.2 × 1.5 mm. Cymes solitary, 1-forked, subsessile; bracts broadly ovate, ± 1.5 × 1.5 mm. Cyathia ± 1.5 × 3.5 mm., with broadly cup-shaped involucres; glands transversely oblong, ± 0.7 × 1.8 mm., with undulate margin, not touching, yellow; lobes subcircular, ± 1 mm. in diameter. Male flowers: bracteoles plumose; stamens 3.5 mm. long. Female flower: ovary subsessile, deeply 3-lobed; styles 3 mm. long, joined for 0.8 mm., apices distinctly bifid, slightly thickened. Capsules and seeds not seen.

TANZANIA. Kilosa District: Ruaha Gorge, 23 km. E. of Mbuyuni, 9 Feb. 1974, *Bally & Carter* 16424! & 30 km. E. of Mbuyuni, 30 Nov. 1974, *Balslev* 369!

DISTR. **T** 6; known only from the Ruaha Gorge

HAB. Sandy stony soil on steep slopes, with dense deciduous bushland; 500 m.

SYN. [*E. stuhlmannii* sensu Pax in E.J. 28: 421 (1900); Goetze & Engl., Veg. Deutsch Ost-Afr.: 19, 20, t. 13, 15 (1902) *nec* Pax (1896), *non* Volk. (1899)]

E. platyacantha Pax in E.J. 34: 84 (1904); N.E. Br. in F.T.A. 6(1): 599 (1912); T.T.C.L.: 214 (1949), *nom. illegit., non* Drake (1903). Type as for *E. dumeticola.*

98. E. quinquecostata *Volk.* in N.B.G.B. 2: 266 (1899); N.E. Br. in F.T.A. 6(1): 579 (1912); T.T.C.L.: 213 (1949); K.T.S.: 200 (1961); S. Carter in K.B. 42: 673 (1987). Type: Tanzania, Moshi District, Lake Chala, *Volkens* 407 (B, holo.†, K, drawing of holo.!)

Tree to 10(–15) m. high, with a simple or sometimes branching trunk, to 50 cm. in diameter, marked with ± loosely spiralled rows of persistent spines and deep pits resulting from the scars of fallen branches; bark flaking, grey-brown; branching dense to form a rounded slightly flattened crown; primary branches 2–4 m. long, spreading outwards and upwards, with whorls of secondary branches which occasionally branch again on larger trees, the lower ones deciduous; terminal branchlets fleshy, (3–)5(–6)-angled, 2–4(–7) cm. thick, ± constricted at intervals of 5–15 cm.; angles straight to sinuately toothed, with teeth 5–15 mm. apart. Spine-shields oblong-triangular, to 1.5 × 2 mm. above the spines, narrowly decurrent to the flowering eye below and soon contiguous. Spines 2–8(–10) mm. long; prickles vestigial or absent. Leaves deltoid, ± 2 × 2 mm. Cymes clustered at the branch apices, solitary or more usually 3 in a horizontal row, 1-forked, with thickened peduncles 2–3 mm. long arranged in a vertical plane; bracts deltoid, ± 1.8 × 2.5 mm. Cyathia ± 2.2 × 4 mm., with cup-shaped involucres; glands transversely oblong, ± 1 × 2 mm., contiguous, golden yellow; lobes circular, 1 mm. in diameter. Male flowers: bracteoles fan-shaped, deeply toothed, plumose; stamens 4.2 mm. long. Female flower: styles 1.2 mm. long, joined to halfway, apices much thickened, bifid. Capsule exserted on a pedicel to 7 mm. long, deeply 3-lobed, with truncate base, ± 4.5 × 7 mm. Seeds subglobose, 2.4 × 2.2 mm., grey-mottled, smooth. Fig. 93/4, 5, p. 496.

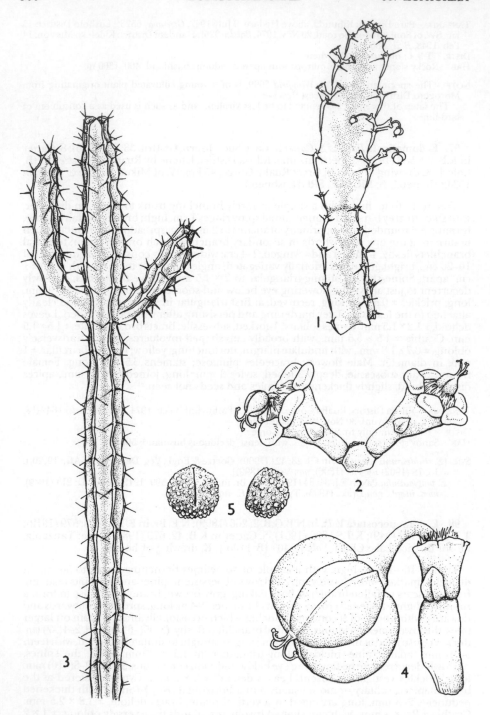

FIG. 94. *EUPHORBIA HETEROCHROMA* subsp. *HETEROCHROMA* — **1**, fruiting branch, × ⅔; **2**, cyathia, × 4. *E. SCARLATINA* — **3**, branch detail, × ⅔; **4**, capsule, × 6; **5**, seeds, × 6. 1, from *Leach & Brunton* 10204; 2, from *Drummond & Hemsley* 3224; 3, from *Glover & Samuel* 2804; 4, 5, from *Verdcourt* 3465. Drawn by Christine Grey-Wilson.

KENYA. Kitui District: Mutomo Hill, 15 Mar. 1968, *Bally* 13107!; Teita District: Voi, Mazinga Hill, 11 Jan. 1964, *Verdcourt* 3891!
TANZANIA. Moshi District: Lake Chala, 28 Oct. 1974, *Balslev* 51!; Pare District: Kisiwani, 24 June 1942, *Greenway* 6489!; Lushoto District: Mbalu Hill, 25 Jan. 1948, *Bally* 5759!
DISTR. K 4, 7; T 2, 3, 6; not known elsewhere
HAB. Rocky hillsides, usually the dominant tree in mixed deciduous woodland; 600–1220 m.

SYN. *E. intercedens* Pax in E.J. 34: 75 (1904); N.E. Br. in F.T.A. 6(1): 590 (1912); T.T.C.L.: 213 (1949). Type: Tanzania, Lushoto District, Mbalu, *Engler* 1472b (B, holo.†, K, drawing from photo. of holo.!)

VAR. The spines of specimens from localities in Kenya, Kitui District, are consistently longer at 5–10 mm. The specimen from Mutha Hill, *MacArthur* in C.M. 7426, appears to be from a seedling plant, comparing well with the seedling on the Kew sheet of *Balslev* 51 from Lake Chala.

NOTE. A possibly distinct subspecies occurs in the Northern Frontier Province of Kenya. Two sterile specimens are known, *Joy Adamson* in *Bally* 3659 from southwest of Maralal at Ngare Narok, and *Gilbert et al.* 5619 from the Ndoto Mts. above Ngoronit. It is also recorded from the Mathews Range at Kichich. It appears to be a shorter tree to about 10 m. high, with 6 vertical rows of branch scars on the trunk and with the terminal branches 3–5-angled.

99. E. classenii *Bally & S. Carter* in K.B. 29: 507 (1974). Type: Kenya, Teita District, Mt. Kasigau, *Classen* 70 in *Bally* 12285 (K, holo.!)

Succulent shrub, densely branching from the base, erect to 1 m. high; branches sparsely rebranched, prominently 6–8-angled, 1–3 cm. thick, bright green; angles sinuately and shallowly toothed, with teeth 6–12 mm. apart. Spinescence grey; spine-shields oblong-triangular, to 2 × 2 mm. above the spines and often extended to surround the flowering eye, decurrent to 4 mm. and halfway to the flowering eye below; spines to 8 mm. long; prickles absent or rudimentary. Leaves acutely deltoid, ± 1 × 1 mm. Cymes solitary, 1(–2)-forked, subsessile; bracts subquadrate, ± 2 × 2 mm. Cyathia ± 3 × 4 mm., with cup-shaped involucres; glands transversely oblong, ± 1 × 2 mm., just touching, golden yellow; lobes subcircular, ± 1.3 × 1.3 mm. Male flowers: bracteoles fan-shaped, laciniate; stamens 3 mm. long. Female flower: styles ± 2 mm. long, joined to almost halfway, apices thickened, rugulose. Capsule exserted on a reflexed pedicel to 4 mm. long, obtusely 3-lobed, ± 4 × 5.5 mm., buff. Seeds ovoid, 2 × 1.5 mm., reddish brown, smooth.

KENYA. Teita District: Mt. Kasigau, 4 June 1960, *Classen* 70 in *Bally* 12285, & cult. Mar. 1971 in *Bally* 14129!
DISTR. K 7; known only from Mt. Kasigau
HAB. Exposed rock faces with deciduous woodland; 900–1200 m.

100. E. heterochroma *Pax* in P.O.A. C: 242 (1895); N.E. Br. in F.T.A. 6(1): 572 (1911), excl. *E. stapfii* Berger; T.T.C.L.: 212 (1949); S. Carter & M.G. Gilbert in K.B. 42: 387 (1987). Type: Tanzania, Moshi District, Himo R., *Volkens* 1759 (B, holo.†, BM, K, iso.!)

Straggly succulent shrub with erect or decumbent branches rooting where they touch the ground, to 2 m. high, sparingly rebranched above; branches 4(–5)-angled, to ± 2 cm. thick, green with usually regular darker patches along the angles; angles straight to shallowly undulate. Spine-shields elongated, to 5 × 2 mm. above the spines, narrowly decurrent to 15 mm. and 1–7 mm. above the flowering eye below, or forming a continuous horny margin with spine-pairs to 2.5 cm. apart; spines 1–6 mm. long, occasionally absent; prickles vestigial. Leaves deltoid, ± 1.5 × 1.5 mm., margin very obscurely toothed. Cymes solitary, 1-forked with stout peduncles and cyme-branches 2–3 mm. long, yellow or red; bracts triangular, ± 1.5 × 2 mm. Cyathia ± 3 × 6 mm., with cup-shaped involucres; glands transversely oblong, ± 1.4 × 3 mm., contiguous; lobes transversely elliptic, 1.2 × 2 mm. Male flowers: bracteoles fan-shaped, deeply laciniate, fimbriate; stamens 3.8 mm. long. Female flower: styles 1.8 mm. long, joined at the base, apices much thickened, rugose. Capsule exserted on a reflexed pedicel ± 5 mm. long, sharply 3-lobed, ± 3 × 5.5 mm. Seeds ovoid, 2.2 × 2 mm., grey, shallowly tuberculate.

subsp. heterochroma

Branches distinctly variegated; spine-shields separate, or sometimes contiguous on older branches; all parts of the cyathia greenish yellow; capsule buff coloured. Fig. 94/1, 2.

TANZANIA. Moshi District: Himo R., 25 Jan. 1936, *Greenway* 4499!; Pare District: Mkomazi–Gonja road, 9 Jan. 1967, *Richards* 21926!; Lushoto District: Kamba–Mnazi, 12 Jan. 1930, *Greenway* 2050!
DISTR. T 2, 3; probably occurring also just over the border in K 7, Teita District
HAB. Sandy stony soils with *Acacia-Commiphora* bushland; 450–1300 m.

SYN. *E. stuhlmannii* Volk. in N.B.G.B. 2: 267 (1899), *nom. illegit., non* Pax. Type: Tanzania, Lushoto
District, coastal steppe, *Volkens* (B, holo.†)
E. mitis Pax in E.J. 34: 70 (1904). Type: Tanzania, Pare District, between Maji-ya-Juu and
Sengina, *Engler* 1613 (B, holo.†, K, drawing of holo.!, K, iso.!)
E. impervia Berger, Sukk. Euph.: 64 (1907). Type as for *E. stuhlmannii* Volk.
E. heterochroma Pax var. *mitis* (Pax) N.E. Br. in F.T.A. 6(1): 572 (1911); T.T.C.L.: 212 (1949)

VAR. The intensity of variegation on the branches varies considerably, as does spine-length even on
one plant. With such variation there is no justification for upholding *E. mitis* even at varietal level.

NOTE. *E. heterochroma* should not be confused with *E. stapfii* and other closely related species
occurring in Uganda, Kenya and Ethiopia. A mixture of *E. heterospina* with its subspecies, and *E.
scarlatina* occur in the distributions cited in K.T.S.: 198 (1961) and U.K.W.F.: 223 (1974).

subsp. **tsavoensis** *S. Carter* in K.B. 42: 387 (1987). Type: Kenya, Machakos District, Kanga, *Verdcourt*
2385 (K, holo.!, EA, iso.!)

As for subsp. *heterochroma*, but the branch variegation less distinct; spine-shields contiguous on all
but young growth; cyathia red, with glands reddish yellow becoming dark red; capsule flushed red.

KENYA. Machakos District: near Kibwezi, 14 July 1960, *Leach & Bayliss* 10248!; Masai District: 4 km. W.
of Kitani Lodge, 13 Aug. 1965, *Gillett* 16839!; Teita District: Tsavo National Park, Mazinga Hill, 8
Aug. 1969, *Bally* 13351!
DISTR. **K** 4, 6, 7; not known elsewhere
HAB. On rock outcrops with deciduous bushland; 550–760 m.

SYN. [*E. heterochroma* sensu Dale & Greenway, K.T.S.: 198 (1961), quoad *Uhlig*; Agnew, U.K.W.F.: 223
(1974), quoad distr. Machakos, *non* Pax]

101. E. stapfii *Berger,* Sukk. Euph.: 59 (1907); S. Carter & M.G. Gilbert in K.B. 42: 389
(1987). Type: Uganda, Mengo District, cult. in Entebbe Botanic Garden, *Dawe* in *E. Brown*
227 (K, iso.!)

Much-branched succulent shrub to 4 m. high; branches 5-angled, ± 1.5 cm. thick,
uniformly green; angles straight to shallowly undulate. Spine-shields oblong-triangular,
with truncate base, to ± 3 × 2 mm. above the spines, decurrent to the flowering eye below
on the youngest growth, otherwise contiguous, with spine-pairs 5–10 mm. apart; spines to
6 mm. long; prickles ± 0.5 mm. long. Leaves deltoid, ± 1.75 × 1.5 mm. Cymes solitary,
1-forked, subsessile; bracts deltoid, ± 1.2 × 1.5 mm. Cyathia ± 2.5 × 5 mm., with cup-shaped
involucres; glands transversely oblong, ± 1.5 × 2.5 mm., contiguous, yellow; lobes
transversely elliptic, ± 1 × 1.5 mm. Male flowers: bracteoles fan-shaped, deeply laciniate,
fimbriate; stamens 4 mm long. Female flower: ovary pedicellate; styles 1.8 mm. long,
joined for 0.8 mm., apices thickened, rugose. Capsules and seeds not seen.

UGANDA. Toro District: ?W. side of Bwamba Pass, c. 27 Jan. 1935, *G. Taylor* 3281!; Mengo District:
Kipayo, Dec. 1913–Jan. 1914, *Dummer* 703! & Gambala, 28 Nov. 1936, *Tothill* 2644!
DISTR. **U** ?3 (see note), 4; not known elsewhere
HAB. Collected so far only as cultivated material usually from hedges; ± 1200 m.

NOTE. This species has been reported as occurring commonly around Entebbe, but so far has not
been collected from this area as authentic wild material. *Snowden* 1046, an isolated specimen from
Mbale District, Bukonde, with truncate bases to the spine-shields and regularly spaced spines of
equal length, may possibly represent a wild population. *G. Taylor* 3281 is without notes and the
locality is uncertain. It may be of a wild population, and is certainly the best representative of the
species. However, until more gatherings and more information are available, *E. stapfii* should be
regarded as a poorly known species which is, nevertheless, distinct from *E. heterochroma* Pax with
which it has long been confused.

102. E. petraea *S. Carter* in Hook., Ic. Pl. 39, t. 3871 (1982). Type: Uganda, Acholi
District, Paimol, *Osmaston* 500 (K, holo.!)

Decumbent spreading succulent shrub, densely branched from the base, sparsely
branched above, to 60 cm. high; branches 4(–5)-angled, 1–2 cm. thick; angles sinuately
toothed, with teeth 8–15 mm. apart. Spine-shields oblong with a truncate base, to 3 × 3
mm. above the spines, decurrent to the flowering eye below, occasionally surrounding it
to form a continuous horny margin; spines to 8 mm. long; prickles rudimentary. Leaves
ovate, ± 1 × 1 mm. Cymes solitary, 1–?-forked, with stout peduncle and cyme-branches 2–4
mm. long; bracts deltoid, ± 1.25 × 1.25 mm. Cyathia ± 2.5 × 5 mm., with cup-shaped
involucres; glands transversely oblong; ± 1.25 × 2.75 mm., contiguous, yellow; lobes
transversely elliptic, ± 1 × 1.25 mm. Male flowers: bracteoles fan-shaped, deeply and

irregularly laciniate; stamens 4.5 mm. long, red. Female flower: styles 1.25 mm. long, joined at the base, apices thickened, rugose. Capsule exserted on a reflexed pedicel to 1 cm. long, obtusely 3-lobed, ± 4 × 5.5 mm., buff. Seeds ovoid, 2.5 × 2 mm., grey, shallowly tuberculate.

UGANDA. Karamoja District: Napak Mt., 23 Jan. 1957, *Dyson-Hudson* 131! & Toror Hills, Moroto–Kotido, 8 Oct. 1952, *Verdcourt* 794!
DISTR. U 1; not known elsewhere
HAB. Exposed rocks and boulders, with sparse deciduous bushland; 950–1850 m.

103. E. heterospina *S. Carter* in K.B. 42: 389 (1987). Type: Kenya, W. Suk District, Kapenguria, *M. Hale* 45 in *Bally* 4042 (K, holo.!)

Erect or subscandent succulent shrub to 3.5 m. high; branches 4–5-angled, 1–2 cm. thick, slightly constricted at irregular intervals of 20–60 cm., uniformly green or indistinctly variegated; angles straight to shallowly toothed, with teeth 0.5–3 cm. apart. Spine-shields to 5 × 3 mm. above the spines, decurrent to shortly above the flowering eye below or forming a continuous horny margin 1–3 mm. wide on older branches; spines to 1 cm. long at the base of the plant, becoming progressively shorter above, to rudimentary or obsolete on upper branches (above ± 1.5 m. high); prickles rudimentary to 1 mm. long. Leaves deltoid, ± 1.5 × 1.5 mm. Cymes solitary, 1-forked, with stout peduncles and cyme-branches 1–3 mm. long; bracts ± subquadrate, to 1.2 × 1.5 mm. Cyathia ± 2.5 × 5 mm., with cup-shaped involucres, all parts yellow, red-tinged or entirely red; glands transversely oblong, ± 1.3 × 2.8 mm., contiguous; lobes transversely elliptic, ± 1 × 1.5 mm. Male flowers: bracteoles fan-shaped, deeply laciniate, plumose; stamens 4 mm. long. Female flower: styles 1.8 mm. long, joined at the base, apices much thickened, rugose, often ± bifid. Capsule exserted on a reflexed pedicel to 6 mm. long, deeply 3-lobed, ± 3.5 × 6 mm., flushed red or dark red. Seeds subglobose, ± 2 mm. in diameter, buff with darker longitudinal streaks, shallowly tuberculate.

subsp. **heterospina**

Laxly-branched shrub erect to 1 m. or subscandent to 3.5 m. high; branches green, sometimes indistinctly variegated; all parts of the cyathia yellow, or occasionally the involucre tinged reddish, with red lobes and orange glands; capsule buff becoming red.

UGANDA. Karamoja District: Moroto R., 1936, *Eggeling* 2954! & Nakiloro, *J. Wilson* 1499! & 32 km. S. of Moroto, 13 Sept. 1956, *Bally* 10811!
KENYA. Turkana District: Lokibennit, Nov. 1939, *Tweedie* 491!; W. Suk District: Suk country, *Tweedie* 72!
DISTR. U 1; K 2, 3; not known elsewhere
HAB. Sandy stony soil with deciduous woodland; 1000–1500 m.
SYN. [*E. heterochroma* sensu Dale & Greenway, K.T.S.: 198 (1961), quoad *M. Hale* 45, *non* Pax]
     [*E. stapfii* sensu Lind & Tallantire in F.P.U.: 112 (1962), *non* Berger]
NOTE. A poor specimen of *Law* 42 from K 5, S. Kavirondo District, Kisii, possibly belongs here.

subsp. **baringoensis** *S. Carter* in K.B. 42: 390 (1987). Type: Kenya, Baringo District, Maji-yo-Moto, *Bally* 4583 (K, holo.!)

As for subsp. *heterospina* but usually no more than 2 m. high; branches usually uniformly bluish green; cyathia red; glands at first yellow flushed with red, becoming entirely red; capsule dark red.

KENYA. Baringo District: Lake Baringo, Ol Kokwa I., 15 Oct. 1978, *Gilbert* 5061A! & Tangulbei, 17 Nov. 1964, *Leippert* 5300!
DISTR. K 3; not known elsewhere
HAB. Rocky slopes with low herbs and sparse deciduous woodland; 550–1600 m.
SYN. [*E. heterochroma* sensu Agnew, U.K.W.F.: 223 (1974), quoad distr. Baringo, *non* Pax]
NOTE. Although no herbarium specimens appear to have been made, this subspecies is known to occur in Laikipia District.

104. E. borenensis *M. Gilbert* in K.B. 42: 391 (1987). Type: Ethiopia, Sidamo, W. of Tertale, *Rippstein* 1978 (K, holo.!)

Sparsely branched succulent shrub erect to 3 m.; branches 4-angled, 1–2 cm. thick, uniformly green; angles straight or shallowly toothed at the base of the plant, with teeth ± 1 cm. apart. Spine-shields to 4 × 2.5 mm. above the spines, decurrent to just above the flowering eye below on young growth, soon forming a continuous horny margin; spines to 8 mm. long below, obsolete on upper branches; prickles vestigial. Leaves deltoid, ± 2 × 1.5 mm. Cymes solitary, 1-forked, with cyme-branches ± 2 mm. long; bracts deltoid, ± 1.2 × 1.5 mm. Cyathia ± 3 × 6 mm., with cup-shaped involucres, all parts yellow; glands transversely oblong, ± 1.5 × 3 mm. contiguous, yellow; lobes subcircular, ± 1.5 mm. in diameter. Male flowers: bracteoles fan-shaped, fimbriate; stamens 4 mm. long. Female flower: styles not seen. Capsule exserted on a reflexed pedicel to 4 mm. long, obtusely 3-lobed, ± 2.5 × 4 mm., red flushed. Seeds subglobose, 1.5 mm. in diameter, grey, shallowly tuberculate.

KENYA. Northern Frontier Province: 30 km. from Ramu on Malka Mari road; 8 May 1978, *Gilbert & Thulin* 1578!
DISTR. **K** 1 (known in East Africa from this collection only); Ethiopia
HAB. Rocky limestone valley with *Acacia-Commiphora* woodland; 400 m.

105. E. vulcanorum S. *Carter* in Hook., Ic. Pl. 39, t. 3872 (1982). Type: Kenya, Northern Frontier Province, N. of Marsabit, Gof Choba, *Carter & Stannard* 663 (K, holo.!)

Densely branching succulent shrub to 1 m. high and 1.2 m. in diameter; branches to 30 cm. long, (4–)5-angled, 1–2 cm. thick, uniformly green; angles very shallowly toothed, with teeth 5–15 mm. apart. Spine-shields forming a continuous horny margin along the angles 2–3 mm. wide; spines stout, to 8 mm. long; prickles vestigial. Leaves ovate, ± 1.75 × 1.25 mm. Cymes solitary, 1-forked, subsessile, all parts yellow; bracts deltoid, ± 1 × 1.5 mm. Cyathia ± 2.5 × 5 mm., with cup-shaped involucres; glands transversely oblong, ± 1.5 × 3 mm., contiguous; lobes transversely elliptic, ± 1 × 1.5 mm. Male flowers: bracteoles deeply laciniate; stamens 4 mm. long. Female flower: styles 1.25 mm. long, joined at the base, apices thickened, rugulose, minutely bifid. Capsule exserted on a reflexed pedicel to 5.5 mm. long, obtusely 3-lobed, ± 3 × 4 mm., buff flushed with red. Seeds ovoid, 2.25 × 1.75 mm., grey, minutely and shallowly tuberculate.

KENYA. Northern Frontier Province: Marsabit, Aug. 1942, *Mrs J. Bally* in *Bally* E.275! & Gof Choba, 11 km. N. of Marsabit, 24 Nov. 1977, *Carter & Stannard* 663!
DISTR. **K** 1; apparently limited to the Marsabit massif
HAB. Lava-strewn rocky slopes with sparse deciduous bushland; 1000–1200 m.

106. E. scarlatina S. *Carter* in K.B. 42: 392 (1987). Type: Kenya, Masai District, 8 km. S. of Magadi, *Verdcourt* 3465 (K, holo.!)

Much-branched succulent shrub erect to 3 m.; branches (4–)5–6-angled, 1.5–2.5 cm. thick, ± constricted at intervals of 10–30 cm., uniformly green; angles straight to sinuately toothed, with teeth 1–2(–3) cm. apart. Spine-shields to 5 × 3 mm. above the spines and forming a continuous horny margin along the angles 2–3 mm. wide; spines to 1 cm. long, reduced to 2–3 mm. long on the uppermost branches; prickles rudimentary to 1 mm. long. Leaves deltoid, ± 2 × 1.5 mm. Cymes solitary, subsessile, 1-forked, with stout cyme-branches 1–3 mm. long; bracts deltoid, ± 1.5 × 1.5 mm., denticulate. Cyathia ± 2.5 × 5 mm., with cup-shaped involucres, all parts bright red; glands transversely oblong, ± 1.3 × 2.8 mm., contiguous, yellowish becoming bright red; lobes transversely elliptic, ± 1.2 × 1.5 mm. Male flowers: bracteoles fan-shaped, deeply laciniate, plumose; stamens 4 mm. long. Female flower: styles 1.8 mm. long, joined at the base, apices much thickened, rugose, slightly bifid. Capsule exserted on a reflexed pedicel ± 5 mm. long, deeply 3-lobed, ± 3.5 × 5.5 mm., dark red. Seeds subglobose, ± 2 mm. in diameter, buff, shallowly and closely tuberculate. Fig. 94/3–5, p. 500.

KENYA. Naivasha District: Kedong Escarpment, Sept. 1933, *Napier* 3019!; Masai District: Mt. Suswa summit, 4 Aug. 1952, *Verdcourt* 703! & Olekejo-Ngiro, opposite Olorgasailie, 16 June 1962, *Glover & Samuel* 2804!
DISTR. **K** 3, 6; confined to the Rift Valley S. of Lake Naivasha
HAB. Rocky slopes and lava ridges with open deciduous bushland; 600–2000 m.

NOTE. This species is included in the distribution of *E. heterochroma* by Dale & Greenway, K.T.S.: 198 (1961) at Magadi, and by Agnew, U.K.W.F.: 223 (1974) in Magadi and Kajiado districts.

107. E. atroflora S. *Carter* in K.B. 42: 393 (1987). Type: Kenya, Northern Frontier Province, Ndoto Mts., Ngurunit, *Gilbert et al.* 5630 (K, holo.!, EA, iso.!)

Sturdy much-branched succulent shrub erect to 2.5 m. high; branches 4–5(–6)-angled, to 2–3 cm. thick, uniformly green or sometimes faintly variegated with darker blotches around the spine-shields, ± constricted at intervals of 10–20 cm.; angles straight to sinuately toothed, with teeth 1–2 cm. apart. Spine-shields to 5 × 3 mm. above the spines and forming a continuous horny margin along the angles ± 3 mm. wide; spines stout, to 1 cm. long, reduced to 2–3 mm. long on the uppermost branches of taller plants; prickles rudimentary to 1 mm. long. Leaves deltoid, ± 2 × 1.5 mm. Cymes solitary, 1-forked, subsessile, with stout cyme-branches ± 1 mm. long; bracts deltoid, ± 1.5 × 1.5 mm. Cyathia ± 2.5 × 5 mm., with cup-shaped involucres, all parts eventually dark red; glands transversely oblong, ± 1.3 × 2.8 mm., contiguous, yellowish red quickly becoming crimson; lobes transversely elliptic, ± 1.2 × 1.5 mm. Male flowers: bracteoles fan-shaped, deeply laciniate, plumose; stamens 4 mm. long. Female flower: styles 1.8 mm. long, joined at the base, apices much thickened, rugose, slightly bifid. Capsule exserted on a reflexed pedicel ± 5 mm. long, deeply 3-lobed, ± 3.5 × 5.5 mm., reddish black. Seeds subglobose, ± 2 mm. in diameter, buff, shallowly and closely tuberculate.

KENYA. Northern Frontier Province: Ndoto Mts., Ndigri-Alori, 2 Jan. 1959, *Newbould* 3457! & 38 km. N. of Maralal on Baragoi road, 2 June 1979, *Gilbert et al.* 5469!; Meru District: 1 km. W. of Isiolo, 14 June 1979, *Gilbert et al.* 5665!
DISTR. **K** 1, 4, 7; not known elsewhere
HAB. Stony soil, usually on rocky slopes, with open deciduous bushland; 900–1760 m.

NOTE. *Robertson* 1709 from **K** 4, Embu District, Kanjiro Hills possibly belongs here, but if a continuous distribution of *E. heterochroma* subsp. *tsavoensis* is eventually found to exist between Kibwezi and Embu, it more probably belongs to this inhabitant of granite outcrops. The species has also been recorded from **K** 7 in the Kora National Reserve.

108. E. tescorum *S. Carter* in Hook., Ic. Pl. 39, t. 3873 (1982). Type: Kenya, Turkana District, junction to Kakuma on Lodwar–Lokitaung road, *Carter & Stannard* 219 (K, holo.!, EA, iso.!)

Sturdy succulent shrub with numerous branches spreading upwards from the base, sparsely rebranched above to 1.5(–2) m. high; branches (4–)5–6(–8)-angled, to 4 cm. thick, ± constricted at intervals of 10–30 cm., usually green-variegated with darker markings around the teeth; angles usually distinctly toothed, with teeth to 1.8 cm. apart. Spine-shields to 3 × 4 mm. above the spines, forming a continuous horny margin 2–4 mm. wide; spines very strong, often somewhat variable in length, to 1.5 cm. long, shorter at the constrictions but rarely reduced to 1–2 mm. long on branches above 1.5 m. high; prickles rudimentary. Leaves deltoid, 2 × 1.5 mm. Cymes solitary, 1-forked, with peduncle and cyme-branches ± 1.5 mm. long; bracts subquadrate, ± 1.25 × 1 mm. Cyathia ± 2.5 × 4.5 mm., with cup-shaped involucres, golden-yellow; glands transversely oblong, ± 1.25 × 2.5 mm., contiguous, orange-yellow, rarely reddish; lobes transversely elliptic, ± 1 × 1.5 mm. Male flowers: bracteoles fan-shaped, deeply laciniate; stamens 3.75 mm. long. Female flower: styles 1.5 mm. long, joined at the base, apices much thickened, rugose. Capsule exserted on a reflexed pedicel to 5 mm. long, obtusely 3-lobed, ± 3.5 × 5.5 m., rosy red. Seeds ovoid, 2.25 × 1.75 mm., grey, minutely tuberculate.

UGANDA. Karamoja District: Loyoru, 3 Nov. 1939, *A.S. Thomas* 3146!
KENYA. Northern Frontier Province: Sololo, Burroli Mt., 4 Sept. 1952, *Gillett* 13778! & Loriu Plateau, 3 June 1970, *Mathew* 6615!; Turkana District: Lokitaung, 22 May 1953, *Padwa* 200!
DISTR. **U** 1; **K** 1, 2; southern Ethiopia
HAB. Rocky ground, often on lava, with very open *Acacia-Commiphora* bushland; 400–1500 m.

VAR. Plants from northern Turkana, notably from the Lokitaung Gorge, are generally shorter (1–1.5 m.) than those east of the lake, but are particularly robust with stouter branches, more prominent toothing of the branch angles, more intense variegation and longer spines. The eastern populations, on the laval plains surrounding the Marsabit massif, are rather more densely branched, with less noticeable branch constrictions and spines of a more even length. These two forms may prove to merit separation at subspecific level. Several different evidently local forms within what has been known as the *E. stapfii-heterochroma* complex are known to occur in **K** 1. Until much more is known about their differences and distribution patterns, these can only be regarded as variations of the better-known taxa described here.

109. E. baioensis *S. Carter* in Hook., Ic. Pl. 39, t. 3870 (1982). Type: Northern Frontier Province, Baio Mt., *Powys & Evans* 76/16 cult. *Carter* (K, holo.!)

Spreading succulent perennial, densely branching from the base; branches shortly decumbent or erect to 30 cm. high, cylindrical to 2 cm. thick, greyish green, with 8–10 longitudinal ribs (angles), without teeth. Spinescence red when young, becoming black with spine-pairs to 8 mm. apart; spine-shields oblong, ± 2.5 mm. wide, to 2.5 mm. long above the spines and almost touching the flowering eye below; spines very slender, to 1 cm. long; prickles absent or rudimentary. Leaves acutely deltoid, ± 1 × 0.7 mm., often persistent and hardening to form a minute prickle. Cymes solitary, 1-forked, subsessile; bracts subquadrate, ± 1.5 × 1.5 mm. Cyathia ± 2 × 3.5 mm., with funnel-shaped involucres; glands transversely elliptic, ± 1 × 1.5 mm., just touching, yellow; lobes subcircular, 1.5 mm. in diameter. Male flowers: bracteoles fan-shaped, laciniate; stamens well exserted, 3.75 mm. long. Female flower: styles 2.5 mm. long, joined for 0.5 mm., apices thickened and minutely bifid. Capsule exserted laterally on a reflexed pedicel to 3 mm. long, deeply 3-lobed ± 2.5 × 4 mm., reddish. Mature seeds not seen.

KENYA. Northern Frontier Province: Baio Mt., 38 km. NW. of Laisamis, 7 Jan. 1976, *Powys & Evans 76/16* cult. *Carter*! (distributed in cultivation under *Lavranos 12532* and *Bally 16966*)
DISTR. **K** 1; known only from the type collection and its clones in cultivation
HAB. On exposed granite rock; 1700 m.

NOTE. Since the original description of this unique species was made, persistence and hardening of the leaf has been noted. Material cultivated in California, with fruits, has been seen, the capsules being exserted horizontally from the lateral cyathia, a feature typical of species related to *E. heterochroma* Pax.

110. **E. colubrina** *Bally & S. Carter* in Hook., Ic. Pl. 39, t. 3868 (1982). Type: Northern Frontier Province, Kubaru, 25 km. E. of Ramu, *Bally & Carter 16587* (K, holo.!, EA, iso.!)

Densely branched succulent perennial, spreading to form tangled masses to 15 cm. high and 50 cm. in diameter; branches 4-angled, to 25 cm. long and 8 mm. thick, with branchlets spreading at right-angles, dark glossy green with a paler longitudinal stripe down each face; angles shallowly toothed with teeth 5–10 mm. apart. Spinescence dark brown; spine-shields oblong, ± 2 mm. wide, to 1.5 mm. long above the spines and decurrent to 5 mm. and to shortly above the flowering eye below; spines to 18 mm. long; prickles absent or rudimentary. Leaves acutely deltoid, ± 1 × 0.7 mm., obscurely denticulate. Cymes solitary, 1-forked, with peduncle and cyme-branches 1–2 mm. long; bracts ovate, 1 × 0.75 mm. Cyathia ± 2 × 2.5 mm., with cup-shaped involucres; glands transversely oblong, ± 0.75 × 1.25 mm., contiguous, brownish yellow; lobes subcircular, ± 0.75 mm. long. Male flowers: bracteoles deeply laciniate; stamens 2 mm. long. Female flower: styles 1 mm. long, joined only at the base, apices scarcely thickend. Capsule exserted on a reflexed pedicel to 4 mm. long, deeply 3-lobed, ± 2 × 2.5 mm., buff with purple-marked sutures. Seeds ovoid, 1.25 × 1 mm., red-brown, minutely and densely rugulose.

KENYA. Northern Frontier Province: 18 km. SW. of Mandera, 12 Dec. 1971, *Bally & Smith 14577*! & Awal Abdulla, 16 km. E. of Ramu, 2 Mar. 1974, *Bally & Carter 16585*!
DISTR. **K** 1; southern Ethiopia; restricted to the NE. corner of Kenya and a corresponding limited area in Ethiopia
HAB. Rocky gravelly limestone slopes with open *Acacia* bushland; 250–300 m.

111. **E. quadrispina** *S. Carter* in Hook., Ic. Pl. 39, t. 3869 (1982). Type: Kenya, Northern Frontier Province, 35 km. NW. of Ramu, *Bally & Smith 14928A* (K, holo.!)

Densely branched succulent perennial spreading to form tangled masses ± 10 cm. high and 30 cm. in diameter; branches cylindrical, to 20 cm. long and 9 mm. thick, dark green blotched with lighter green, obscurely toothed with teeth 5–10 mm. apart in 5 loosely spiralled series. Spinescence black; spine-shields obtusely triangular, 1.5–5 × 1.5–3 mm.; spines slender, 3–20 mm. long, very variable in length on each branch; prickles 2–10 mm. long. Leaves deltoid, ± 1.5 × 1 mm., shallowly toothed. Cymes solitary, 1-forked, with peduncle ± 1 mm. long and cyme-branches ± 2 mm. long; bracts subquadrate, ± 1 × 0.75 mm. Cyathia ± 2.5 × 3.5 mm., with funnel-shaped involucres; glands transversely oblong, ± 0.75 × 1.5 mm., contiguous, pinkish orange; lobes subcircular, ± 0.75 mm. in diameter. Male flowers: bracteoles spathulate, laciniate; stamens 2.5 mm. long. Female flower: ovary deeply 3-lobed, pedicellate; styles 1.5 mm. long, joined for 0.5 mm., apices scarcely thickend. Capsules and seeds not seen.

KENYA. Northern Frontier Province: 35 km. NW. of Ramu, 21 Jan. 1972, *Bally & Smith 14928A*; 30 km. from Ramu on Malka Mari road, 8 May 1978, *Gilbert & Thulin 1579*!
DISTR. **K** 1; known only from between Ramu and Malka Mari

HAB.   Rocky limestone slopes with *Acacia-Commiphora* bushland; 400–450 m.

112. E. ellenbeckii *Pax* in E.J. 33: 285 (1903); N.E. Br. in F.T.A. 6(1): 576 (1911); E.P.A.: 446 (1958). Type: Ethiopia, Sidamo Region, *Ellenbeck* 2099 (B, holo.†, K, drawing of holo.!)

Densely tufted succulent perennial, branching from the base to 10(–15) cm. high and 30(–45) cm. in diameter; branches simple, cylindrical, 8–14 mm. thick, 10–20 cm. long, surface tessellated, with divisions to 8 × 6 mm. arranged in 5 loosely spiralled series and each surmounted by a spine-shield, bright green, paler in the grooves. Spinescence buff; spine-shields broadly triangular with corners rounded and often enlarged, 2–6 × 2.5–5 mm., the base of larger shields sometimes produced into a stout spine to 1.5 mm. long; spines sturdy, 1–22 mm. long, extremely variable in length on each branch; prickles 0.5–2 mm. long. Leaves deltoid, ± 1 × 1 mm., denticulate. Cymes solitary, 1-forked, with peduncles and cyme-branches ± 2 mm. long; bracts oblong, ± 1 × 0.7 mm. Cyathia ± 1.5 × 2 mm., with cup-shaped involucres; glands transversely rectangular, ± 1 × 2 mm., contiguous to form an undivided circle, yellowish pink; lobes subcircular, ± 0.7 mm. long. Male flowers: bracteoles deeply and finely laciniate; stamens 2.5 mm. long. Female flower: styles 2 mm. long, joined for 0.5 mm., apices thickened. Capsule exserted on a recurved pedicel to 5 mm. long, deeply and acutely 3-lobed with a truncate base, ± 2 × 2.7 mm. Mature seeds not seen.

KENYA.   Northern Frontier Province: 35 km. WNW. of Ramu, 21 Jan. 1972, *Bally & Smith* 14928! & Kubaru, 25 km. E. of Ramu, 3 Mar. 1974, *Bally & Carter* 16586!
DISTR.   **K** 1, limited to the NE. corner; southern Ethiopia and SW. Somalia
HAB.   On stony limestone slopes with open *Acacia* bushland; 250–750 m.

VAR.   The form occurring in Kenya appears to be more sturdy, with a greater proportion of longer spines, than the form so far collected in Ethiopia and Somalia. This may be due to the proximity of the Kenya plants to the seasonal Dawa Parma River, with available moisture rarely far below the surface.

113. E. proballyana *Leach* in Journ. S. Afr. Bot. 34: 289 (1968). Type: Tanzania, Kilosa District, Great Ruaha Gorge, 48 km. W. of Mikumi, *Leach & Brunton* 10138 (PRE, holo., BR, EA, G, K!, LISC, SRGH, ZSS, iso.)

Densely branching succulent shrub to 1.3(–2) m. high and 1.5 m. in diameter; branches 4-angled, 1–3 cm. thick, uniformly greyish green; angles sinuately and shallowly toothed, with teeth 1–1.5 cm. apart. Spinescence greyish buff; spine-shields oblong-triangular, to 2 × 3 mm. above the spines, broadly decurrent from half-way to shortly above the flowering eye below; spines to 13 mm. long; prickles to 3.5 mm. long. Leaves acutely deltoid, ± 2 × 2 mm. Cymes solitary producing 2–4 subsessile lateral cyathia; bracts subquadrate, ± 1.8 × 2 mm. Cyathia ± 2.5 × 5 mm., with cup-shaped involucres; glands transversely oblong, ± 0.8 × 2 mm., scarcely spreading, contiguous, dull red; lobes transversely elliptic, ± 0.8 × 1.3 mm. Male flowers: bracteoles laciniate; stamens 3 mm. long. Female flower: styles 1.8 mm. long, apices thickened and distinctly bifid. Capsule sessile, obtusely 3-lobed, ± 3.5 × 4 mm. Seeds subglobose, 1.8 × 1.7 mm., brown, tuberculate.

TANZANIA.   Dodoma District: 104 km. N. of Iringa on Dodoma road, 5 Dec. 1974, *Balslev* 399!; Kilosa District: Great Ruaha valley, 12 Sept. 1971, *Bally & Gilbert* 14252!
DISTR.   **T** 5–7; apparently limited to the Great Ruaha valley
HAB.   Rocky slopes with sandy soil and open *Acacia* woodland; 500–750 m.

VAR.   The spinescence of some plants, apparently growing in shade, is weaker with spines ± 5 mm. long. The number of lateral cyathia produced is by no means constant, even on one plant. Specimens from the Iringa–Dodoma road area often produce the normal 2, while those near Mbuyuni (including the type) usually but not always produce 3.

114. E. reclinata *Bally & S. Carter* in K.B. 40: 120 (1985). Type: Tanzania, Mpwapwa, *Greenway* 6653 (K, holo.!)

Procumbent, occasionally scandent succulent perennial, much branched from the base; branches sparsely rebranched, 4-angled, to 1 m. long and 1 cm. thick, dull greenish brown; angles shallowly toothed, with teeth to 1.5 cm. apart. Spine-shields oblong-triangular, ± 1 × 2 mm. above the spines, decurrent to shortly above the flowering eye below; spines to 12 mm. long; prickles to 2.5 mm. long. Leaves acutely deltoid, ± 1.25 × 1 mm. Cymes solitary, 1-forked, subsessile, but cyme-branches to 2.5 mm. long; bracts subquadrate ± 1.5 × 1.5 mm. Cyathia ± 2.5 × 4 mm., with cup-shaped involucres; glands transversely oblong, ± 0.7 × 1.7 mm., just touching, reddish yellow; lobes transversely elliptic, ± 0.75 × 1.3 mm. Male flowers: bracteoles spathulate, laciniate; stamens 2.5 mm. long. Female flower: styles 2 mm. long, joined at the base, apices distinctly bifid. Capsule sessile, obtusely 3-lobed, ± 2.5 × 3.2 mm. Seeds ovoid, 1.5 × 1.3 mm, buff, tuberculate.

TANZANIA. Mpwapwa District: Gulwe, 19 Aug. 1930, *Greenway* 2409! & Mpwapwa, 27 Oct. 1957, *Hornby* 850!
DISTR. T 5, not known elsewhere
HAB. On rocks in *Acacia-Commiphora* bushland; 850–1100 m.
SYN. *E. hepatica* Bally & S. Carter in Hook., Ic. Pl. 39, t. 3867 (1982), *non* Urb. & Ekm. (1929). Type as for *E. reclinata*
NOTE. Two other gatherings are tentatively identified as this species: *Bally* 7883, from Dodoma, has much smaller involucral bracts and slightly larger more spreading glands; *B.D. Burtt* 839 from Simbo, SW. of Kondoa, appears to produce a more cushion-like habit with prominently toothed branches. *Greenway* 7944 probably also belongs here, but its precise locality from the northern escarpment of the Ruaha River is not clear.

115. E. quadrilatera *Leach* in Journ. S. Afr. Bot. 46: 318 (1980). Type: Tanzania, Njombe District, Makambako, *Leach & Brunton* 10350 (SRGH, holo., K, iso.!)

Laxly branched succulent shrub erect to 2 m. high, or more densely branched when damaged; stem 2–5 cm. thick near the base; branches 4-angled, 1–2.5 cm. thick, green with darker mottling especially along the angles; angles sinuately-toothed, with teeth 1–2.5 cm. apart. Spinescence brownish grey; spine-shields elongated, ± 3 × 2 mm. above the spines, decurrent to just above the flowering eye below, often with a short spine to 1.5 mm. long at the base, eventually forming a continuous horny margin; spines 5–10 mm. long; prickles 0.5–2 mm. long. Leaves deltoid, ± 1.5 × 1.5 mm. Cymes solitary, 1-forked, subsessile; bracts broadly deltoid, ± 1.75 × 2 mm. Cyathia ± 3 × 8 mm., with broadly funnel-shaped involucres; glands transversely rectangular, ± 2 × 4.5 mm., contiguous, deep yellow with a red margin; lobes transversely elliptic, ± 1.5 × 2.5 mm. Male flowers: bracteoles spathulate, deeply dentate; stamens 4.5 mm. long. Female flower: styles 3.5 mm. long, joined at the base, apices thickened, rugose. Capsule sessile, obtusely 3-lobed, ± 3.5 × 6 mm. Seeds ovoid, 2.3 × 2 mm., dark brownish grey, minutely and shallowly tuberculate.

TANZANIA. Chunya District: Ntande Hill, 97 km. N. of Chunya, 8 Feb. 1974, *Bally & Carter* 16498!; Iringa District: Mufindi, below Lake Ngwazi Dam, 12 Aug. 1971, *Perdue & Kibuwa* 11046! & Iringa, 4 Dec. 1974, *Balslev* 397!
DISTR. T 7; not known elsewhere
HAB. Among rocks in *Brachystegia* woodland; 1500–2100 m.

116. E. quadrangularis *Pax* in E.J.: 119 (1894); N.E. Br. in F.T.A. 6(1): 574 (1911); T.T.C.L.: 212 (1949); Cribb & Leedal, Mount. Fl. S. Tanz.: 77 (1982). Type: Tanzania, Kondoa District, Irangi [Itarige], *Fischer* 519 (B, holo.†, K, fragment & drawing of holo.!)

Very sparsely branched erect to subscandent succulent perennial to 3.5 m. high, or sometimes with the stem apex damaged producing a more densely branched habit to ± 1 m. high; stem and branches 4-angled, stem to 5 cm. thick near the base, branches 1–2 cm. thick, pale greyish green mottled with darker bluish green especially along the angles; angles straight to shallowly toothed, with teeth 1–3 cm. apart. Spinescence greyish buff; spine-shields elongated, ± 3 × 2 mm. above the spines, decurrent to just above the flowering eye below, eventually forming a continuous horny margin; spines 5–8(–10) mm. long; prickles 1–2 mm. long. Leaves ovate, ± 7 × 2.5 mm. Cymes solitary, 1-forked, subsessile, but cyme-branches to 4 mm. long; bracts broadly deltoid, ± 2.5 × 2 mm. Cyathia ± 3 × 9 mm., with broadly funnel-shaped involucres; glands transversely rectangular, ± 2 × 4.5 mm., contiguous, yellow-green and granular, with a distinct smooth red margin 1 mm. wide; lobes transversely elliptic, ± 1.7 × 3 mm. Male flowers: bracteoles spathulate, dentate; stamens 4.5 mm. long. Female flower: styles 3 mm. long, joined for 0.7 mm., apices thickened. Capsule subsessile, obtusely 3-lobed, ± 5 × 7 mm. Seeds subglobose, 2.2 mm. in diameter, greyish brown, shallowly tuberculate. Fig.95/1–3.

TANZANIA. Maswa District: 80 km. W. of Endulen, 27 July 1957, *Bally* 11606!; Kilosa District: N. of Malolo near Ruaha R., 1 Dec. 1974, *Balslev* 374!; Iringa District: Ruaha National Park, Maganga-Madung'u, Great Ruaha R., 19 Aug. 1969, *Greenway & Kanuri* 13762!

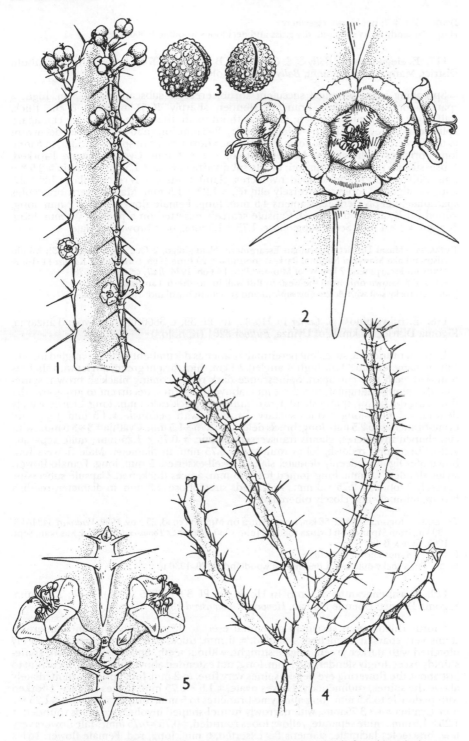

FIG. 95. *EUPHORBIA QUADRANGULARIS* — **1**, flowering and fruiting branch, × ⅔; **2**, cyathia and spinescence, × 3; **3**, seeds, × 6. *E. TETRACANTHOIDES* — **4**, sterile branch, × ⅔; **5**, cyathia and spinescence, × 3. 1, from *Balslev* 374; 2, from *Bally* 11606; 3, from *Thulin & Mhoro* 631; 4, 5, from *Bally & Carter* 16463. Drawn by Christine Grey-Wilson.

DISTR. T 1, 4–7; not known elsewhere
HAB.   In sandy rocky soils with dry grass and deciduous woodland; 500–1720 m.

117. E. elegantissima *Bally & S. Carter* in K.B. 29: 507 (1974). Type: Tanzania, Mbulu District, Manyara Escarpment, *Bally* E.10 (K, holo.!)

Spreading many-stemmed succulent shrub, erect or subscandent to 3 m. high, ± sparsely branched above; branches slender, sharply 4(–5)-angled, 1–2 cm. thick, uniformly dark glaucous green; angles without teeth, the spine-pairs 1–1.5 cm. apart. Spinescence grey-black; spine-shields slender, 2–3 mm. long above the spines, decurrent to the flowering eye below to form a continuous horny margin; spines slender, to 8 mm. long; prickles rudimentary. Leaves deltoid, ± 1 × 1 mm. Cymes solitary, 1-forked, subsessile, all parts dark crimson; bracts subquadrate, ± 1.5 × 1.5 mm. Cyathia ± 2.5 × 5 mm., with broadly funnel-shaped involucres; glands transversely rectangular, ± 1.25 × 2.5 mm., contiguous; lobes transversely elliptic, ± 1.25 × 1.5 mm. Male flowers: bracteoles spathulate, deeply dentate; stamens 4.5 mm. long. Female flowers: styles 2 mm. long, joined for ⅓, apices thickened. Capsule scarcely exserted on a pedicel ± 2 mm. long, 3-lobed, ± 3 × 4 mm. Seeds subglobose, 1.75 × 1.5 mm., pale brown, areolate.

TANZANIA. Masai District: Nguruman Escarpment, Naidigidigo, 2 Oct. 1944, *Bally* 3863!; Mbulu District: Lake Manyara National Park, W. boundary, 29 June 1965, *Greenway & Kanuri* 11918! & Manyara Escarpment, 7 km. W. of Mto-wa-Mbu, 13 Feb. 1974, *Bally & Carter* 16528!
DISTR. T 2; known only from the western Rift wall in northern Tanzania
HAB.   In rocky soil with *Acacia-Commiphora* and succulent bushland; 1370–1550 m.

118. E. rubrispinosa *S. Carter* in Hook., Ic. Pl. 39, t. 3866 (1982). Type: Tanzania, Kigoma District, 57 km. S. of Uvinza, *Bullock* 3261 (K, holo.!)

Densely branching succulent perennial; branches decumbent forming tangled mats to 1 m. in diameter and 30 cm. high, 4-angled, ± 1 cm. thick, bright green; angles with shallow rounded teeth to 1 cm. apart. Spinescence dark red becoming blackish brown; spine-shields oblong-triangular, to ± 2 × 2 mm. above the spines, decurrent to just above the flowering eye below; spines slender, to 6 mm. long; prickles to 1 mm. long. Leaves acutely deltoid, ± 1 mm. long. Cymes solitary, 1-forked, with peduncle ± 1.5 mm. long and cyme-branches to 2.5 mm. long; bracts deltoid, ± 1.75 × 1.5 mm. Cyathia ± 3 × 3.5 mm., with cup-shaped involucres; glands transversely elliptic, ± 0.75 × 1.25 mm., quite separate, yellow becoming reddish; lobes rounded, ± 0.75 mm. in diameter. Male flowers few: bracteoles ligulate, deeply dentate; stamens well-exserted, 5 mm. long. Female flower: styles slender, 4.5 mm. long, joined for 0.75 mm. apices thickened. Capsule subsessile, obtusely 3-lobed, ± 2.75 × 3 mm., red. Seeds subglobose, 1.5 mm. in diameter, reddish brown, minutely and closely tuberculate.

TANZANIA. Kigoma District: 56 km. from Uvinza on Mpanda road, 23 Nov. 1962, *Verdcourt* 3434A! & 189 km. from Mpanda on Uvinza road, 12 June 1980, *Hooper & Townsend* 1959B! & Kasakati, Sept. 1965, *Suzuki* B.48!
DISTR. T 4; not known elsewhere
HAB.   On a rocky outcrop in *Brachystegia* woodland; 1370–1730 m.

119. E. asthenacantha *S. Carter* in Hook., Ic. Pl. 39, t. 3865 (1982). Type: Tanzania, Kigoma District: Ubende Plateau, *Hooper & Townsend* 1957B (K, holo.!)

Shortly rhizomatous succulent perennial erect to 15 cm. high; rhizomes tuberous; stems very sparsely branched, 4-angled, ± 6 mm. thick, pale greyish green obscurely blotched with darker green; angles straight, without teeth. Spinescence brown; spine-shields exceedingly slender, to 11 mm. long, not extended above the spines, decurrent to just above the flowering eye below; spines very fine, to 2 mm. long; prickles immediately above the spines, rudimentary. Leaves ovate, ± 1.5 × 0.75 mm. Cymes solitary, 1-forked with peduncle to 3.5 mm. long and cyme-branches to 5 mm. long; bracts ovate, ± 1.25 × 1 mm. Cyathia ± 4 × 3.75 mm., with narrowly funnel-shaped involucres; glands rounded, ± 1.25 × 1.5 mm., quite separate, yellow; lobes rounded, ± 0.75 × 0.75 mm. Male flowers very few: bracteoles laciniate; stamens far exserted, 6 mm. long, red. Female flower: styles slender, 5 mm. long, joined at the base, the apices scarcely thickened. Capsule subsessile, obtusely 3-lobed, ± 3 × 4 mm., reddish. Seeds subglobose, 2 × 1.8 mm., brown, minutely and shallowly tuberculate.

TANZANIA.   Kigoma District: Ubende Plateau, 86 km. N. of Mpanda on Uvinza road, 12 June 1980, *Hooper & Townsend* 1957B! & June 1979, *Classen* 106!
DISTR.   T 4; known only from these two collections
HAB.   In depressions of rocky outcrop; 1500 m.

120. E. torta *Pax & K. Hoffm.* in E.J. 45: 240 (1910); N.E. Br. in F.T.A. 6(1): 568 (1911). Type: Tanzania, Tabora, *von Trotha* 167 (B, holo.†, K, drawing of holo.!)

Cushion-forming succulent perennial ± 15 cm. high, with a thick woody rhizomatous rootstock and stems branching densely from the base; branches to 15(–25) cm. long, 4-angled, 4–5 mm. thick; angles usually prominently toothed, with teeth to 1.5 cm. apart. Spinescence brown; spine-shields elongated triangular, to 1 × 2 mm. above the spines, decurrent to shortly above the flowering eye below; spines to 3 mm. long; prickles to 2.5 mm. long. Leaves acutely deltoid, ± 1.5 × 1 mm. Cymes solitary, 1-forked, with peduncles and cyme-branches ± 3 mm. long; bracts ovate, ± 1.5 × 1.25 mm. Cyathia ± 3 × 4.5 mm., with cup-shaped involucres; glands transversely rectangular, ± 1 × 2 mm., not quite touching; lobes rounded, ± 1 × 1 mm. Male flowers few: bracteoles spathulate, deeply dentate; stamens 4.5 mm. long. Female flower: styles 3.5 mm. long, joined for 1.5 mm., apices slightly thickened. Capsule and seeds not seen.

TANZANIA.   Tabora District: Lyela, Sept. 1951, *Groome* 12! & without precise locality or date, *Swynnerton* 1399!; Iringa District: Ruaha National Park, 21 km. N. of Msembe on Ikuka track, 5 Dec. 1972, *Bjørnstad* 2028!
DISTR.   T 4, 7; Zambia (Mpulungu Escarpment)
HAB.   *Brachystegia* woodland; 1050–1500 m.

NOTE.   The wide distribution of the cited localities, the only ones known, suggest this small plant is more common than it appears to be from the collections so far seen. The one from the Mpulungu Escarpment (*B.D. Burtt* 6199) has longer (25 cm.) more prominently toothed branches than the Tanzanian specimens, but otherwise appears to be identical.

121. E. greenwayi *Bally & S. Carter* in K.B. 29: 512 (1974). Type: Tanzania, Iringa Escarpment, *Greenway & Kanuri* 13812 (K, holo.!, EA, iso.!)

Branching succulent perennial to 1.2 m. high; branches sharply 4-angled, bluish green usually with darker blotches along the angles; angles with shallow to prominent teeth 1–1.5 cm. apart. Spine-shields elongated, to 2 × 1 mm. above the spines, decurrent to, or just above the flowering eye below; spines slender, to 1 cm. long; prickles 0.5–5 mm. long. Leaves acutely deltoid, ± 1.5 × 1 mm. Cymes solitary, 1-forked, with peduncles and cyme-branches ± 3 mm. long; bracts ovate, ± 1.5 × 1.25 mm. Cyathia ± 3 × 4.5 mm., with cup-shaped involucres; glands transversely rectangular, ± 1 × 2 mm., not quite touching; lobes rounded, ± 1 × 1 mm. Male flowers few: bracteoles spathulate, deeply dentate; stamens 4.5 mm. long. Female flower: styles 3.5 mm. long, joined for 1.5 mm., apices slightly thickened. Capsule and seeds not seen.

subsp. **greenwayi**

Procumbent to 30 cm. high; stems and branches markedly variegated with darker blue-green along the prominently and sharply-toothed angles, and with teeth to 1 cm. apart; spine-shields decurrent to the flowering eye below and often forming a continuous horny margin; spines 8–10 mm. long; prickles 3–5 mm. long; cyme peduncles to 5 mm. long.

TANZANIA.   Iringa District: Iringa Escarpment ± 65 km. N. of Iringa on Dodoma road, 28 Aug. 1909, *Greenway & Kanuri* 13812!
DISTR.   T 7; known only from the type collection
HAB.   Rocky scarp with *Brachystegia* open woodland; 1000 m.

subsp. **breviaculeata** *S. Carter* in K.B. 42: 381 (1987). Type: Kahama, *Bullock* 3036 (K, holo.!)

Much-branched shrub to 1.2 m. high; stems and branches sometimes with darker markings along the sinuately toothed angles, and with teeth to 1.5 cm. apart; spine-shields decurrent to just above the flowering eye below; spines 5–8 mm. long; prickles 0.5–1.5 mm. long; cyme peduncles 2 mm. long.

TANZANIA.   Kahama, 27 July 1950, *Bullock* 3036!; Iringa District: Kalenga, 35 km. W. of Iringa, 25 June 1960, *Leach & Brunton* 10123!
DISTR.   T 4, 7; not known elsewhere
HAB.   Among rocks in open deciduous woodland; 1300–1500 m.

SYN.   [*E. greenwayi* sensu Bally & S. Carter in K.B. 29: 512 (1974) excl. typo]

NOTE.   Both of the above collections differ in some degree from each other, as well as from the type of the species. *Leach & Brunton* 10123 is densely-branched and bushy with narrow unmarked bluish branches and a sturdy spinescence; another collection, *Leach* 10347, from a cultivated plant purported to originate from 'around Iringa', shows a more laxly branched habit, with wider pale bluish green faintly variegated branches and a slender spinescence, and is probably identifiable with *Redmayney* in *E.A.H.* 13266 from Mazombe. *Bullock* 3036, the type of this subspecies, was collected from over 500 km. to the north-west and most nearly matches Leach's cultivated specimen but with more distinctly marked branches. Apart from habit, the typical subspecies differs most noticeably in the intense variegation of its branches and in its longer spines, with much longer prickles. The distinctive inflorescence characters serve to unite all the gatherings of both subspecies closely, with further separation unjustified before more material can be examined.

122. E. angustiflora *Pax* in E.J. 34: 82 (1904); N.E. Br. in F.T.A. 6(1): 568 (1911). Type: Tanzania, Chunya District, Ilunga Mts., *Goetze* 1095 (B, holo.†, K, fragment of holo.!, BM, iso.!)

Densely branched succulent perennial forming clumps to ± 20 cm. high and 50 cm. in diameter; branches 4-angled, to 1 cm. thick, glaucous; angles prominently sinuate-toothed, with teeth 1–1.5 cm. apart. Spinescence purplish black; spine-shields oblong-triangular, to ± 2 × 2.5 mm. above the spines, decurrent to 8 mm. and halfway to ⅔ towards the flowering eye below; spines to 8 mm. long; prickles to 2 mm. long. Leaves acutely deltoid, ± 1.5 × 1 mm. Cymes solitary, 1-forked, with peduncles and cyme-branches ± 3 mm. long; bracts ovate, ± 2 × 1.5 mm. Cyathia ± 4 × 3 mm., with elongated barrel-shaped involucres; glands subcircular, ± 1 mm. in diameter, quite separate and facing inwards; lobes rounded, ± 0.75 × 0.75 mm., deeply toothed. Male flowers: bracteoles spathulate, deeply dentate; stamens far exserted, 8 mm. long, red. Female flower: styles 7 mm. long, joined for 2.5 mm., apices scarcely thickened, red. Capsule subsessile, obtusely 3-lobed, ± 3.5 × 3.5 mm. Mature seeds not seen.

TANZANIA.   Chunya District: Iloma Mt., 3 Sept. 1942, *Bally* E.125! & Rukwa, Kwimba Mt., 27 Sept. 1976, *Leedal* 3822!; Mbeya District: Madibira, 12 July 1959, *Anderson* 1243!
DISTR.   T 7; not known elsewhere
HAB.   On rocky outcrops in *Brachystegia* woodland; 1100–1500 m.
SYN.   [*E. schinzii* sensu Pax in E.J. 30: 341 (1901), *non* Pax (1898)]

123. E. tetracanthoides *Pax* in E.J. 30: 342 (1901); N.E. Br. in F.T.A. 6(1): 569 (1911). Type: Tanzania, Njombe District, Tyuni [Ujuni] Mt., *Goetze* 1004 (B, holo.†, K, drawing of holo.!)

Densely branching succulent perennial forming tangled masses 20–30 cm. high and to 1.5 m. in diameter; branches trailing to 30 cm. long, 1–1.3 cm. thick, 4-angled, uniformly dull green; angles ± distinctly toothed, with rounded teeth 1–1.5 cm. apart. Spinescence grey; spine-shields triangular, ± 2 mm. wide and 2–4 mm. long above the spines, 2–6 mm. decurrent and about halfway to the flowering eye below; spines to 8 mm. long; prickles 0.5–2.5 mm long. Leaves deltoid, ± 1.5 × 1 mm. Cymes solitary, 1-forked, subsessile; bracts subquadrate, ± 2 × 2 mm. Cyathia ± 3 × 6.5 mm., with broadly funnel-shaped involucres; glands transversely rectangular, ± 1.8 × 3.5 mm., contiguous, yellow; lobes transversely elliptic, ± 1.5 × 2.5 mm. Male flowers: bracteoles fan-shaped, dentate; stamens 4.5 mm. long. Female flower: styles 4 mm. long, joined at the base, apices thickened. Capsule sessile, obtusely 3-lobed, ± 4 × 5.5 mm. Seeds ovoid, 2.5 × 2 mm., pale brown, shallowly tuberculate. Fig. 95/4, 5, p. 509.

TANZANIA.   Njombe District: Poroto Mts., 4 km. up track above Chimala, 6 Feb. 1974, *Bally & Carter* 16463! & Ujuni Mt., 1 June 1982, *Leedal* 7004! & 3 km. NW. of Njombe, 11 July 1958, *Milne-Redhead & Taylor* 11059!
DISTR.   T 7; not known elsewhere
HAB.   Amongst rocks usually on steep slopes with scattered deciduous woodland; 1750–2200 m.

124. E. discrepans *S. Carter* in K.B. 42: 380 (1987). Type: Tanzania, Njombe District, Kimani R., 16 km. E. of Chimala, *Leach* 10352 (K, holo.!)

Shrubby succulent perennial, densely branching, spreading-erect to 1.5 m. high; branches 4-angled, ± 1 cm. thick; angles shallowly toothed, with teeth to 1 cm. apart. Spine-shields oblong-triangular, ± 3 × 2 mm. above the spines, decurrent to just above the flowering eye below; spines slender to 6 mm. long; prickles to 1 mm. long. Leaves deltoid,

1 mm. long. Cymes solitary. 1-forked, subsessile; bracts subquadrate, ± 1.5 × 1.5 mm. Cyathia ± 3 × 5.5 mm., with funnel-shaped involucres; glands transversely rectangular, ± 1.5 × 2.8 mm, contiguous, yellow; lobes transversely elliptic, ± 1 × 2 mm. Male flowers: bracteoles fan-shaped, dentate; stamens 3.5 mm. long. Female flower: styles 2.3 mm. long, joined at the base, apices thickened. Capsule sessile, obtusely 3-lobed, ± 3.5 × 5 mm. Seeds ovoid, 2.5 × 2 mm., brown, shallowly tuberculate.

TANZANIA. Mbeya District: Songwe Bat Caves, 3 Oct. 1976, *Leedal* 3872!; Njombe District: Ruaha R., Usalimwani, 8 km. E. of Chimala, 11 Aug. 1979, *Leedal* 5569!
DISTR. T 7; known only from these 3 collections
HAB. Rocky ground at the foot of the north-facing scarp; 1340–1370 m.
NOTE. More material of this species is needed before the extent of variation, especially in habit can be accurately assessed.

125. E. nyassae *Pax* in E.J. 34: 375 (1904); N.E. Br. in F.T.A. 6(1): 575 (1911); T.T.C.L.: 212 (1949). Type: Tanzania, Mbeya District, Manayeme's village, *Goetze* 1443 (B, holo.†, K, fragments and drawing of holo.!)

Shrubby or semi-prostrate succulent perennial 25–100 cm. high with branching stems sometimes trailing; branches 4–5-angled, 1–1.5 cm. thick; angles shallowly to prominently toothed, with rounded teeth 1–1.5 cm. apart. Spine-shields triangular or oblong-triangular, to 3 × 3 mm. above the spines, decurrent to 5(–8) mm. and ½–¾ towards the flowering eye below; spines to 8 mm. long; prickles 1–2 mm. long. Leaves deltoid, ± 1 × 1 mm. Cymes solitary, 1-forked, with peduncles 0.5–1.5 mm. long; bracts subquadrate, ± 1.4 × 1.4 mm. Cyathia to ± 2.5 × 5 mm., with funnel-shaped involucres; glands transversely rectangular, to 1.25 × 2.5 mm., contiguous; lobes transversely elliptic, to 1 × 1.5 mm. Male flowers: bracteoles fan-shaped, dentate, stamens 3 mm. long. Female flower: styles 2–2.2 mm. long, joined at the base, apices thickened. Capsule sessile, obtusely 3-lobed, ± 2.5 × 4 mm. Seeds ovoid, 1.8 × 1.5 mm., brown, shallowly tuberculate.

subsp. **nyassae**

Bushy shrub 50 cm. high; branches ± 1 cm. thick, apparently shallowly toothed; spine-shields to 2 × 2 mm. above the spines, decurrent to 4 mm., spines to 4 mm. long; prickles to 1.5 mm. long. Cyathia funnel-shaped; glands yellowish red.

TANZANIA. Mbeya District: Unyika [Unyiha], Manayeme's village [± 9°15'S, 33°07'E], Nov. 1899, *Goetze* 1443!
DISTR. T 7; known with certainty only from the type collection
HAB. Amongst rocks on a plateau; 1500 m.

SYN. *E. tetracantha* Pax in E.J. 30: 341 (1901), *non* Rendle. Type as for *E. nyassae* Pax

NOTE. This description has been taken from Goetze's notes and the fragments, in Kew, of his collection, including all details of the cyathia, capsule and seeds. N.E. Brown's description of dark purple glands is incorrect, Goetze describing them as yellowish red.

subsp. **mentiens** *S. Carter* in K.B. 42: 379 (1987). Type: Zambia, Mbala District, Sunzu Mt., *Richards* 13168 (K, holo.!)

Shrubby or semi-prostrate, 25–100 cm. high, with trailing stems to 1 m. long; branches ± prominently toothed. Cyathia with broadly funnel-shaped involucres; glands bright yellow.

TANZANIA. Ufipa District: near Zambia border on new Sumbawanga–Mbala road, 4 Dec. 1960, *Richards* 13644! & 8 km. N. of border on old Mbala–Sumbawanga road, 16 June 1960, *Leach & Brunton* 10052!
DISTR. T 4; northern Zambia (Mbala region)
HAB. Amongst rocks, in the open or in light shade; 1700–1740 m.

NOTE. Subsp. *mentiens* closely resembles *E. nyassae* as represented by Goetze's notes and the remaining fragments of his collection and may well represent the true species. *Richards* 13644 is especially similar, apparently differing only in the colour of the cyathial glands. However, until adequate material has been collected from Goetze's locality for comparison, it is best to regard the two taxa as distinct.

E. whellanii Leach, in Journ. S. Afr. Bot. 33: 247 (1967), is a closely related dwarf, densely tufted species so far known only at its type locality near Kawimbe, Zambia, close to the Tanzanian border, which suggests that it could occur in Ufipa District. It grows in rock crevices in damp shady situations. Its branches are uniformly dark green, subcylindrical to ± 17 cm. long and 8 mm. thick, usually 6-angled with pale grey spine-shields ± 5 mm. apart; spine-shields obtusely triangular, not decurrent, ± 2 × 2 mm.; spines and prickles subequal in length 2.5–3.5 mm. long, very slender, arranged in an X; cyathia subsessile, ± 2 × 3.5 mm., all parts yellow; capsule subsessile, ± 2.5 × 4 mm., flushed with red; seeds subglobose, 1.75 × 1.5 mm., pale grey and mottled, densely tuberculate.

126. E. exilispina *S. Carter* in K.B. 42: 382 (1987). Type: Tanzania, Njombe District, Lumbila, *Gilli* 280 (W, holo.!)

Branching succulent perennial erect to 1 m. high; branches 4-angled, 5–8 mm. thick; angles very shallowly toothed with teeth 5–8 mm. apart. Spine-shields oblong-triangular, ± 1 × 1 mm. above the spines, decurrent to ± 2.5 mm. and half-way to the flowering eye below; spines very slender, to 6 mm. long; prickles 1–2 mm. long. Leaves acutely deltoid, ± 1 × 1 mm. Cymes solitary, 1-forked, sessile; bracts subquadrate, ± 1.2 × 1.2 mm. Cyathia ± 2 × 4.5 mm., with funnel-shaped involucres; glands transversely rectangular, ± 1 × 2 mm., contiguous, yellow; lobes transversely elliptic, ± 0.8 × 1 mm. Male flowers: bracteoles fan-shaped, dentate; stamens 3.5 mm. long. Female flower: styles 3 mm. long, joined at the base for 0.5 mm., apices thickened. Capsule sessile, obtusely 3-lobed, ± 2.5 × 4 mm. Seeds ovoid, 1.8 × 1.5 mm., buff, shallowly tuberculate.

TANZANIA. Njombe District: Lumbila, Aug. 1958, *Gilli* 280!
DISTR. T 7; known only from the type collection
HAB. Sandy stony soil in dry woodland; 530 m.

SYN. [*E. tenuispinosa* sensu Gilli in Ann. Nat. Mus. Wien 78: 167 (1974), pro parte, quoad *Gilli* 280, *non* Gilli sensu stricto]

127. E. isacantha *Pax* in E.J. 34: 82 (1904); N.E. Br. in F.T.A. 6(1): 575 (1911); T.T.C.L.: 212 (1949). Type: Tanzania, Songea District, Ngaka valley, *Busse* 949 (B, holo.†, K, drawing of holo.!)

Succulent perennial densely branching from the base forming tangled masses ± 25 cm. high and 1 m. in diameter; branches decumbent to ± 50 cm. long, sparsely rebranched, sharply 4-angled, to ± 1 cm. thick, uniformly bright green; angles ± without teeth. Spinescence brown, the spine-pairs 5–15 mm. apart; spine-shields oblong-triangular, to 2–3 × 2 mm. above the spines, decurrent to 8 mm. and about halfway to 2 mm. above the flowering eye below; spines and prickles very slender, subequal in length or the prickles slightly longer, to 5 mm. long, spreading to form an X. Leaves deltoid, ± 1 × 1 mm. Cymes solitary, 1-forked, with peduncles ± 2 mm. long; bracts ovate, ± 1.75 × 1.5 mm. Cyathia ± 3 × 5 mm., with funnel-shaped involucres; glands transversely rectangular, ± 1.2 × 2.7 mm., contiguous, pinkish yellow; lobes transversely elliptic, 1 × 1.5 mm. red. Male flowers: bracteoles fan-shaped, dentate; mature stamens not seen. Female flower: ovary sessile; styles 4 mm. long, joined at the base, apices thickened. Capsule and seeds not seen.

TANZANIA. Songea District: Matengo Highlands, near Miyau, 2 Mar. 1956, *Milne-Redhead & Taylor* 8643A! & Kwamponjore valley, 9.5 km. WSW. of Songea, 19 June 1956, *Milne-Redhead & Taylor* 8643B!
DISTR. T 8; Malawi (Northern Province)
HAB. On rocks in the shade of *Brachystegia* woodland; 990–1015 m.

128. E. cataractarum *S. Carter* in K.B. 42: 379 (1987). Type: Tanzania, Ufipa District, Kalambo Falls, *Bullock* 1081 (K, holo.!)

Shrubby succulent perennial, sparingly branched, erect to 1.5 m. high; branches 4-angled, ± 1 cm. thick, bright green; angles shallowly toothed with teeth to 1.5 cm. apart. Spinescence pale grey; spine-shields elongated, to ± 3 × 2 mm. above the spines, decurrent to shortly above the flowering eye below; spines to 6 mm. long; prickles to 1.5 mm. long. Leaves deltoid, ± 1.5 × 1.5 mm. Cymes solitary, subsessile, 1-forked; bracts subquadrate, ± 1.5 × 1.5 mm. Cyathia ± 2.5 × 5 mm., with cup-shaped involucres; glands transversely rectangular, ± 1 × 2.5 mm., contiguous, yellowish becoming red; lobes transversely elliptic, ± 1 × 1.5 mm. Male flowers: bracteoles fan-shaped, dentate; stamens 3 mm. long. Female flower: styles 2.5 mm. long, joined at the base, apices thickened. Capsule sessile, obtusely 3-lobed, ± 2.8 × 4.5 mm. Seeds ovoid, 2.2 × 1.8 mm., brown, minutely tuberculate.

TANZANIA. Ufipa District: Kawa Falls, 1 Oct. 1956, *Richards* 6345! & Kalambo gorge, 15 Sept. 1959, *Richards* 11447!
DISTR. T 4; northern Zambia around the southern end of Lake Tanganyika
HAB. Amongst rocks in river gorges, characteristically near waterfalls; 1200–1290 m.

129. E. eyassiana *Bally & S. Carter* in Hook., Ic. Pl. 39, t. 3864 (1982). Type: Tanzania, Mbulu District, between Oldeani and Lake Eyassi, *Bally* 10615 (K, holo.!)

Fibrous-rooted succulent perennial, with shortly rhizomatous stems densely branching at the base, sparsely so above, erect to 80 cm.; stems and branches 4-(5)-angled, to 1 cm. thick, greyish green often with a purplish tinge, darker along the angles; angles with shallow teeth 1-2 cm. apart. Spinescence grey; spine-shields oblong-triangular, fairly slender, ± 2 mm. long above the spines, decurrent to 8 mm. and about halfway to the flowering eye below; spines slender, to 15 mm. long; prickles to 3.5 mm. long. Leaves acutely deltoid, ± 1.5 × 1 mm. Cymes solitary, 1-forked, subsessile; bracts deltoid, ± 1.5 × 1.5 mm. Cyathia ± 2.5 × 4.5 mm., with funnel-shaped involucres; glands transversely rectangular, ± 1.25 × 2.5 mm., contiguous, yellowish brown; lobes subcircular, 1 mm. in diameter. Male flowers: bracteoles fan-shaped, deeply dentate; stamens 3.25 mm. long. Female flower: styles 2.75 mm. long, joined for ⅓, apices thickened. Capsule sessile, obtusely 3-lobed, ± 2.5 × 3 mm. Seeds ovoid, 1.75 × 1.5 mm., pale brown, shallowly tuberculate.

TANZANIA. Musoma District: 28 km. from Seronera on Soitayai road, 10 June 1961, *Greenway* 10373!; Masai District: Olduvai, 27 Sept. 1977, *Raynal* 19322!; Mbulu District: Mangola, 24 Sept. 1977, *Raynal* 19268!
DISTR. T 1, 2; not known elsewhere
HAB. Stony soils with sparse grassland and open dry bushland; 1000-1800 m.

VAR. Specimens from the east of the Serengeti have slightly more toothed stems and stronger spinescence.

130. E. furcata *N.E. Br.* in F.T.A. 6(1): 566 (1911). Type: Kenya, Kwale District, Inepanga, *Kassner* [mixed with *Kassner* 430, *E. tenuispinosa* Gilli] (BM, holo.!, K, drawing of holo.!)

Succulent perennial with densely tufted stems from a thick fleshy root, branching from the base, 4-angled, to 15 cm. long and ± 7 mm. thick, greyish to brownish green; angles with laterally compressed prominent very sharply pointed teeth to 1 cm. long and ± 1.5 cm. apart. Spinescence grey; spine-shields to 8 mm. long above the spines, occasionally with 2 scarcely diverging arms to 1.5 mm. long, decurrent to 3-5 mm. above the flowering eye below; spines 2-12 mm. long, often joined at the base for up to 3 mm,; prickles to 1 mm. long. Leaves deltoid, ± 1 × 1 mm. Cymes 1-2 in a vertical line developing successively, 1-forked, with branches to 5 mm. long; bracts subquadrate, ± 1.5 × 1.25 mm. Cyathia ± 2.5 × 4.5 mm., with funnel-shaped involucres; glands transversely elliptic, ± 1 × 2 mm., contiguous, pinkish brown; lobes subcircular, 1 mm. in diameter. Male flowers: bracteoles deeply dentate; stamens 3.25 mm. long. Female flower: styles 2 mm. long, joined at the base, apices thickened, rugulose. Capsule subsessile, obtusely 3-lobed, ± 2.5 × 3 mm. Mature seeds not seen. Fig. 96/1-3, p. 516.

KENYA. Kilifi District: 46.5 km. from Mombasa on Voi road, 14 Apr. 1969, *Bally* 13261!
TANZANIA. Moshi District: Lake Chala-Taveta, 21 Jan. 1936, *Greenway* 4472!; Pare District: Mwembe valley, 30 June 1942, *Greenway* 6509!
DISTR. K 7; T 2, 3; known only from a small area of extreme SE. Kenya and NE. Tanzania
HAB. In dry rocky sandy soils with *Acacia* bushland; 300-915 m.

NOTE. The specimen used by N.E. Brown as the type was mixed with a Kassner collection of *E. tenuispinosa* Gilli (*E. taitensis* Pax, *nom. illegit.*). There is no reason to suppose that they were not collected at the same locality near Inepanga at the northern part of the Shimba Hills in Kwale District.

131. E. uhligiana *Pax* in E.J. 43: 86 (1909); N.E. Br. in F.T.A. 6(1): 567 (1911), pro parte, excl. *Scott-Elliot* 6271 & *Uhlig* 88; U.K.W.F.: 223 (1924). Type: Tanzania, Masai District, Engaruka, *Uhlig* 227 (B, holo.†, K, drawing of holo.! & iso.!)

Succulent perennial with densely tufted stems from a thick fleshy root, branching from the base, to 30(-100) cm. long, 4-angled, ± 1 cm. thick, dark green, often lightly variegated with paler green and the surface shallowly fluted; angles with prominent sharply pointed teeth to 7 mm. long and to 1.5 cm. apart. Spinescence pale grey; spine-shields to 2 cm. long, to 5 mm. long above the spines and with 2 widely diverging arms 2-5 mm. long, decurrent from halfway to just above the flowering eye below, with the base sometimes free and forming a hook to 1 mm. long; spines to 1.3 cm. long, often joined at the base for up to 2 mm.; prickles to 2.5 mm. long. Leaves deltoid, ± 1.5 × 1.5 mm. Cymes 1-3 in a vertical line, developing successively, 1-forked, with branches 1-2.5 mm. long; bracts subquadrate, ± 1.5 × 1.25 mm. Cyathia ± 2.5 × 4.5 mm., with broadly funnel-shaped

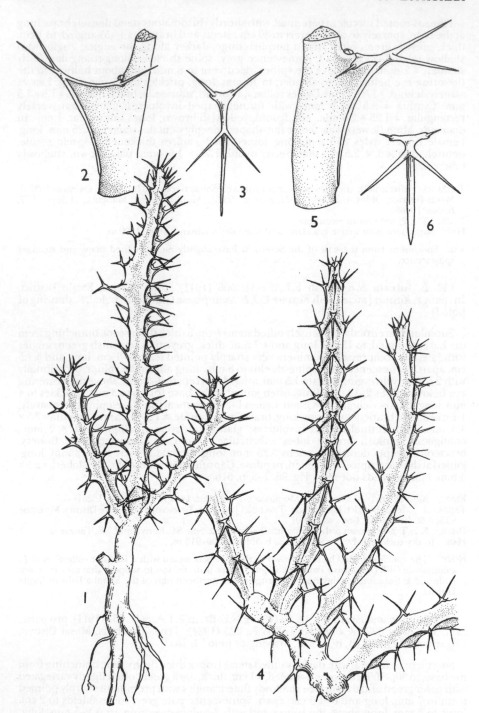

FIG. 96. *EUPHORBIA FURCATA* — 1, portion of plant showing branch, × ²⁄₅; 2, branch-tooth and spinescence, × 2; 3, spine-shield, × 2. *E. UHLIGIANA* — 4, branching stem, × ²⁄₅; 5, branch-tooth and spinescence, × 2; 6, spine-shield, × 2. 1–3. from *Greenway* 11473; 4–6 from *Bally* 10666. Drawn by Christine Grey-Wilson.

involucres; glands spreading, transversely elliptic; ± 1 × 2.5 mm., contiguous, yellow; lobes subcircular, 1 mm. long. Male flowers: bracteoles 2 mm. long, laciniate, deeply dentate; stamens 3 mm. long. Female flower: styles ± 1.75 mm. long, joined at the base, apices thickened. Capsule sessile, obtusely 3-lobed, ± 3.25 × 4.5 mm. Seeds ovoid with truncate base, 2 × 1.5 mm., greyish brown, shallowly and minutely tuberculate. Fig. 96/4–6.

KENYA. Masai District: SE. foot of Mt. Suswa, 3 Feb. 1964, *Glover & Oledonet* 4066! & Ngong Hills on road to Magadi, Feb. 1972, *Greenway* in *Bally* 15006! & 9 km. S. of Ewaso Nyiro, S. of Masandare R., 11 Jan. 1981, *Kuchar* 13886!
TANZANIA. Masai District: Engaruka, 3 Feb. 1932, *St. Clair-Thompson* 300! & 9 July 1956, *Bally* 10666!; Lushoto District: Lake Manka, 12 Jan. 1967, *Richards* 21955!
DISTR. K 6; T 2, 3; not known elsewhere
HAB. Stony sandy soils in open deciduous bushland, usually on exposed slopes; 435–1550 m.

VAR. Populations from the Narok area in Kenya possess spine-shields with the upper arms more widely diverging and with shorter prickles than those from the type locality. They also appear to form larger clumps than populations elsewhere, with branches to 1 m. long.

132. E. petricola *Bally & S. Carter* in Hook., Ic. Pl. 39, t. 3856 (1982): Type: Kenya, Machakos District, 64 km. NW. of Voi, *Greenway* 9831 (K, holo.!, EA, iso.!)

Succulent perennial with stems densely tufted, sparsely branched, subcylindrical, to 30(–75) cm. long and 8 mm. thick, uniformly green; branches with ± shallow teeth to 2 cm. apart in 4 longitudinal series. Spinescence brownish grey, black when young; spine-shields elongated to 13 mm., divergent above the spines into 2 arms to 2 mm. long with expanded ends, decurrent to 1 cm.; spines very slender, to 12 mm. long, sometimes joined at the base; prickles very fine to 5 mm. long. Leaves deltoid, ± 0.75 × 0.75 mm. Cymes solitary, 1-forked, with cyme-branches ± 1.25 mm. long, all parts indistinctly papillose; bracts subquadrate, ± 1.25 × 1.25 mm. Cyathia ± 2.25 × 3.5 mm., with broadly funnel-shaped involucres; glands transversely rectangular, ± 0.75 × 1.75 mm., contiguous, yellow; lobes transversely elliptic, ± 1 × 1.5 mm. Male flowers: bracteoles spathulate, dentate; stamens 4 mm. long. Female flowers: styles 1.75 mm. long, free to the base, apices thickened. Capsule sessile, obtusely 3-lobed, ± 2.5 × 3.25 mm., minutely papillose. Seeds ovoid, with truncate base, 1.75 × 1.5 mm., grey, shallowly and minutely tuberculate.

KENYA. Machakos District: Ngomeni, 20 Nov. 1893, *Scott-Elliot* 6271!; Kitui District: Mutomo Hill, 15 Mar. 1968, *Bally* 13108!; Teita District: Mudanda Rock, 3 Feb. 1953, *Bally* 8680A!
DISTR. K 4, 7; not known elsewhere
HAB. In crevices of rocky outcrops with *Xerophyta* and grass tufts; 500–1000 m.

SYN. [*E. uhligiana* sensu N.E. Br. in F.T.A. 6(1): 567 (1911) pro parte, quoad *Scott-Elliot* 6271, *non* Pax]

133. E. marsabitensis *S. Carter* in Hook., Ic. Pl. 39, t. 3857 (1982). Kenya, Northern Frontier Province, Marsabit, *Carter & Stannard* 651 (K, holo.!, EA, iso.!)

Succulent perennial with stems densely tufted, procumbent usually to one side, sparsely branched, to 30 cm. long and 8 mm. thick, subcylindrical, green strongly variegated with yellow-green; branches with shallow teeth to 1.5 cm. apart in 4 longitudinal series. Spinescence brownish grey; spine-shields elongated to 1 cm., divergent above the spines into 2 arms to 2.5 mm. long, decurrent to 8 mm.; spines slender, to 18 mm. long; prickles to 2 mm. long. Leaves deltoid, ± 1 × 1 mm. Cymes 1–2 in a vertical line, 1-forked, subsessile, all parts minutely papillose; bracts subquadrate, ± 1 × 1 mm. Cyathia ± 2 × 4 mm., with funnel-shaped involucres; glands spreading, transversely rectangular, ± 1 × 1.75 mm., contiguous, yellow; lobes subcircular, ± 1 mm. in diameter. Male flowers: bracteoles spathulate, dentate; stamens 3.2 mm. long. Female flower: ovary subsessile; styles 1.8 mm. long, joined at the base, apices thickened, rugulose. Capsule 3-lobed, sessile, ± 3 × 4 mm. Seeds ovoid, 2.2 × 1.5 mm., minutely tuberculate.

KENYA. Northern Frontier Province: Marsabit, Gof Choba, 1 Mar. 1963, *Bally* 12567! & 12 km. SW. of Marsabit, Karsadera, 24 Nov. 1977, *Carter & Stannard* 651!
DISTR. K 1; apparently restricted to the Marsabit massif
HAB. In rocky soils amongst grass and low open bushland; ± 1200 m.

134. E. vittata *S. Carter* in Hook., Ic. Pl. 39, t. 3858 (1982). Type: Kenya, Northern Frontier Province, SW. of L. Turkana, Mugurr, *Mathew* 6660 (K, holo.!)

Tufted succulent perennial; stems semi-prostrate, sparsely branching, to 30 cm. long and 2 cm. thick, 4–5-angled, darkish green longitudinally streaked with brighter yellow-green; angles with distinct teeth to 1.5 cm. apart. Spinescence brownish grey; spine-shields elongated to 15 mm. long, divergent above the spines into 2 arms to 3 mm. long, decurrent almost to the flowering eye below; spines to 2 cm.long; prickles to 2.5 mm. long. Leaves deltoid, ± 1 × 1.25 mm. Cymes 1–3 in a vertical line, maturing successively, 1-forked, with cyme-branches ± 1.5 mm. long, all parts papillose; bracts subquadrate, ± 1.25 × 1.25 mm. Cyathia ± 2 × 3 mm., with funnel-shaped involucres; glands transversely rectangular, ± 0.7 × 1.5 mm., yellow; lobes subcircular, ± 1 mm. in diameter. Male flowers: bracteoles spathulate, deeply dentate; stamens 3 mm. long. Female flower: styles 1.5 mm. long, free to the base, apices thickened. Capsule sessile, obtusely 3-lobed, ± 2.5 × 3.2 mm. Seeds ovoid, 1.7 × 1.4 mm., reddish brown, minutely tuberculate.

KENYA. Northern Frontier Province: Mt. Kulal, 22 Nov. 1978, *Hepper & Jaeger* 6996! & Archers Post by Uaso Nyiro, 14 June 1979, *Gilbert et al.* 5657!; Tana River District: Kora Game Reserve, Kamunyu Hill, 24 May 1983, *Mungai et al.* 367!
DISTR. K 1, 7; not known elsewhere
HAB. In gravelly soils and amongst rocks with open deciduous bushland; 700–1200 m.

VAR. The most common form, from which the inflorescence measurements have been taken, occurs east of the Mathews Range. It has usually thinner 4-angled branches which are longer and more distinctively striped and toothed than the form of the type from the southern end of Lake Turkana. Branches on these latter plants are usually 5-angled and subcylindrical to ± 2 cm. thick and 15 cm. long. The two forms may well prove to represent distinct species.

135. E. turkanensis *S. Carter* in Hook., Ic. Pl. 39, t. 3859 (1982). Type: Kenya, Northern Frontier Province, 1.5 km. SW. of Lokichar, *Carter & Stannard* 304 (K, holo.!, EA, iso.!)

Succulent perennial, spreading-erect to 45 cm. high and 1.25 m. in diameter; stems crowded, sparsely branching, to 1.5 cm. thick, 4-angled, dark green blotched with brown; angles with prominent teeth 1–3 cm. apart. Spinescence very sturdy, grey, dark red when young; spine-shields elongated to 23 mm. long, diverging above the spines into 2 arms to 5 mm. long, decurrent to 20 mm.; spines to 3 cm. long occasionally joined at the base; prickles to 8 mm. long. Leaves deltoid, 2.7 × 2.7 mm. Cymes 1–5 in a vertical line, maturing successively, 1-forked, with cyme-branches ± 2 mm. long, all parts densely papillose; bracts deltoid, ± 1.25 × 1.25 mm. Cyathia ± 2.5 × 4.5 mm., with funnel-shaped involucres; glands transversely rectangular, ± 1 × 2 mm., contiguous, brownish pink; lobes transversely elliptic, ± 1.25 × 1.5 mm. Male flowers: bracteoles spathulate, deeply dentate; stamens 3 mm. long. Female flower: styles 2.5 mm. long, joined at the base, lightly papillose, apices thickened. Capsule subsessile, obtusely 3-lobed, ± 3.5 × 4 mm., densely papillose, purple. Seeds ovoid with truncate base, 2 × 1.5 mm., grey, minutely tuberculate.

KENYA. Northern Frontier Province: Lokori, 27 July 1968, *Mwangangi & Gwynne* 1041!; Turkana Province: Kokidodoka, Mar. 1965, *Newbould* 7323!
DISTR. K 1, 2; limited to a small area of NW. Kenya
HAB. In gravelly soils and crevices of lava rocks with very open deciduous bushland; 550–800 m.

136. E. samburuensis *Bally & S. Carter* in Hook., Ic. Pl. 39, t. 3860 (1982). Type: Kenya, Northern Frontier Province, 15 km. E. of Baragoi, *Carter & Stannard* 510 (K, holo.!, EA, iso.!)

Loosely tufted succulent perennial with a thick fleshy root; stems sparsely branched, to 90 cm. long, 1–2 cm. thick, subquadrangular, greyish green, darker along the angles; angles with ± prominent teeth 1–2.5 cm. apart. Spinescence red when young, becoming pale grey; spine-shields oblong-triangular, 2–4 × 2–3.5 mm. above the spines, decurrent to 10 mm. and 5–10 mm. above the flowering eye below; spines stout, to 2.5 cm. long, sometimes slightly curved outwards and joined at the base; prickles minute or rudimentary. Leaves deltoid, 1.5 mm. long. Cymes solitary, 1-forked, with peduncles and cyme-branches ± 2.5 mm. long; bracts subquadrate, ± 1.5 × 1.5 mm. Cyathia ± 2.5 × 6 mm., with broadly funnel-shaped involucres; glands transversely elliptic, ± 1.25 × 2.75 mm., just touching, bright yellow; lobes transversely elliptic, ± 1 × 2 mm. Male flowers: bracteoles fan-shaped, deeply dentate; stamens 3.75 mm. long. Female flower: styles 2 mm. long, apices thickened. Capsule subsessile, obtusely 3-lobed, ± 3.75 × 4 mm., purplish. Seeds ovoid, 2.75 × 1.5 mm., grey, tuberculate

KENYA. Northern Frontier Province: El Barta Plains, 20 km. S. of Baragoi, 23 Feb. 1974, *Bally & Carter* 16550! & base of Bartagwet, NE. of Baragoi, 19 Nov. 1977, *Carter & Stannard* 560! & 43 km. N. of

Maralal, 3 June 1979, *Gilbert, Kanuri & Mungai* 5471!
DISTR. **K** 1; limited in distribution to the El Barta Plains and surrounding foothills
HAB. Sandy, stony soil with low deciduous bushland; 1200–1710 m.

137. E. septentrionalis *Bally & S. Carter* in K.B. 29: 514 (1974). Type: Kenya, Laikipia
District, Uaso Nyiro, 32 km. N. of Nanyuki, *Ritchie* in *Bally* E. 84 (K, holo.!)

Densely tufted succulent perennial with a thick fleshy root; stems densely branching
from the base, sparsely so above, erect to 15 cm. or decumbent and occasionally
stoloniferous, to 50(–100) cm. long, 5–8 mm. thick, cylindrical, usually greyish green,
often with darker longitudinal stripes; branches with shallow teeth 1–2 cm. apart in 4
longitudinal series. Spinescence pale grey; spine-shields triangular, 1–2 × 1–1.5 mm.
above the spines, decurrent for 1–4(–6) mm.; spines 5–15 mm. long; prickles less than 1
mm. long. Leaves ovate, 1.5 × 1 mm. Cymes solitary, 1-forked, with peduncles and
cyme-branches ± 2.5 mm. long; bracts subquadrangular, ± 1 mm. long. Cyathia ± 2.5 × 5
mm., with broadly funnel-shaped involucres; glands transversely elliptic, ± 1 × 2.5 mm.,
just touching, bright yellow; lobes transversely elliptic, ± 1 × 1.5 mm. Male flowers:
bracteoles fan-shaped, deeply dentate; stamens 3.25 mm. long. Female flower: styles 2
mm. long, apices thickened, rugulose. Capsule subsessile, obtusely 3-lobed, ± 3.25 × 4 mm.,
purplish. Seeds ovoid, 1.75 × 1.5 mm., brown, tuberculate.

UGANDA. Karamoja District: Amudat, Dec. 1957, *Tweedie* 1469 in *Bally* 11963!
KENYA. Turkana District: 1 km. NE. of Loiya on Lodwar road, 8 Nov. 1977, *Carter & Stannard* 268!;
   Laikipia District: Seya R., 30 km. SE. of Maralal, 25 Feb. 1974, *Bally & Carter* 16556!
DISTR. **U** 1; **K** 1–3; not known elsewhere
HAB. In sandy rocky soils with *Acacia* bushland, usually in the open; 1075–1850 m.

SYN. *E. sp. B* sensu Agnew, U.K.W.F.: 223 (1974)

VAR. The two geographically separated populations of *E. septentrionalis*, one on each side of the Rift
Valley, may eventually prove to be distinct, at least at subspecific level. The western one occurs at a
lower altitude than the one from Laikipia District and the southern end of the Mathews Range,
and appears, from the few good specimens available, to produce usually slightly more slender,
virtually toothless branches with longer finer spines. Branches of plants from the region of the
type locality are also more distinctly striped with a brighter green than those of plants from north
of Rumuruti.

NOTE. The Ethiopian specimens cited in the original description appear to belong to an
undescribed species which does not occur in the Flora area.

138. E. dichroa *S. Carter* in Hook., Ic. Pl. 39: t. 3861 (1982). Type: Uganda, Karamoja
District, Kanamugit, *Eggeling* 2983 (K, holo.!)

Tufted succulent perennial with a fleshy rootstock; stems branching densely from the
base, unbranched above; branches to 15 cm. long and 5–8 mm. thick, subcylindrical,
bright yellow-green, conspicuously striped longitudinally with darker or sometimes
purplish green; branches with prominent teeth 5–10 mm. apart in 4 longitudinal series.
Spinescence brown; spine-shields triangular, 1–2 × 2–2.5 mm. above the spine-shields,
decurrent 1–5.5 mm.; spines to 10 mm. long; prickles 1–2 mm. long. Leaves deltoid, 1 mm.
long. Cymes solitary, 1-forked, with peduncles and cyme-branches ± 2.5 mm. long; bracts
subquadrate, ± 1.5 × 1 mm. Cyathia ± 2.5 × 4 mm., with funnel-shaped involucres; glands
transversely elliptic, ± 1 × 1.75 mm., just touching, yellow becoming reddish; lobes
subcircular, ± 1.25 mm. in diameter, red. Male flowers: bracteoles fan-shaped, deeply
dentate; stamens 3.5 mm. long. Female flower: styles 2 mm. long, apices thickened.
Capsule subsessile, obtusely 3-lobed, ± 2.5 × 3 mm., reddish. Seeds ovoid, 2.25 × 1.25 mm.,
grey, tuberculate.

UGANDA. Acholi District: Chua, Amiel, 1936, *Eggeling* 2339!; Karamoja District: Kanamugit, 1937,
   Eggeling 2983!; Kotido, Nov. 1945, *Tweedie* 661!
DISTR. **U** 1; known only from a small area of NE. Uganda
HAB. Rocky outcrops in dry wooded grassland; 1350–1500 m.

139. E. cuprispina *S. Carter* in K.B. 42: 378 (1987). Type: Kenya, Northern Frontier
Province, 8 km. from South Horr on Ilaut track, *Gilbert, Kanuri & Mungai* 5511 (K, holo.!,
EA, iso.!)

Succulent perennial with a fleshy rootstock, densely branched forming tufts erect to 20
cm. high, ± 30 cm. in diameter, sometimes sprawling, rarely stoloniferous; branches

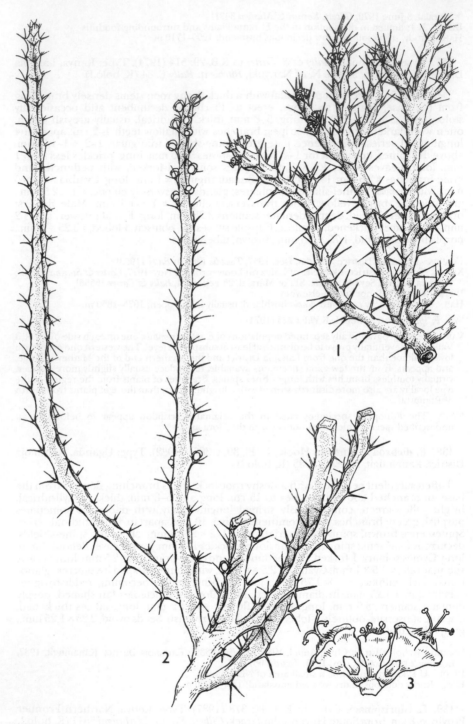

FIG. 97. *EUPHORBIA CUPRISPINA* — 1, branching stem, in flower, × ⅔. *E. TENUISPINOSA* var. *TENUISPINOSA* — 2, branching stem, in fruit, × ⅔; 3, cyme, × 4. 1, from *Gilbert et al.* 5182; 2, from *Bally* 13415; 3, from *Bally* E.9. Drawn by Christine Grey-Wilson.

cylindrical, ± 8 mm. thick, greyish green with a pale longitudinal stripe between the spine-shields; branches with very shallow teeth to 1.5 cm. apart, in 4 longitudinal series. Spinescence yellowish when young becoming bronze or copper-coloured; spine-shields oblong-triangular, ± 1.5 × 1.5 mm. above the spines, decurrent for 2–8 mm.; spines 10–17 mm. long; prickles 0.5–1.5 mm. long. Leaves deltoid, ± 1.25 × 1 mm. Cymes solitary, 1-forked, with peduncles and cyme-branches 1–2 mm. long; bracts subquadrate, ± 1.25 × 1 mm. Cyathia ± 2.5 × 5.5 mm., with funnel-shaped involucres; glands transversely elliptic, ± 1.25 × 2 mm., just touching, bright yellow; lobes transversely elliptic, ± 1 × 1.5 mm. Male flowers: bracteoles spathulate, deeply dentate; stamens 3.5 mm. long. Female flower: styles ± 1.5 mm. long, joined at the base, apices thickened. Capsule subsessile, obtusely 3-lobed, ± 3.25 × 4.5 mm., purplish. Seeds ovoid, 2.2 × 1.5 mm., brown, lightly tuberculate. Fig. 97/1.

KENYA. Northern Frontier Province: Kowop, SSW. of Mt. Nyiru, 28 Oct. 1978, *Gilbert, Gachathi & Gatheri* 5182! & 28 km. N. of Baragoi, Oct. 1979, *Lavranos & Newton* 17715!
DISTR. **K** 1; known only from the southern end of the Nyiru Range
HAB. Rocky soil in dry open deciduous bushland; 950–1500 m.

140. E. subscandens *Bally & S. Carter* in Hook., Ic. Pl. 39, t. 3862 (1982). Type: Kenya, Northern Frontier Province, Ngoronit, *J. Adamson* 365 in *Bally* E. 291 (K, holo.!, EA, iso.!)

Shortly rhizomatous succulent perennial, with a thickened rootstock; stems branching densely from the base, erect to ± 15 cm., or to 1 m. long and arching over, sprawling and rooting to form new plants; branches 7–10 mm. thick, subcylindrical, dark greyish green occasionally with paler streaks, sparsely rebranched; branches with teeth absent or very shallow and 1–2 cm. apart in 4 longitudinal series. Spinescence dark reddish brown; spine-shields oblong-triangular, 1–2 × 1–2 mm. above the spines, decurrent to 8 mm. and ± halfway to the flowering eye below; spines slender, to 15 mm. long; prickles fine, 1–3 mm. long. Leaves deltoid, ± 1 mm. long. Cymes solitary, 1-forked, with peduncles and cyme-branches to 8 mm. long; bracts deltoid, ± 1.25 × 1 mm. Cyathia ± 3.5 × 5.5 mm. with funnel-shaped involucres; glands rounded, ± 1.5 × 2 mm., completely separate, pinkish yellow; lobes transversely rectangular, ± 1 × 1.5 mm. Male flowers: bracteoles few, filamentous; stamens well exserted, 5.5 mm. long, red. Female flower: styles slender, to 4 mm. long, joined only at the base, apices thickened. Capsule shortly exserted on an erect pedicel ± 2.5 mm. long, obtusely 3-lobed, ± 3 × 4.5 mm. Seeds ovoid, 2 × 1.6 mm., pale brown, with small widely spaced tubercles.

KENYA. Northern Frontier Province: 10 km. E. of Ngoronit, 1 Nov. 1978, *Gilbert, Gachathi & Gatheri* 5271! & Ngoronit, 3 Dec. 1978, *Hepper & Jaeger* 7211!
DISTR. **K** 1; known only from the eastern side of the Ndoto Mts. and Mathews Range
HAB. Amongst rocks with low vegetation and very open deciduous bushland; 750–1350 m.

VAR. Individual collections differ in sturdiness of the spines, length of prickles, shallowness of the stem-teeth, stem variegation and length. *Gilbert et al.* 5521 from 1 km. W. of Ilaut appears to be an extreme form, with subquadrangular sprawling stems, very long spines and cymes apparently with shorter peduncles than is typical.

141. E. gemmea *Bally & S. Carter* in Hook., Ic. Pl. 39, t. 3863 (1982). Type: Kenya, Northern Frontier Province, Dandu, *Gillett* 12633 (K, holo.!)

Fibrous rooted succulent perennial, with stems shortly rhizomatous, decumbent and weakly erect to 45 cm. high, sparsely branching; stems and branches sharply 4-angled, 5–10 mm. thick, irregularly blotched with pale and darker green; angles with shallow teeth 1–1.5 cm. apart. Spinescence reddish brown becoming grey; spine-shields very slender, ± 2 mm. long above the spines, decurrent up to 10 mm. and halfway to 5 mm. above the flowering eye below; spines slender, to 12 mm. long; prickles rudimentary, to 0.5 mm. long. Leaves deltoid, ± 1.5 × 1.5 mm. Cymes solitary, 1-forked, with cyme-branches ± 2 mm. long; bracts subquadrate, ± 1.25 × 1 mm. Cyathia ± 2.5 × 5 mm., with broadly funnel-shaped involucres; glands transversely rectangular, ± 1.5 × 3 mm., contiguous, ruby-red; lobes transversely elliptic, ± 1 × 1.5 mm. Male flowers: bracteoles ligulate, dentate; stamens 3.5 mm. long. Female flower: styles 1.5 mm. long, free, apices thickened. Capsule shortly exserted on a pedicel 2 mm. long, obtusely 3-lobed, ± 2.5 × 3.5 mm., dark red. Seeds ovoid, 1.8 × 1.4 mm., pale greyish-brown, tuberculate-rugose.

KENYA. Northern Frontier Province: Dandu, 24 Mar. & 18 June 1952, *Gillett* 12633!; 160 km. E. of Moyale, Takabba Hill, 1 Mar. 1974, *Bally & Carter* 16577!

DISTR. **K** 1; known only from a small area east of Moyale
HAB. Shallow soils of rocky outcrops in the shade of *Acacia-Commiphora* bushland; 700–1000 m.

SYN. [*E. saxorum* sensu Bally & S. Carter in K.B. 29: 510 (1974), pro parte (*Gillett* 12633, in error as 13633), *non* Bally & S. Carter sensu stricto]

142. E. saxorum *Bally & S. Carter* in K.B. 29: 510 (1974), excl. *Gillett* 12633 (as 13633). Type: Kenya, Fort Hall District, Maboloni Rock, *Bally* 8383 (K, holo.!)

Fibrous rooted succulent perennial, with numerous semi-prostrate and stoloniferous stems to 45 cm. long, sparsely branching; stems and branches sharply 4-angled, 5–8 mm. thick, dark green irregularly blotched with dark purplish green; angles with shallow teeth to 1.3 cm. apart. Spinescence black; spine-shields slender, 1.5–3 mm. long above the spines, decurrent to 10 mm. and to just above or touching the flowering eye below, often forming a continuous horny margin; spines slender to 10 mm. long; prickles 0.5–2.5 mm. long. Leaves deltoid, 1.5 mm. long, purplish. Cymes solitary, simple and bisexual or 1-forked, subsessile, all parts deep crimson; bracts subquadrate ± 1.5 mm. long. Cyathia ± 2.5 × 6.5 mm., with broadly funnel-shaped involucres; glands transversely rectangular, ± 1.5 × 3 mm., contiguous; lobes transversely elliptic, 1.5 × 2.5 mm. Male flowers: bracteoles spathulate, dentate; stamens 3 mm. long. Female flower: styles 2.5 mm. long, joined at the base, apices thickened, rugulose. Capsule sessile, obtusely 3-lobed, ± 3 × 4 mm. Seeds subglobose, 2 × 1.75 mm., brownish grey, surface shallowly tuberculate.

KENYA. Kitui District: 1.5 km. N. of Kangondi on Embu road, 8 May 1960, *Napper* 1658! & 84 km. from Embu on Mwingi road, 15 Nov. 1979, *Gatheri, Mungai & Kibui* 79/22!
DISTR. **K** 4; known only from near Kangondi, Nzukini and Yatta Hill
HAB. Exposed places in rocky clefts with very shallow soil; 1200 m.

SYN. *E. sp. A* sensu Agnew, U.K.W.F.: 223 (1974)

143. E. taruensis *S. Carter* in K.B. 42: 377 (1987). Type: Kenya, Kwale District, Taru Hill, *Lavranos & Newton* 12315 (K, holo.!)

Succulent perennial with rhizomatous stems becoming erect to 30(–45) cm. high, branched only from the base; branches sharply 4-angled, 5–8 mm. thick, variegated with mid- and pale-green; angles straight, without teeth. Spinescence black, extremely slender; spine-shields 5–15 mm. long, decurrent to just above the flowering eye below; spines rarely developed, or not more than 1 mm. long and situated 1–2 mm. below the apex of the spine-shield; prickles 1–3 mm. long, extremely fine, clasping the stem. Leaves acutely deltoid, ± 1 × 1 mm. Cymes solitary, subsessile, 1-forked; bracts deltoid, ± 1.25 × 1 mm. Cyathia ± 2.5 × 4.5 mm., with funnel-shaped involucres; glands transversely oblong, ± 1 × 2.5 mm., brownish yellow; lobes transversely elliptic, ± 1.5 × 2 mm. Male flowers: bracteoles fan-shaped, deeply dentate; stamens 3.5 mm. long. Mature ♀ flower, capsule and seeds not seen.

KENYA. Kwale District: Taru Hill, 1975, *Lavranos & Newton* 12315 cult. *La Fon!*
DISTR. **K** 7; known only from the type locality and seen only as cultivated material
HAB. Leaf-litter amongst rocks beneath trees; 150–480 m.

144. E. tenuispinosa *Gilli* in Ann. Nat. Mus. Wien. 78: 167 (1974); S. Carter in K.B. 32: 84 (1977). Type: Kenya, Teita District, Buchuma, *Hildebrandt* 2859 (B, holo.†, K, drawing of holo.!)

Tuberous-rooted shrubby succulent perennial, erect to 1 m. or subscandent to 1.8 m.; branches 4-angled, 5–10(–15) mm. thick, olive-green, mottled or variegated and darker along the angles; angles with very shallow to prominent teeth 1–2.5(–3) cm. apart. Spinescence blackish; spine-shields slender, 3–6 mm. long above the spines, decurrent to shortly above or touching the flowering eye below; spines 2–12(–15) mm. long; prickles 2–6 mm. long. Cymes solitary, 1-forked, subsessile; bracts subquadrate, 1.5 mm. long. Cyathia 2–3 × 4.5–6 mm., with funnel-shaped involucres; glands transversely rectangular, ± 1 × 2.5 mm., contiguous, yellowish brown; lobes transversely elliptic, ± 1 × 1.5 mm. Male flowers bracteoles spathulate, dentate; stamens 4 mm. long. Female flower: styles 1.8 mm. long, joined at the base, apices thickened, rugulose. Capsule just exserted on a pedicel 2 mm. long, obtusely 3-lobed, ± 3.5 × 5.5 mm. Seeds subglobose, 2 mm. in diameter, brown, tuberculate.

var. tenuispinosa

Branches 5–8 mm. thick, with usually very shallow teeth; spines fine, to 5 mm. long; prickles very fine, to 3 mm. long, often ± the same length as the spines; cyathia ± 2.5 × 5 mm. Fig. 97/2, 3, p. 520.

KENYA. Kitui District: Tulima, Jan. 1940, *Ritchie* in *Bally* E.100!; Teita District: plain W. of Maktao Hill, 17 Aug. 1969, *Bally* 13415!; Kwale District: Taru, 5 Sept. 1953, *Drummond & Hemsley* 4182!
DISTR. K 4, 7; not known elsewhere
HAB. Amongst grass usually in *Acacia-Commiphora* bushland, occasionally in open evergreen forest (*Bally* 8535); 150–1100 m.

SYN. *E. taitensis* Pax in E.J. 34: 83 (1904); N.E. Br. in F.T.A. 6(1): 571 (1911), *non* Boiss. (1860). Type as for *E. tenuispinosa* Gilli
E. ndurumensis Bally in Candollea 20: 39 (1965), *nom. inval.*

NOTE. There appears to be only the one herbarium specimen from K 4, but photographic records and living plants which I have seen originate from Kitui and Meru Districts.

var. robusta *Bally & Carter* in K.B. 42: 377 (1987). Type: Kenya, Teita District, Tsavo Station, *Sheldrick* in *Bally* 8572 (K, holo.!)

Branches to 1(–1.5) cm. thick, with teeth ± obvious or prominent on particularly robust specimens; spines sturdy to very robust, 5–12(–15) mm. long, often joined at the base for up to 2 mm.; prickles fine, 2–6 mm. long, always much shorter than the spines; cyathia ± 3 × 6 mm.

KENYA. Kitui District: 200 km. from Nairobi on Garissa road, Mar. 1941, *Powell* in *Bally* E.151!; Teita District: 48 km. N. of Voi on Nairobi road, 5 Feb. 1961, *Greenway* 9830! & Sobo Rocks, 67 km. NE. of Voi Gate, 3 Jan. 1967, *Greenway & Kanuri* 12956!
DISTR. K 1, 4, 7; not known elsewhere
HAB. Rocky sandy soils in dry deciduous bushland; 200–900 m.

NOTE. Distribution overlaps that of var. *tenuispinosa* in Kitui District and northern Teita District. Sturdiness of plants appears to be dependent upon environmental conditions, extreme forms of the variety (stout stems, branches with prominent teeth and robust spination) occurring in drier habitats. A shrubby form from NE. of Isiolo, near Kom (*Powys* 644, cultivated material only seen), with a stout main stem to 4 cm. thick at the base, may prove to be distinct.

145. E. odontophora *S. Carter* in Hook., Ic. Pl. 39, t. 3855 (1982). Type: Kenya, Northern Frontier Province, War Gedud, *Gilbert & Thulin* 1299 (K, holo.!, EA, iso.)

Succulent perennial with a fleshy root and spreading tangled branches to 40 cm. high; branches 4-angled, to ± 1 cm. thick, bright glossy green with a darker stripe along the angles; angles with prominent teeth ± 1.5 cm. apart. Spinescence dark reddish brown, yellowish when young; spine-shields elongated, contiguous; spines to 2 cm. long and often joined at the base for up to 5 mm.; prickles to 3 mm. long. Leaves deltoid ± 1.5 mm. long. Cymes solitary, 1-forked, with peduncles ± 1.5 mm. long; bracts oblong, ± 1.25 × 1 mm. Cyathia ± 2.5 × 3.5 mm., with broadly funnel-shaped involucres; glands transversely rectangular, ± 1 × 2 mm., contiguous, brick-red; lobes subcircular, 1 mm. in diameter. Male flowers: bracteoles ligulate, dentate; stamens 3 mm. long. Female flower: styles 1.75 mm. long, joined at the base, apices thickened. Capsule just exserted on a pedicel 2 mm. long, obtusely 3-lobed, 2 × 3.5 mm., speckled red along the sutures. Seeds ovoid, 1.5 × 1.25 mm., grey, areolate.

KENYA. Northern Frontier Province: N. of El Wak at War Gedud, 1 May 1978, *Gilbert & Thulin* 1299!
DISTR. K 1; known only from near El Wak
HAB. Sandstone ridges with rich *Acacia-Commiphora* woodland; 450–500 m.

NOTE. Other collections from the area have been made but these are known only as living plants (*Lavranos & Newton* 12238!)
A distinct species, of which no herbarium material is yet available for examination but which could possibly be related to *E. odontophora*, occurs at the southern foot of Mt. Kulal. It is a shrubby plant with erect 4-angled branching stems ± 1 m. high and to 1 cm. thick, and with stout spines conspicuously varying in length on each branch from ± 5–25 mm.

146. E. dauana *S. Carter* in K.B. 42: 376 (1987). Type: Kenya, Northern Frontier Province, 30 km. on the Ramu–Malka Mari road, *Gilbert & Thulin* 1537 (K, holo.!, EA, iso.!)

Succulent perennial with fibrous roots, spreading-erect or subscandent to ± 1 m. high, laxly branched; branches 4-angled, to 1 cm. thick, conspicuously blotched with dark and paler green; angles with prominent teeth 1–2 cm. apart. Spinescence dark brown, dark red when young; spine-shields longitudinal to 1.5 cm. long, 2.5 mm. wide above the spines, decurrent up to 10 mm., usually separated up to 6 mm. or occasionally confluent;

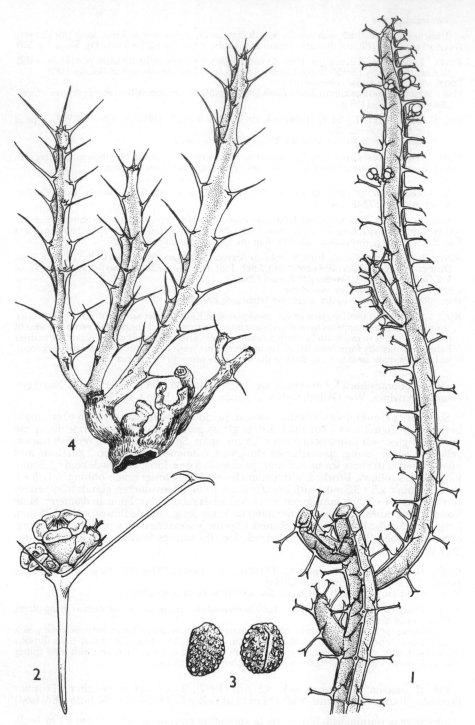

FIG. 98. *EUPHORBIA GLOCHIDIATA* — 1, fruiting branch, × ²⁄₃; 2, cyme and spine-shield, × 3; 3, seeds, × 6. *E. GRACILIRAMEA* — 4, portion of plant showing branching, × ²⁄₃. 1, 3, from *Bally & Carter* 16580; 2, from *Bally* 9440; 4, from *Raynal* 19266. Drawn by Christine Grey-Wilson.

spines to 15 mm. long; prickles fine, to 5 mm. long. Leaves deltoid, ± 1.5 × 1.5 mm. Cymes solitary, 1-forked, with peduncles ± 2.5 mm. long; bracts deltoid, ± 1.2 × 1.2 mm. Cyathia ± 2.5 × 5.5 mm., with funnel-shaped involucres; glands transversely oblong, ± 1.5 × 2.8 mm., contiguous, yellow; lobes transversely elliptic, ± 1.3 × 1.5 mm. Male flowers: bracteoles fan-shaped, deeply dentate; stamens 3.8 mm. long. Female flower: styles 2 mm. long, joined at the base, apices thickened, rugulose. Capsule and seeds not seen.

KENYA. Northern Frontier Province: 30 km. on the Ramu–Malka Mari road, 6 May 1978, *Gilbert & Thulin* 1537! & 35 km. NW. of Ramu on the Malka Mari track, Aug. 1975, *Lavranos & Newton* 12219!
DISTR. K 1; known only from the one localised area
HAB. Rocky limestone slopes with *Acacia-Commiphora* woodland; ± 400 m.

147. E. glochidiata *Pax* in Ann. Ist. Bot. Roma 6: 187 (1897); N.E. Br. in F.T.A. 6(1): 566 (1911); E.P.A.: 446 (1958). Type: Ethiopia, Harerghie Region, Web Karanle, *Ruspoli & Riva* 1122 [336] 1018 (B, holo.†, FT, iso.!, K, drawing of holo.!)

Succulent perennial with a fleshy root, ± sparsely branched, shrubby to ± 1.5 m. or occasionally subscandent to 2 m. high; stems and branches 4(–5)-angled, to 1.5 cm. thick, variously vareigated, mottled, or striped darker green along the angles; angles with shallow teeth to 3 cm. apart. Spinescence greyish brown; spine-shields longitudinal and contiguous or decurrent to just above the flowering eye below; spines single, to 2 cm. long terminating in forked tips to 5 mm. long; prickles to 5 mm. long. Leaves deltoid, ± 3 × 2 mm. Cymes solitary, 1-forked, subsessile; bracts deltoid, ± 1.5 × 1.5 mm. Cyathia ± 3 × 5 mm., with funnel-shaped involucres; glands transversely rectangular, ± 1 × 2.5 mm., contiguous, yellow to reddish; lobes transversely elliptic, ± 1 × 1.5 mm. Male flowers: bracteoles spathulate, laciniate, deeply dentate; stamens 4.25 mm. long. Female flower: styles 2 mm. long, joined at the base, apices thickened, rugose. Capsule subsessile, obtusely 3-lobed, ± 3.5 × 5 mm., speckled purplish. Seeds ovoid, 2 × 1.5 mm., grey, closely tuberculate. Fig. 98/1–3.

KENYA. Northern Frontier Province: near El Wak, 26 May 1952, *Gillett* 13335! & Lugh Olla, 68 km. W. of Ramu, 2 Mar. 1974, *Bally & Carter* 16580! & Dadaab–Wajir road, 38 km. N. of Sabule airstrip, 29 Nov. 1978, *Brenan, Gillett et al.* 14838!
DISTR. K 1, 4, 7; S. and SE. Ethiopia and southern Somalia
HAB. Sandy soils usually overlying limestone, with open *Acacia-Commiphora* bushland; 190–750 m.

VAR. Stem markings vary considerably, as does spination. The forked tips of spines on specimens from nearest the Somalia border show an increasing tendency to diverge to about 180°. Some specimens from south-west of Ramu produce a particularly sturdy spinescence, but with the forked tips much shorter at about 2 mm. long. Specimens from west of Garissa (Kitui and Tana River Districts on Nairobi road, and near Garba Tula) have the longest spines but with the forked tips divergent at about 90° or less.

148. E. fluminis *S. Carter* in Hook., Ic. Pl. 39, t. 3854 (1982). Type: Kenya, Tana River District, 73 km. N. of Malindi on Garsen road, *Gillett* 19534 (K, holo.!, EA, iso.)

Succulent perennial with a fleshy root, subscandent to 2 m. high; stems and branches 4-angled, to ± 1 cm. thick, variously mottled, variegated or striped; angles with shallow to subprominent teeth to ± 3 cm. apart. Spinescence brownish grey, the spine-shields elongated to 2.5 cm. long, decurrent to 2 cm. rarely reaching the flowering eye below; spines single to 1.5 cm. long, terminating in forked tips to 3 mm. long, weakly divergent; prickles to 2 mm. long. Leaves deltoid, ± 1.25 × 1.25 mm. Cymes solitary, 1-forked, subsessile; bracts deltoid, ± 1.5 × 1.5 mm. Cyathia ± 3 × 4.5 mm., with funnel-shaped involucres; glands transversely rectangular, ± 1 × 2.5 mm., contiguous, reddish; lobes transversely elliptic, ± 1 × 1.5 mm. Male flowers: bracteoles spathulate, laciniate, deeply dentate; stamens 4.25 mm. long. Female flower: ovary subsessile; styles 2 mm. long, joined at the base, apices thickened, rugose. Entire capsule not seen, probably ± 2.5 mm. in diameter. Seeds not seen.

KENYA. Tana River District: Pump Station, 20 Dec. 1964, *Gillett* 16478! & Galole–Garsen road, 8 km. towards Garsen from Wenjie turn-off, 11–12 July 1974, *Faden* 74/1055!
DISTR. K 7; not known elsewhere
HAB. Alluvial soils in dry deciduous woodland; 15–70 m.

VAR. Herbarium specimens I have seen of this species are easily distinguished from those of *E. glochidiata*, principally by the separated spine-shields and shorter spines with weakly divergent tips. Separation at specific level is therefore upheld until more investigation can show a possible

gradation between this and the weakest forms of *E. glochidiata.*
A strongly variegated population with very short spines occurs in the Galana River area and may prove to be distinct.

149. E. meridionalis *Bally & S. Carter* in Hook., Ic. Pl. 39, t. 3853 (1982). Type: Kenya, Machakos District, near Athi, *Verdcourt* 2358 (K, holo.!, EA, iso.)

Succulent perennial, with a thick fleshy root, branching from the base and weakly erect to 1 m., or grazed to ± 25 cm. high and then usually more densely branched; branches 4-angled, to 1.5 cm. thick, glaucous with darker stripes along the angles, becoming purplish; angles with prominent teeth to ± 2(–3.5) cm. apart. Spinescence grey, rather weak; spine-shields elongated up to 2 cm. long, decurrent to 1.5 cm.; spines single to 2 cm. long terminating in forked tips to 5 mm. long, weakly divergent; prickles to 2.5 mm. long. Leaves deltoid, ± 1.5 × 1.5 mm. Cymes solitary, 1-forked, with vivid carmine subsessile cyathia; bracts deltoid, ± 2 × 2 mm. Cyathia ± 3 × 5.5 mm., with funnel-shaped involucres; glands transversely elliptic, ± 1.25 × 1.75 mm., contiguous; lobes transversely elliptic, ± 1.5 × 2.5 mm. Male flowers: bracteoles spathulate, deeply dentate; stamens 3.5 mm. long. Female flower: ovary subsessile; styles 2.5 mm. long, joined at the base, apices rugose. Entire capsule not seen, probably ± 4 mm. in diameter. Seeds globose, 1.75 mm. in diameter, dark brownish red, indistinctly alveolate.

KENYA. Nairobi District: 1 km. N. of Marimbeti Railway Station, 19 Feb. 1972, *Mwangangi & Gillett* 1963!; Machakos District: Stony Athi, 24 Feb. 1973, *Hansen* 859!
TANZANIA. Masai District: Longido–Merkerstein, 13 Jan. 1936, *Greenway* 4352!
DISTR. K 4; T 2; probably also occurs in K 6, Masai District around Kajiado; not known elsewhere
HAB. In sandy stony soils amongst grass in dry open bushland; 1200–1750 m.

SYN. [*E. glochidiata* sensu Jex-Blake, W.F.K.: 38 (1948); U.K.W.F.: 223 (1974), *non* Pax]

150. E. erlangeri *Pax* in E.J. 33: 286 (1903); N.E. Br. in F.T.A. 6(1): 575 (1911); E.P.A.: 446 (1958). Type: Somalia, Gedo Region, Gara Libin, *Ellenbeck* 2099 (B, holo.†, K, drawing of holo.!)

Tuberous-rooted perennial succulent, scandent to 3 m., sparsely branched; branches cylindrical, 5–10 mm. thick, with 3–6 ribs (angles) without teeth. Spinescence purplish brown; spine-shields 1–3 cm. long, contiguous and forming longitudinal strips 1–3 mm. wide, separated by grooves ± 1 mm. wide; spines 1–4 mm. long, recurved, immediately below the flowering eye; prickles vestigial. Leaves lanceolate, ± 1.5 mm. long. Cymes solitary, 1-forked, sessile; bracts subcircular, ± 1.25 mm. in diameter. Cyathia ± 2 × 4 mm., with broadly funnel-shaped involucres; glands transversely rectangular, ± 1 × 2.5 mm., contiguous, brownish red; lobes transversely elliptic, ± 1 × 1.5 mm. Male flowers: bracteoles laciniate; stamens 3.5 mm. long. Female flower: styles 1 mm. long, joined at the base, apices thickened, rugulose, minutely bifid. Capsule sessile, obtusely 3-lobed, ± 2.5 × 3.5 mm. Seeds subglobose, 1.5 mm. in diameter, pale brown, tuberculate.

KENYA. Northern Frontier Province: Lag Ola, 45 km. W. of Ramu on Banessa road, 1 June 1952, *Gillett* 13412! & 48 km. from Ramu on El Wak road, 9–10 May 1978, *Gilbert & Thulin* 1599!
DISTR. K 1 (confined to the north-east corner of the Northern Frontier Province); southern Somalia
HAB. Red sandy soils usually overlying limestone, with open *Acacia-Commiphora* bushland; 450–650 m.

151. E. cryptospinosa *Bally* in Candollea 18: 351 (1962); E.P.A.: 444 (1958). Type: Kenya, Tana River District, Garissa, *Bally* E. 159 (K, holo.!, EA, iso.)

Perennial succulent with a small tuberous root, much branched and erect from 20 cm. high or subscandent to 3(–5) m. and sparsely branched; branches cylindrical, 3–10 mm. thick, with 5–10 ribs (angles). Spinescence greyish brown covered in minute bristly hairs on young growth; spine-shields 1–3 cm. long, contiguous and forming longitudinal strips ± 2 mm. wide, completely covering the stem, separated only by very narrow grooves; spines very fine, 1–3 mm. long, clasping the stem immediately below the flowering eye, present only on young growth; prickles absent. Leaves linear, 2–3 mm. long, minutely hispid. Cymes solitary, 1-forked, sessile, with all parts bright crimson; bracts subcircular, ± 1.25 mm. long. Cyathia ± 2 × 3.5 mm., with broadly funnel-shaped involucres; glands transversely rectangular, ± 0.75 × 1.75 mm., contiguous; lobes transversely elliptic, ± 0.75 × 1.25 mm. Male flowers: bracteoles laciniate; stamens 3 mm. long. Female flower: styles 1.25 mm. long, joined at the base, apices thickened, rugulose, minutely bifid. Capsule sessile, obtusely 3-lobed, ± 2.5 × 3.5 mm. Seeds subglobose, 1.5 mm. in diameter, grey, tuberculate.

KENYA. Northern Frontier Province: Garissa–Liboi road, 24 km. E. of Mado Gashi junction, 28 Jan. 1972, *Bally & Smith* 14981! & 24 km. from Baragoi–South Horr road on road to Ilaut, 4 June 1979, *Gilbert, Kibui & Mungai* 5517!; Teita District: Tsavo National Park East, Voi camp-site, 11 Jan. 1967, *Greenway & Kanuri* 13021!
DISTR. **K** 1, 4, 7; southern Ethiopia, southern Somalia
HAB. Sandy gravelly soils with open dry deciduous bushland; 380–1350 m.

VAR. Usually found scrambling through bushes and small trees, with branches considerably lengthened; when growing in the open without support plants form tightly branched shrublets, presenting a completely different appearance.

152. E. ballyana *Rauh* in Kakt. & And. Sukk. 17: 46 (1966). Type: Kenya, Masai District, 80 km. from Nairobi on Magadi road, *Rauh* Ke. 105 (HEID, holo.)

Succulent perennial branching from the base, with a thick fleshy root; branches erect to 30(–50) cm. high, 7–10 mm. thick, sparsely rebranching, cylindrical, greyish green with darker stripes between the spine-shields, teeth very obscure, ± 2 cm. apart. Spine-shields in 4 longitudinal series, elongated, ± 17 mm. long, decurrent to 15 mm.; spines single to 15 mm. long; prickles to 3 mm. long. Leaves deltoid, 1.5 mm. long. Cymes solitary, 1-forked, subsessile; bracts deltoid, ± 1.5 × 1.5 mm. Cyathia ± 3 × 5 mm., with funnel-shaped involucres; glands transversely rectangular, ± 1 × 2 mm., contiguous, dull green; lobes subcircular, ± 1.5 mm. long. Male flowers: bracteoles fan-shaped, ± 2 mm. long, deeply dentate; stamens 3.5 mm. long. Female flower: styles ± 1.5 mm. long, joined at the base, spreading, apices thickened, rugose. Capsule subsessile, obtusely 3-lobed, ± 3.5 × 4 mm., buff. Seeds ovoid, 2.2 × 1.8 mm., grey, surface alveolate.

KENYA. Masai District: 80 km. from Nairobi on Magadi road, Feb. 1960, *Rauh* Ke. 105! (cult. material preserved in K)
DISTR. **K** 6, known only from this one collection
HAB. In dry *Acacia-Commiphora* bushland; 900 m.

NOTE. Although the type locality is easily accessible from Nairobi, this plant has not been found again.

153. E. similiramea *S. Carter* in K.B. 42: 371 (1987). Type: Tanzania, Masai District, 38 km. S. of Longido on Namanga–Arusha road, *Polhill & Paulo* 1029 (K, holo.!, EA, iso.!)

Loosely tufted succulent perennial to 30 cm. high and ± 50 cm. in diameter, branching densely at the base from a thick fleshy root; branches to 30 cm. long, 1.5 cm. thick, cylindrical, sparsely rebranched, with usually prominent teeth, ± 2 cm. apart, in 4–5 longitudinal ± spiral series. Spinescence grey; spine-shields oblong-triangular, to 10 × 3 mm., decurrent to ± 8 mm. and ± halfway to the shield below; spines single, to ± 2 cm. long; prickles rudimentary to 1.5 mm. long. Leaves deltoid, ± 2.5 × 1.5 mm. Cymes solitary, 1-forked, with peduncles and cyme-branches 1–3 mm. long; bracts deltoid, ± 1.5 × 1.5 mm. Cyathia ± 2.5 × 5 mm., with funnel-shaped involucres; glands transversely rectangular, ± 1 × 2.5 mm., contiguous, greenish yellow; lobes transversely elliptic, ± 1 × 2 mm. Male flowers: bracteoles fan-shaped, denticulate; stamens 4.5 mm. long. Female flower: styles 1.5 mm. long, joined at the base, spreading, with thickened rugose apices. Capsule subsessile, obtusely 3-lobed, ± 3 × 4 mm., buff. Mature seeds not seen.

KENYA. Naivasha District: NE. foot of Suswa, 27 Jan. 1964, *Glover & Oledonet* 4065!; Machakos District: Athi R. at Athi, 12 Mar. 1960, *Polhill* 210!; Masai District: 72 km. from Namanga on Kajiado road, 20 Nov. 1960, *Verdcourt* 3014!
TANZANIA. Masai District: Longido, 13 Jan. 1936, *Greenway* 4350!; Arusha District: Ngare Nanyuki, 29 Dec. 1970, *Richards & Arasululu* 26588!
DISTR. **K** 3, 4, 6; **T** 2; not known elsewhere
HAB. On rocky usually volcanic soils in wooded grassland; 1200–1780 m.

SYN. [*E. graciliramea* sensu Jex-Blake, W.K.F.: 38 (1948); U.K.W.F.: 223 (1974), pro parte, quoad *Hanid & Kiniaruh* 897, *non* Pax]

NOTE. This species, from the eastern side of the Rift, has been previously identified with the more westerly *E. graciliramea* Pax. It can be distinguished from the latter by its usually thicker, longer, more sharply toothed branches (although this feature can vary considerably), with the teeth in usually 5 ± spiral series bearing spine-shields which are decurrent, often halfway to the shield below. When the teeth are in 4 series, these tend towards a spiral arrangement and are not all strictly opposite as they are with *E. graciliramea*.

154. E. graciliramea *Pax* in E.J. 34: 78 (1904); N.E. Br. in F.T.A. 6(1): 564; U.K.W.F.: 223 (1974), pro parte; Blundell, Wild Fl. Kenya: 52, t. 160 (1982), excl. distr. (see Note). Type: Tanzania, Musoma District, [possibly north of Nata] *Fischer* 522 (B, holo.†, K, fragment & drawing of holo.!)

Tufted succulent perennial to 15 cm. high and 30(–60) cm. in diameter, branching densely from the base at the apex of a thick fleshy root; branches to 25 cm. long and 5–10 mm. thick, cylindrical, seldom rebranched except at the base, with teeth shallow to ± prominent, ± 2 cm. apart in 4 longitudinal series, and in strictly opposite pairs. Spinescence sturdy, grey; spine-shields triangular, T-shaped above the spines, to 8 × 4 mm., shortly decurrent to 4(–6) mm.; spines single to ± 2 cm. long; prickles rudimentary, to ± 1 mm. long. Leaves deltoid, 1 × 1 mm. Cymes solitary, 1-forked, with peduncles and cyme-branches 1(–5) mm. long; bracts obtusely triangular, ± 1.5 × 2 mm. Cyathia ± 2.5 × 6 mm., with broadly funnel-shaped involucres; glands transversely rectangular, ± 1 × 3 mm., contiguous, yellow occasionally becoming reddish; lobes transversely elliptic, ± 1 × 1.5 mm. Male flowers: bracteoles fan-shaped, finely dentate; stamens 4 mm. long. Female flower: styles 2 mm. long, free to the base, spreading, with thickened rugose apices. Capsule subsessile, obtusely 3-lobed, ± 2.5 × 3 mm., buff. Mature seeds not seen. Fig. 98/4, p. 524.

KENYA. Masai District: Ololua, NW. of Narosura, 18 Jan. 1981, *Kuchar* 14319! & Nguruman Escarpment, Lenyora, 27 Sept. 1944, *Bally* 3884! & Ngorengore, 12 Dec. 1963, *Verdcourt* 3819A!
TANZANIA. Shinyanga District: Seseku, 13 June 1931, *B.D. Burtt* 3442!; Masai District: Meserani Dam, 21 June 1946, *Greenway* 7800!; Mbulu District: Mangola, 24 Sept. 1977, *Raynal* 19266!
DISTR. **K** 6; **T** 1, 2, 5; not kwown elsewhere
HAB. Stony soils in grassland and dry open deciduous bushland; 700–2025 m.

VAR. A species which appears to be confined to the western side of the Rift. The strictly opposite branch-teeth are usually fairly prominent, especially on dried specimens, but are often shallow on very turgid actively growing plants. Collections from the Narok area in Kenya indicate the occurrence of a possibly smaller form with a less sturdy spinescence, which may, however, be the result of overgrazing and more exposed habitat conditions.

NOTE. Blundell's distribution for *E. graciliramea* and Agnew's in U.K.W.F. covers also that of the two related species: *E. graciliramea* occurs in Narok and Masai Mara, *E. similiramea* in Rift Valley, Magadi, Nairobi and Kajiado, and *E. laikipiensis* in Nanyuki areas.

155. E. laikipiensis *S. Carter* in K.B. 42: 373 (1987). Type: Kenya, Laikipia District, 40 km. N. of Rumuruti on Maralal road, *Gilbert et al.* 5082 (K, holo.!, EA, iso.)

Tufted succulent perennial to 10 cm. high and 20 cm. in diameter, branching densely from the base, with a thick fleshy root; branches to 10 cm. long and 8–10 mm. thick, cylindrical, not rebranching, with shallow teeth 1–1.5 cm. apart in 4 longitudinal series. Spinescence whitish; spine-shields oblong-triangular, to 6 × 2 mm., decurrent to 4 mm.; spines single, to ± 12 mm. long; prickles absent or occasionally rudimentary. Leaves sharply deltoid, ± 1 × 1 mm. Cymes solitary, 1-forked, subsessile; bracts deltoid, ± 1.5 × 1.5 mm. Cyathia ± 2.5 × 4 mm., with funnel-shaped involucres; glands transversely rectangular, ± 1 × 2 mm., contiguous, yellow; lobes subcircular, ± 1 mm. in diameter. Male flowers: bracteoles fan-shaped, finely dentate; stamens 3.5 mm. long. Female flower: styles 2 mm. long, joined at the base, spreading, with thickened rugose apices. Capsule subsessile, obtusely 3-lobed, ± 3 × 4 mm., buff. Seeds ovoid, 2 × 1.6 mm., pale brown, surface wrinkled.

KENYA. Laikipia District: 13 km. NE. of Rumuruti, 22 Feb. 1974, *Bally & Carter* 16535!
DISTR. **K** 3; not known outside Laikipia District
HAB. In rocky soil in grassland, with deciduous shrubs; 1700–1850 m.

SYN. [*E. graciliramea* sensu Agnew, U.K.W.F.: 223 (1974), pro parte; Blundell, Wild Fl. Kenya: 52 (1982), pro parte, *non* Pax]

NOTE. This species can be distinguished from *E. graciliramea* and *E. similiramea* by its shorter branches which do not rebranch, by the shallower branch-teeth, by shorter spine-shields and almost complete absence of prickles.

156. E. schizacantha *Pax* in Ann. Ist. Bot. Roma 6: 187 (1894); N.E. Br. in F.T.A. 6(1): 566 (1911); E.P.A.: 457 (1958); S. Carter in K.B. 42: 375 (1987). Type: Ethiopia, Bale Region, Web Kuspoll, Elba, *Riwa* 764 [544] (B, holo.†, K, drawing of holo.!, ?FT, iso.)

Densely branched succulent perennial to 60 cm. high, with fibrous roots and a main stem to 50 cm. high and 6 cm. thick; branches to 40 cm. long, erect at first then drooping and trailing, usually unbranched, cylindrical, to 12 mm. thick, longitudinally striped with dark and light green; teeth very shallow, to 1.5 cm. apart in 3–5 ± spirally arranged series. Spinescence greyish brown; spine-shields longitudinal, to 11 × 3 mm., decurrent to 10 mm.; spines single, to 2.5 cm. long terminating in forked tips to 1 mm. long with one of these slightly shorter and weaker than the other; prickles to 4 mm. long. Leaves deltoid, ± 2 × 1.5 mm. Cymes solitary, 1-forked, subsessile; bracts oblong, ± 1.5 × 1 mm. Cyathia ± 2 × 5.5 mm., with broadly funnel-shaped involucres; glands transversely rectangular, ± 1.5 × 3 mm., contiguous, dark red; lobes subcircular, ± 1.5 mm. in diameter, deeply toothed. Male flowers: bracteoles fan-shaped, laciniate, deeply dentate; stamens 4 mm. long. Female flower: styles slender, 2 mm. long, free to the base, apices thickened, rugulose. Capsule subsessile, obtusely 3-lobed, ± 3.5 × 5 mm., purplish. Mature seeds not seen.

KENYA. Northern Frontier Province: 20 km. S. of Mandera, 12 Dec. 1971, *Bally & Smith* 14578! & Ramu–Banissa road, 10 km. from turning to Banissa, 4 May 1978, *Gilbert & Thulin* 1426!
DISTR. K 1 (known only from the extreme north-east); S. and SE. Ethiopia, southern Somalia
HAB. Stony soils with *Acacia-Commiphora* bushland; 380–520 m.

SYN. *E. chamaecormos* Chiov., Fl. Somal. 1: 303 (1929); E.P.A.: 444 (1958). Types: Somalia, Shabeellaha Dhexe Region, inland from Hariri, *Puccioni & Stefanini* 747 [825] & Bari Region, between Dhur and Hossa Uein, *Puccioni & Stefanini* 700 [774] (both FT syn.!, K, photos of syn.!)

157. E. kalisana *S. Carter* in Hook., Ic. Pl. 89, t. 3852 (1982). Type: Turkana District, Lokitaung–Lodwar road, *Carter & Stannard* 198 (K, holo.!)

Sprawling succulent perennial with main stem to 7.5 cm. high and 20 cm. in diameter, the apex usually ± at ground-level from a short thick fleshy root; branches produced successively from the apical growing point, spreading, to 1 m. long and 2 cm. thick, sparsely rebranched when mature, cylindrical, green with yellow-green longitudinal stripes and obscure teeth to 4 cm. apart in 5 longitudinal series. Spinescence very robust, pale grey; spine-shields oblong-triangular, to ± 2.5 × 1 cm., decurrent to 2 cm.; spines single, to 7 cm. long; prickles to 8 mm. long. Leaves ovate, ± 3.5 × 2.5 mm. Cymes solitary, 1-forked, with peduncles and cyme-branches ± 1.5 mm. long; bracts deltoid, ± 1.75 × 1.75 mm. Cyathia ± 3 × 7.5 mm., with broadly funnel-shaped involucres; glands transversely rectangular, ± 1.5 × 4 mm., contiguous, yellow; lobes transversely elliptic, ± 1.5 × 2.5 mm. Male flowers: bracteoles fan-shaped, dentate; stamens 5 mm. long. Female flower: styles 2 mm. long, free to the base, apices thickened. Capsule on a pedicel 2 mm. long, deeply 3-lobed, ± 3.25 × 5 mm., purplish. Seeds globose, 2 mm. in diameter, pale brown, lightly areolate.

KENYA. Northern Frontier Province: 12 km. NE. of Habaswein, 6 Mar. 1974, *Bally & Carter* 16592!; Turkana District: Lodwar, Feb. 1965, *Newbould* 6883!; Tana River District: 8.3 km. from Bura turn-off on Garissa–Garsen road, 7–9 July 1974, *Faden* 74/998!
DISTR. K 1, 2, 7; southern Ethiopia
HAB. Gravelly, often laval soils in sparse *Acacia-Commiphora* bushland, 100–1000 m.

SYN. [*E. triaculeata* sensu Jex-Blake, W.F.K.: 38 (1948), *non* Forssk.]

VAR. Plants from the very arid Turkana District tend to be larger and more robust than those from the lower altitude plains of eastern Kenya, but distribution is continuous. Some plants from near the Ethiopian border produce short branches (± 15 cm. long) arising from a distinct stem and may represent a separate taxon.

158. E. actinoclada *S. Carter* in Hook., Ic. Pl. 39, t. 3851 (1982). Type: Kenya, Northern Frontier Province, Dandu, *Gillett* 12631 (K, holo.!)

Tufted succulent perennial with a thick fleshy root and main stem to 5 cm. high and 2 cm. in diameter, the apex just above ground-level; branches produced successively from the apical growing-point, erect then spreading, to 15 cm. long, usually shorter, 1 cm. thick, cylindrical, seldom rebranched, dark green with lighter longitudinal stripes and shallow teeth to 1.5 cm. apart in 5 longitudinal series. Spinescence greyish brown; spine-shields longitudinal, to 9 × 1.5 mm., decurrent to 8 mm.; spines single to 2 cm. long; prickles to 2.5 mm. long. Leaves deltoid, 1 × 1 mm. Cymes solitary, 1-forked, with peduncle and cyme-branches 1 mm. long; bracts oblong, ± 1.75 × 1 mm. Cyathia ± 2 × 4 mm., with funnel-shaped involucres; glands transversely rectangular, ± 1 × 1.75 mm., contiguous, reddish; lobes subcircular, ± 1.25 mm. in diameter, deeply toothed. Male flowers: bracteoles few, fan-shaped, deeply dentate; stamens 3.5 mm. long. Female flower: ovary shortly pedicellate; styles 1.75 mm. long, joined at the base, with thickened rugose apices. Capsule and seeds not seen.

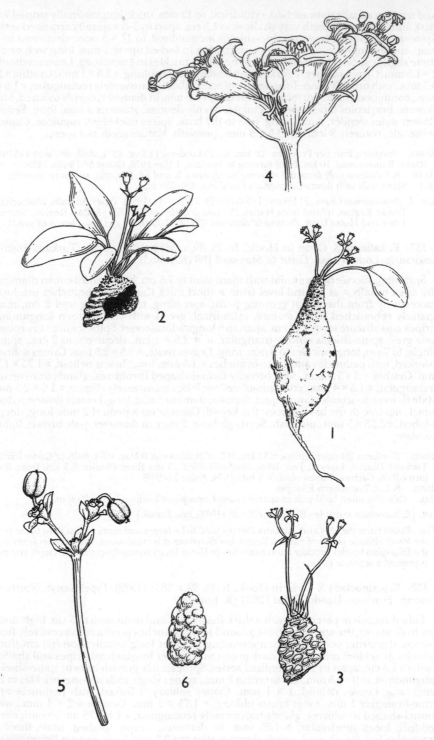

FIG. 99. *EUPHORBIA BRUNELLII* — 1, habit, × ⅔; 2, stem-apex with leaves in full growth × ⅔; 3, stem-apex, showing leaf-scars, × ⅔; 4, cyme, ×8; 5, fruiting cyme, × 2; 6, seed, ×8. 1, 4, from *Gillett* 12860; 2, 6, from *Tweedie* 3608; 3, 5, from *Tweedie* 355. Drawn by Christine Grey-Wilson.

KENYA. Northern Frontier Province: Malka Mari, Aug. 1975, *Lavranos & Newton* 12224! & 48 km. from Ramu on El Wak Road, 9–10 May 1978, *Gilbert & Thulin* 1612!
DISTR. **K** 1; southern Ethiopia
HAB. Gravelly slopes with sparse deciduous bushland; 650–900 m.

159. E. monadenioides *M. Gilbert* in K.B. 42: 233 (1987). Type: Ethiopia, Gojjam Region, *M. & S. Gilbert* 2295 (K, holo.!, ETH, iso.)

Hysteranthous geophyte, with branched rhizomes to 4 cm. thick and ± 15 cm. long; stems erect, below ground, ± 5 cm. high, with horizontally spreading branches and a dark roughened surface. Leaves glabrous, sessile, crowded and spreading in a rosette at soil level from the stem apex, oblanceolate to obovate to 8.5 × 3.2–4 cm., base cuneate, apex obtuse, mucronate, margin entire or minutely crisped, ± fleshy, green streaked with red. Cymes produced before the leaves, simple or sparsely branched, with peduncles 3.3–3.7 cm. long; bracts sessile, deltoid, 3.2–3.5 mm. long. Cyathia sessile, ± 3.5 × 5.2 mm., with cup-shaped involucres; glands transversely elliptic, ± 0.5 × 1.5 mm., contiguous, brownish; lobes 2.2 mm. wide, deeply dentate. Male flowers: bracteoles spathulate, deeply laciniate; stamens ± 4.2 mm. long. Female flower: styles ± 1.9 mm. long, joined at the base, apices thickened. Capsule exserted on a recurved pedicel to 1 cm. long, obtusely 3-lobed, ± 3.3 × 4.2 mm. Seeds not seen.

UGANDA. Teso District: Serere, Mar. 1933, *Chandler* 1133!
DISTR. **U** 3 (known in Uganda from the one specimen only); S. Ethiopia
HAB. Rocky ground with black clay in open deciduous woodland; 1100 m.

NOTE. This specimen is sterile, but the fleshy leaves and thick rhizomes make identification certain. Affinities of the species are obscure, but Gilbert relates it to the West African *E. baga* Chev., and in turn possibly to *E. decidua* Bally & Leach (see Note under 91. *E. venenifica*). With relationships so uncertain, the species is placed here at the end of subgenus *Euphorbia*.
The species is probably more common in Uganda than it appears to be from the one collection.

9. Subgen. **Lacanthis**

Perennial herbs or shrubs, or occasionally hysteranthous geophytes; roots, stems and branches thick and fleshy. Leaves entire, persistent, sessile or shortly petiolate; stipules usually developed into persistent feathery ridges on the stems and branches, or sometimes stiffened into spines, rarely absent (as in East Africa). Cymes axillary usually crowded at the stem and branch apices, branching dichotomously; bracts large, enveloping the cyathium, often brightly coloured. Involucres bisexual; glands 4–5 or apparently 6 (in East Africa). Stamens just exserted from the involucre; bracteoles included. Perianth of the ♀ flower reduced to a rim below the ovary. Capsule subsessile, rarely exserted. Seeds without a caruncle.

160. E. brunellii *Chiov.* in Webbia 8: 234 (1952); E.P.A.: 443 (1958). Type: Ethiopia, Sidamo Region, *Vatova* 1997 (FT, holo.!)

Glabrous hysteranthous geophyte, with a tuberous root ± 3 × 2 cm. tapering abruptly into a long tap-root and with several horizontal to ascending lateral roots; subterranean caudex tapering from the root, ± 3 × 1 cm., closely covered with tessellated leaf-scars. Leaves oblong, to ± 5.5 × 3.5 cm., spreading above ground; petiole to 2.5 cm. long. Cymes appearing before the leaves, on peduncles to 2 cm. long, 2–4(–7)-forked, with cyme-branches to 1 cm. long; bracts ovate, ± 3 × 2 mm., pinkish buff. Cyathia ± 2 × 3 mm., with cup-shaped involucres; lobes subquadrate, 0.5 mm. long, deeply 3–4-toothed; glands 5, with 4 transversely elliptic, ± 0.75 × 1 mm., plus 1 divided into subcircular halves 0.5 mm. in diameter, brownish yellow, all quite separate. Male flowers: bracteoles very few, linear; stamens 2.5 mm. long. Female flowers: styles 1 mm. long, joined at the base, spreading, apices distinctly bifid, thickened. Capsule 3-lobed, exserted on a pedicel to 5 mm. long, ± 2.5 × 3 mm. Seeds ovoid, 1.5 × 1.2 mm., brown, tuberculate. Fig. 99.

UGANDA. Mbale District: Tororo Hill, 25 Apr. 1952, *Bally* E.375!
KENYA. Northern Frontier Province: Dandu, 26 Mar. 1952, *Gillett* 12631!; W. Suk District: N. Cherangani Mts., Mar. 1969, *Tweedie* 3608!; N. Kavirondo District: S. Elgon, Dec. 1936, *Tweedie* 355!
DISTR. **U** 3; **K** 1, 2, 5; southern Ethiopia & southern Sudan; certainly more common than the few collections suggest

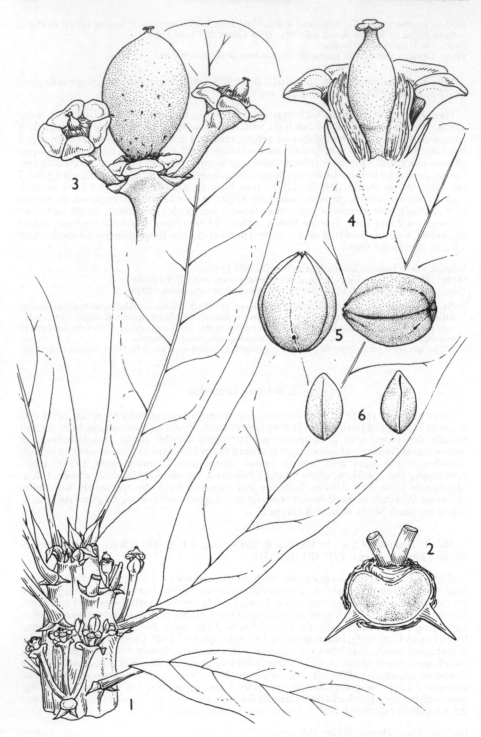

FIG. 100. *ELAEOPHORBIA DRUPIFERA* — 1, flowering branch, × ⅔, 2, spines etc. around leaf-scar and flowering eye, × 2; 3, fruiting cyme, × 2; 4, section of cyathium, showing ovary, × 2; 5, endocarps, × 2; 6, seeds, × 2. 1, from *Johnson* 1053 (Ghana); 2–6, from *Burton* (Nigeria). Drawn by Christine Grey-Wilson.

HAB.   In grassland, usually noticeable only after burning; 1080–2500 m.

SYN.   *E. rubella* Pax var. *brunellii* (Chiov.) Bally in Candollea 22: 262 (1967)
    [*E. rubella* sensu Agnew, U.K.W.F.: 221 (1974), *non* Pax]

NOTE.   Relationships of this species are obscure. Gilbert believes them to lie in a group of predominantly Madagascan species with usually ornately developed stipules and large persistent cyathial bracts (in K.B. 42: 239 (1987)).

69. ELAEOPHORBIA

Stapf in Hook., Ic. Pl. 23, t. 2823 (1906) & in Johnston, Liberia 2: 646 (1906); N.E. Br. in F.T.A. 6(1): 604 (1912); Croizat in B.J.B.B. 15: 109 (1938); Keay in F.W.T.A., ed. 2, 1: 423 (1958)

Trees with fleshy spiny branches and copious caustic milky latex, monoecious. Leaves fleshy, the petiole subtended by a pair of short spines and vestigial stipules. Inflorescence with cyathia in simple or dichotomous axillary cymes; bracts paired, persistent. Involucres with 5 spreading glands around the rim and 5 alternating lobes with denticulate margins. Male flowers in 5 groups, bracteolate. Female flower subsessile or shortly pedicellate in fruit; perianth entirely absent with the ovary merging into the pedicel; ovary 3-locular with thickened fleshy walls and 1 pendulous ovule in each locule; styles very short with thickened bifid stigmas. Fruit drupaceous, with a thick fleshy exocarp and woody endocarp; endocarp with a pore at the base, and each locule marked with a shallow longitudinal groove and a pore between the grooves near the apex. Seeds without a caruncle.

A genus of 4 species: *E. acuta* N.E. Br., reputedly from Swaziland but described from incomplete material of unknown origin; *E. hiernii* Croizat from Angola; *E. grandifolia* (Haw.) Croizat and *E. drupifera* (Thonn.) Stapf both originating from W. Africa, of which *E. drupifera* has the more easterly distribution.

The genus is distinguished from all species of *Euphorbia* by the complete absence of a perianth in the female flower, with the ovary being confluent with the pedicel, and by the large indehiscent drupaceous fruit.

E. drupifera (*Thonn.*) *Stapf* in Hook., Ic. Pl. 23, t. 2823 (1906); N.E. Br. in F.T.A. 6(1): 604 (1912), excl. cult. spec. ex Berlin; Croizat in B.J.B.B. 15: 113 (1938); I.T.U., ed. 2: 125 (1952); Keay in F.W.T.A., ed. 2, 1: 423 (1958). Type: Ghana, *Thonning* 266 (holotype not traced)

Glabrous tree to 22 m. high, with a stout trunk to 60 cm. in diameter branching at ± 1.5–2 m. high and with a rough grey bark; branches spreading, loosely rebranching to form a large rounded crown; terminal branchlets obtusely 5-angled becoming cylindrical, 2–3 cm. thick; angles with spine-pairs ± 1 cm. apart. Spine-shields very obtusely triangular, to 8 × 9 mm. on mature branches but not quite joined below the large semi-circular leaf-scar (± 6 × 7 mm.) on young growth, eventually ± surrounding the flowering eye immediately above; spines stout, conical, to 3 mm. long; prickles vestigial. Leaves persistent at the branch apices, obovate, to 28 × 10 cm., tapering to a petiole to 2.5 cm. long, apex rounded or sometimes slightly emarginate, margin entire, midrib prominent beneath. Cymes (1–)3 in a horizontal line, 1–2-forked, with peduncles to 4.5 cm. long and cyme-branches to 2.5 cm. long; bracts broadly deltoid, ± 3 × 7 mm., apex minutely denticulate. Cyathia sessile, ± 4 × 12 mm., with widely funnel-shaped involucres; glands transversely oblong, ± 2.7 × 6 mm., contiguous, brownish yellow; lobes transversely elliptic, ± 2 × 4 mm., very minutely denticulate. Male flowers: bracteoles fan-shaped, apices deeply dentate; stamens many, well exserted, 4 mm. long. Female flower: styles stout, 1.5 mm. long, joined at the base, with stigmas spreading, thickened, rugose, distinctly bifid. Fruit subsessile, obovoid, to 5 × 3.5 cm., green becoming yellow; endocarp subglobose, ± 13 × 11 mm., with 1–2 locules often arbortive. Seeds ovoid, 7 × 4.5 mm., with a dorsal and ventral ridge and acute apex, greyish brown, smooth. Fig. 100.

UGANDA.   Toro District: Bwamba, Butogo, Oct. 1940, *Eggeling* 4067!; Kigezi District: Malabigambo [Maramagambo] Forest, Sept. 1936, *Eggeling* 3313!
DISTR.   U 2; W. Africa from Ghana eastwards to Cameroon
HAB.   Forest edges; ± 1000 m.

SYN.   *Euphorbia drupifera* Thonn. in Schum. & Thonn., Beskr. Guin. Pl.: 250 (1827)
    E. renouardii Pax in Bull. Mus. Hist. Nat. Paris 8: 61 (1902). Type: Dahomey, *Poisson* (P, holo.)

NOTE. *Euphorbia juvoklantii* Pax (in E.J. 43: 86 (1909)) from Cameroon, included in synonymy by
N.E. Brown, was described from a sterile specimen, *Winkler* 490, which Brown apparently did not
see. With its very long narrow spine-shield, to 3.5 cm. long, it cannot be identified as any known
species of *Elaeophorbia*.

70. SYNADENIUM

Boiss. in DC., Prodr. 15(2): 187 (1860); N.E. Br. in F.T.A. 6(1): 462 (1911)

Shrubs or trees with subfleshy cylindrical branches and copious caustic milky latex,
monoecious. Leaves fleshy, with stipules modified as small but ± conspicuous dark brown
glands. Inflorescence with sessile cyathia in dichotomously branching axillary cymes,
usually crowded into pseudo-umbels (pseudumbels) at the branch apices; bracts paired,
persistent. Involucres with an entire (rarely lobed) spreading furrowed glandular rim and
5 fringed lobes. Male flowers in 5 groups, with stamens shortly exserted from the
involucre, bracteoles included. Female flower shortly pedicellate, the pedicel elongating
slightly in fruit; perianth reduced to a 3-lobed rim below the ovary; ovary 3-locular with 1
pendulous ovule in each locule; styles 3, joined at the base, with bifid stigmas. Fruit a
3-lobed capsule, dehiscent (a regma). Seeds with a sessile, often rudimentary caruncle.

A genus confined to east and southern tropical Africa in which 19 closely-related spieces (12 from
East Africa with 3 varieties) have been described, mostly from single specimens. 9 species with 1
variety are recognized here for our area. Differences between these, at least from herbarium
specimens, appear to be very slight and have been based upon leaf and cyme-branching characters.
Leaf characters especially are undoubtedly important, but emphasis should also be placed upon
habit, colour of the involucral glands and features of the ♀ flower. Relatively little material, especially
at the fruiting stage has been collected of some species which nevertheless appear to be fairly
common. More information about distribution, habit and variation, is required before this account
of the genus in East Africa can be regarded as entirely satisfactory.

1. Small shrub to 1.5 m. high; inflorescence-branches, bracts
   and involucres glabrous or almost so; caruncle of the
   seed obvious . . . . 9. *S. glabratum*
   Shrubs or trees more than 2 m. high; inflorescence-
   branches (at least the uppermost), bracts and involucres
   obviously pubescent; caruncle minute 2
2. Perianth-lobes of ♀ flower elongated into obvious
   filaments 2–4 mm. long at fruiting stage 3
   Perianth of ♀ flower bluntly 3-lobed or reduced to a rim 4
3. Shrub to 4.5 m. high; leaves oblanceolate, to 16 × 4 cm.,
   green-variegated with darker veining; perianth-lobes
   to 2 mm. long . . . . 1. *S. cymosum*
   Tree to 8 m. high; leaves obovate, to 20 × 7 cm., uniformly
   green; perianth-lobes 4 mm. long or more . . . 8. *S. calycinum*
4. Leaves pubescent on both surfaces, densely so when young 5. *S. molle*
   Leaves glabrous or pubescent only on the margin and
   midrib beneath, and sometimes sparsely so on the
   lower surface 5
5. Midrib rounded beneath 6
   Midrib acutely prominent or keeled beneath 7
6. Glandular rim of involucre red; pseudumbel of 2–6-forked
   cymes . . . . 2. *S. grantii*
   Glandular rim of involucre yellow; pseudumbel of solitary
   cyathia . . . . 3. *S. volkensii*
7. Glandular rim red . . . . 4. *S. compactum*
   Glandular rim yellow 8
8. Shrubs 2–3(–5) m. high; leaf-apex usually rounded; midrib
   acutely prominent to slightly keeled beneath . . . 6. *S. pereskiifolium*
   Trees, sometimes shrubby, 3–9 m. high; leaf-apex usually
   obtuse; midrib distinctly keeled beneath . . . . 7. *S. glaucescens*

1. S. cymosum *N.E. Br.* in F.T.A. 6(1). 469 (1911); I.T.U., ed. 2: 142 (1952). Type:
Uganda, Bunyoro District, Butiaba, E. of L. Albert, *Bagshawe* 850 (BM, holo.!, K, fragment
of holo.!)

Much branched shrub to ± 4.5 m. high. Leaves oblanceolate, to ± 16 × 5 cm., base tapering to a petiole to 1 cm. long, apex rounded, lamina sparsely pilose on and near the entire margins when young, becoming ± glabrous at maturity, ± variegated with darker green along the veins, with the midrib acutely prominent and sparsely pilose beneath. Inflorescence lax in pseudumbels of 3–4 cymes on peduncles 1–4.5 cm. long; cymes 3–4-forked, with branches 0.5–1.5 cm. long, shortly pubescent; bracts subquadrate, ± 3.5 × 3.5 mm., with 2–3 small teeth at the apex, pubescent, reddish green. Cyathia ± 2.5 × 5 mm., with broadly funnel-shaped involucres, ± densely pubescent; glandular rim 1.25 mm. wide, dull red; lobes subquadrate, ± 1.5 × 1.5 mm., glabrous above, pubescent beneath, reddish. Male flowers: bracteoles densely plumose; stamens 3 mm. long. Female flower: pedicel and ovary pubescent; perianth 3-lobed, the lobes usually extended into linear filaments to 2 mm. long, often with divided tips, glabrous or sparsely puberulous below; styles ± 1.5 mm. long, joined at the base, with distinctly bifid thickened apices. Capsular pedicel ± 3mm. long, recurved; capsule and seeds not seen.

UGANDA. Bunyoro District: Bugungu, Waiga R. by Lake Albert, Sept. 1937, *Eggeling* 3397! & Biseruka, 21 Sept. 1960, *Lind* 2770!
TANZANIA. Kigoma District: Kalinzi, 4 July 1963, *Carmichael* 919!
DISTR. U 2; T 4; not known elsewhere
HAB. Dry plains and rocky slopes with deciduous woodland; 670–1675 m.

NOTE. Apart from the lanceolate variegated leaves, this species is distinguished from *S. grantii* which occurs in the same area, by the elongated perianth lobes of the ♀ flower.

2. **S. grantii** *Hook. f.* in Bot. Mag. 93, t. 5633 (1867); N.E. Br. in F.T.A. 6(1): 468 (1911); T.T.C.L.: 227 (1949); I.T.U., ed. 2: 142 (1952); F.P.S. 2: 98 (1952); S. Carter in K.B. 42: 669 (1987). Type: Uganda, W. Nile District, Madi, *Grant* 754 (K, holo.!)

Bush or shrubby tree to 10 m. high. Leaves sessile, obovate, to ± 15 × 6 cm., base tapering, apex obtuse, margin entire or ± crimped towards the apex and occasionally forming a few minute teeth, midrib prominent below, rounded and often ± acute towards the apex, lamina very sparsely pilose with long hairs (to 1 mm. long) towards the base, more densely so and extending upwards on young leaves, uniform green or sometimes tinged red beneath. Inflorescence lax in pseudumbels of 2–5 cymes on pubescent peduncles to 5 cm. long; cymes 2–6-forked (occasionally trichotomously), with pubescent branches 1–3 cm. long; bracts subquadrate, ± 4 × 4 mm., entire or occcasionally with a few teeth, pubescent, reddish green. Cyathia ± 3 × 6.5 mm., with broadly funnel-shaped involucres, pubescent towards the base, dull crimson; glandular rim ± 1 mm. wide, deeply furrowed, crimson; lobes subquadrate, ± 1.8 × 2 mm., margin and base minutely puberulous, red. Male flowers: bracteoles linear, laciniate, densely plumose, with red tips; stamens 3.8 mm. long. Female flower: ovary densely pubescent; styles ± 2 mm. long, joined to ± halfway with distinctly bifid ± thickened apices. Capsule exserted on a pubescent pedicel 3–5(–9) mm. long, ± 7 × 8 mm., pubescent, red. Seeds ovoid, slightly compressed laterally, 2.5 × 2 mm., pale brownish grey, shallowly and minutely tuberculate; caruncle rudimentary. Fig. 101/1, 2, p. 536.

UGANDA. Ankole District: Ruizi R., 29 Jan. 1951, *Jarrett* 462!; Mbale District: Bugishu, Lukongi, Aug. 1936, *Tothill* 2605!; Masaka District: Sembabule, 23 July 1946, *A.S. Thomas* 4507!
KENYA. W. Suk District: Mwina, 8 Jan. 1979, *Meyerhoff* 120 M!; S. Kavirondo District, 5 May 1978, *Plaizier* 1302!; Masai District: Endama, 23 June 1961, *Glover, Gwynne, Samuel & Tucker* 1969!
TANZANIA. Ngara District: Bugufi, Murewanza, 20 Jan. 1961, *Tanner* 5795!; Mwanza District: Nyegezi, 8 Mar. 1961, *Leippert* 5619!; Musoma District: Kogatindi, Mara R., 19 Feb. 1968, *Greenway, Kanuri & Turner* 13308!
DISTR. U 1–4; K 2, 3, 5, 6; T 1, 2, 5, 7; Rwanda, Burundi and eastern Zaire
HAB. Rocky slopes with dry open woodland; 950–2100 m.

SYN. *S. umbellatum* Pax in E.J. 19: 125 (1894); N.E. Br. in F.T.A. 6(1): 464 (1911); T.T.C.L.: 227 (1949). Type: Tanzania, Maswa District, Simiu R., *Fischer* 517 (B, holo.†, K, drawing of holo.!)
S. umbellatum Pax var. *puberulum* N.E. Br. in F.T.A. 6(1): 465 (1911); F.P.N.A. 1: 483 (1948); T.T.C.L.: 227 (1949). Type: Tanzania, Mwanza District, Ukerewe, Multanga, *Conrads* 166 (B, holo.†, K, drawing of holo.!, EA, iso.!)

NOTE. The footnote by N.E. Brown in F.T.A., concerning the occurrence of *S. grantii* in the Zambezi valley, refers to *S. kirkii* N.E. Br., with leaves sowewhat pubescent especially on the margins and midrib of the lower surface, with the glandular rim of the involucre yellow, and seeds with an obvious caruncle. This species has so far not been collected in East Africa.

FIG. 101. *SYNADENIUM GRANTII* — 1, flowering branch, × ⅔; 2, section through leaf, × ⅔. *S. COMPACTUM* var. *COMPACTUM* — 3, flowering branch, × ⅔; 4, section through leaf, × ⅔; 5, cyathium, × 4. Var. *RUBRUM* — 6, fruiting inflorescence, × ⅔; 7, seeds, × 4. *S. GLAUCESCENS* — 8, flowering branch, × ⅔; 9, section through leaf, × ⅔. 1, 2, from *A.S. Thomas* 4507; 3, 4, from *Gillett & Mathew* 19063; 5, from *Bally* E.66; 6, 7, from *E. Polhill* 152; 8, 9, from *Faulkner* 1446. Drawn by Christine Grey-Wilson.

3. S. volkensii *Pax* in P.O.A. C: 243 (1895); N.E. Br. in F.T.A. 6(1): 464 (1911); T.T.C.L.: 227 (1949). Type: Tanzania, Moshi District, Kilimanjaro, by Mission Station, *Volkens* 1059a (B, holo.†, K, fragment & drawing of holo.!, BM, iso.!)

Trees to 18 m. high. Leaves subsessile, obovate, to ± 16 × 8 cm., apex obtuse, margin minutely serrate, midrib prominent and rounded beneath or ± acute towards the apex, lamina glabrous except occasionally for a few marginal hairs at the base. Inflorescence in pseudumbels of 2–4 cymes on pubescent peduncles to 5 cm. long; cymes 1-forked or cyathia solitary, with pubescent branches to 1.5 cm. long; bracts subquadrate, ± 4 × 4 mm., indistinctly toothed, pubescent. Cyathia ± 3.5 × 8 mm., with broadly funnel-shaped involucres, pubescent at the base; glandular rim ± 1.8 mm. wide, minutely grooved and wrinkled, yellow; lobes transversely oblong ± 2 × 3 mm., ± pubescent on the margins and at the base. Male flowers: bracteoles fan-shaped, laciniate, plumose; stamens 3.5 mm. long. Female flower: ovary pubescent; styles 2 mm. long, joined at the base, with deeply bifid apices. Capsule exserted on a pedicel ± 4 mm. long, ± 7 × 8 mm., densely pubescent. Seeds ovoid, 3 × 2.5 mm., pale brown, minutely and shallowly tuberculate; caruncle rudimentary.

KENYA. Masai District: between Loitokitok and Useri, *Bally* E. 47!
TANZANIA. Arusha District: Arusha National Park, Engare Nanyuki R., 11 Apr. 1968, *Greenway & Kanuri* 13453!; Moshi District: Kilimanjaro by Mission Station, 21 Jan. 1894, *Volkens* 1059a!
DISTR. K 6; T 2; known only from the lower slopes of Kilimanjaro
HAB. Deciduous thicket; 1000–1700 m.

4. S. compactum *N.E. Br.* in F.T.A. 6(1): 465 (1911); K.T.S.: 221 (1961). Type: Kenya, Machakos, *Kassner* 956 (K, holo.!, BM, iso.!)

Much-branched bush or shrubby tree to 7 m. high. Leaves sessile, obovate, to ± 18 × 9 cm., apex abruptly acute and recurved, margin minutely toothed, midrib acutely prominent to ± keeled beneath, glabrous, glossy green, often flecked or tinged purplish, sometimes entirely dark red. Inflorescences congested at the branch apices, in pseudumbels of up to 6 cymes on pubescent peduncles 2–4 cm. long; cymes 2–3-forked, with usually densely pubescent branches 5–20 mm. long; bracts subquadrate, ± 3 × 3.5mm., pubescent, often densely so, obscurely dentate. Cyathia ± 2.5 × 7 mm., with broadly funnel-shaped involucres, densely pubescent below, often reddish; glandular rim ± 1 mm. wide, entire or often ± lobed by one or more deep notches, shallowly and minutely grooved, red; lobes subquadrate ± 1.5 × 2 mm., pubescent, reddish. Male flowers: bracteoles fan-shaped, laciniate, plumose, reddish; stamens 3.8 mm. long, with pedicels rarely minutely pubescent. Female flower: styles ± 1.8 mm. long, joined to ± halfway, puberulous, with thickened distinctly bifid apices. Capsule exserted on a pubescent pedicel to 4 mm. long, broader above the middle, ± 7.5 × 7.5 mm., pubescent, becoming dark red. Seeds ovoid and obtusely 4-angled, 2.8 × 2.5 mm., pale grey, minutely, densely and shallowly tuberculate; caruncle rudimentary.

var. **compactum**
Leaf-surfaces uniformly green or occasionally flecked with purplish red; bracts, involucral lobes and anthers green, yellow or pink-tinged. Fig. 101/3–5.

KENYA. S. Nyeri District: between Karatina and Sagana, 24 Apr. 1943, *Bally* 2541!; Embu District: Thuchi R. crossing on Embu–Meru road, 4 Apr. 1970, *Gillett & Mathew* 19063!; Kitui District: 16 km. N. of Migwani on Tharaka road, 3 May 1960, *Napper* 1596!
DISTR. K 3, 4, 6, 7; not known elsewhere
HAB. Sandy, gritty soils often on rocky slopes with ± open deciduous bushland; 760–1825 m.
SYN. [*S. grantii* sensu Dale & Greenway, K.T.S.: 222 (1961), quoad *Dale* 2887, *non* Hook. f.]

var. **rubrum** *S. Carter* in K.B. 42: 669 (1987). Type: Kenya, Naivasha District, approaches to lake shore, *E. Polhill* 152 (K, holo.!)

At least the lower leaf-surface entirely purplish red; all parts of the inflorescence dark red. Fig. 101/6, 7.

KENYA. Nairobi District: Choromo Estate, 24 Feb. 1970, *Mathenge* 572!; Embu, 11 Apr. 1932, *Sunman* 2231!
DISTR. K 3, 4; probably occurring elsewhere
HAB. Possibly wild around Embu, otherwise naturalised in some places near habitation; 1700–1830 m.
SYN. [*S. grantii* sensu Jex-Blake, Gard. E. Afr., ed. 4: 188, 351 (1957), *non* Hook. f.]
NOTE. Used for its foliage as an ornamental, especially in Kenya.

5. S. molle *Pax* in E.J. 43: 88 (1909); N.E. Br. in F.T.A. 6(1): 466 (1911); K.T.S.: 222 (1961).
Type: Kenya, Machakos District, Kibwezi, *Scheffler* 137 (B, holo.†, BM, K, iso.!)

Many-branched bush or shrubby tree to 9 m. high. Leaves sessile (rarely petiolate, see note on variation below) obovate, to ± 17 × 7 cm., apex rounded to obtuse with a recurved tip, margin minutely toothed, midrib prominent and ± keeled beneath, both surfaces of the lamina ± densely pubescent especially when young. Inflorescence in pseudumbels of 3–5 cymes on pubescent peduncles to 5 cm. long; cymes 1–2-forked, with densely pubescent branches to 2.5 cm. long; bracts rounded to subquadrate, ± 3.5 × 3.5 mm., entire or with a few teeth, pubescent. Cyathia ± 2.5 × 8 mm., with broadly funnel-shaped involucres, pubescent below; glandular rim ± 1.3 mm. wide, distinctly grooved with a very narrow smooth margin, yellow to greenish red; lobes subquadrate, ± 2 × 2.5 mm., sparsely pubescent. Male flowers: bracteoles fan-shaped, laciniate, plumose; stamens 3.7 mm. long, with pedicels rarely pubescent. Female flower: styles 1.5 mm. long, joined at the base, pubescent, with distinctly bifid thickened apices. Capsule subsessile, widest above the middle, to 10 × 10 mm., densely pubescent. Seeds subglobose, obscurely 4-angled, 2.5 × 2.5 mm., brown, minutely tuberculate; caruncle rudimentary.

KENYA. Kitui District: Mutomo Hill, 2 May 1970, *Gillett* 19141!; Teita District: Mwatate, 29 Apr. 1963, *Bally* 12725!
TANZANIA. Pare District: above Mwembe, 9 Apr. 1972, *Harris* 6342!; Dodoma, 11 Apr. 1945, *Hartnell* H26/45!; Iringa District: Image, 27 Feb. 1962, *Polhill & Paulo* 1618!
DISTR. K 4, 7; T 2, 3, 5, 7, 8; not known elsewhere
HAB. Sandy soils and slopes with open deciduous woodland; 600–1550 m.

VAR. The few good specimens from Tanzania all possess a less compact inflorescence than those from Kenya, including the type. Longer peduncles and larger cyathia which are not so densely pubescent and which appear to lack any reddish colouring, may result from environmental differences, or may be indication that separation is needed. Further well-preserved material is needed before this can be determined. Specimens from Lindi District (*Migeod* 533 & 665 and *Schlieben* 6373, all at BM) are particularly lax with obviously petiolate leaves.

6. S. pereskiifolium (*Baill.*) *Guill.* in Bull. Mus. Hist. Nat. Paris, sér. 2, 7: 135 (1935).
Type: Zanzibar, *Boivin* cult. *Richard* in Réunion [Bourbon] (P, holo.!, K, photo. of holo.!)

Sparsely branching shrub to 5 m. high. Leaves obovate, to ± 19 × 10 cm., base tapering to a winged petiole ± 1 cm. long, apex obtuse or usually rounded, margin entire or ± crimped towards the apex and sometimes forming minute teeth, midrib acutely prominent to keeled beneath, lamina glabrous or rarely the margin and midrib with a few long hairs towards the base. Inflorescence lax in pseudumbels of 2–5 cymes, sometimes also with 1 or 2 cymes produced below, on sparsely pubescent peduncles to 4 cm. long; cymes 2–3-forked, with pubescent branches to 2 cm. long; bracts subquadrate, ± 3.5 × 3 mm., entire or with a few obscure teeth, pilose. Cyathia ± 2.5 × 6 mm., with funnel-shaped involucres, densely pubescent below; glandular rim ± 1.2 mm. wide, deeply furrowed and wrinkled, occasionally with 1 or 2 deep notches, greenish yellow; lobes subquadrate, ± 1.5 × 1.8 mm., ciliate. Male flowers: bracteoles fan-shaped, laciniate, plumose; stamens 3.3 mm. long. Female flower: perianth obtusely 3-lobed, rarely with the lobes elongated into narrow filaments to ± 0.3 mm. long; styles 2 mm. long, pubescent below, joined to halfway, with deeply bifid, scarcely thickened apices. Capsule exserted on a recurved pubescent pedicel to 6 mm. long, ± 7 × 7 mm., pubescent. Seeds ovoid, very obtusely 4-angled, 2.5 × 2.2 mm., buff, shallowly tuberculate; caruncle minute.

KENYA. Kwale District: Tanga–Mombasa road, ± 1.5 km. from border, 14 Aug. 1953, *Drummond & Hemsley* 3759! & Shimba Hills National Park above Sheldrick Falls, 23 Nov. 1977, *Gillett & Stearn* 21608!; Kilifi District: Chasimba, 6 Nov. 1974, *B. Adams* 94!
TANZANIA. Uzaramo District: Masasini, 25 Aug. 1968, *Harris* 2163! & 99 km. from Dar es Salaam on Morogoro road, 21 June 1955, *Welch* 294!; Kilwa District: Selous Game Reserve, 7 km. NNW. of Kingupira, 25 July 1975, *Vollesen* 2580!
DISTR. K 7; T 3, 6, 8; not known elsewhere
HAB. Sandy soil or on rocks in coastal or riverine woodland; 0–250 m.

SYN. *Euphorbia pereskiifolia* Baill., Adansonia 1: 105 (1860)
     Synadenium carinatum Boiss. in DC., Prodr. 15(2): 187 (1862); N.E. Br. in F.T.A. 6(1): 465 (1911); T.T.C.L.: 227 (1949); U.O.P.Z.: 457 (1949) Type as for *S. pereskiifolium*
     S. piscatorium Pax in E.J. 19: 125 (1894). Type: Zanzibar, *Stuhlmann* 460 (B, holo.!, K, drawing of holo.!)
     [*S. glaucescens* sensu Dale & Greenway, K.T.S.: 221 (1961), excl. *Napier* 1322, *non* Pax]

VAR. Specimens collected from around Jilori, W. of Malindi (K 7, Kilifi District) may represent a distinct taxon. Leaves appear to be larger than usual and are described as growing in tufts at the branch apices. The inflorescences also are a little more densely pubescent, with slightly larger cyathia which appear sometimes to be tinged with red.

7. **S. glaucescens** *Pax* in E.J. 33: 289 (1903); N.E. Br. in F.T.A. 6(1): 467 (1911); T.T.C.L.: 227 (1949); S. Carter in K.B. 42: 670 (1987). Types: Tanzania, Lushoto District, Mazinde, *Busse* 1147 (B, syn.†, K, fragment & drawing of syn.!) & Morogoro District, Mkata, *Busse* 165 (B, syn.†)

Bush or shrubby tree to 9 m. high. Leaves oblanceolate to obovate, to ± 18 × 9 cm., base tapering to a winged petiole ± 1 cm. long, apex rounded or usually obtuse, margin entire or minutely toothed towards the apex, midrib distinctly keeled beneath, lamina glabrous or sometimes with a few long hairs at the base and on the midrib beneath. Inflorescence in pseudumbels of 3–5 cymes on sparsely pubescent peduncles to 3 cm. long; cymes 1–3-forked, with pubescent branches to 2 cm. long; bracts subquadrate, ± 3.5 × 3.5 mm., entire, pubescent. Cyathia ± 2.5 × 6 mm., with funnel-shaped involucres, pubescent below; glandular rim, ± 1.5 mm. wide, deeply furrowed, occasionally with 1 or 2 deep notches, yellow; lobes subquadrate, ± 2 × 2 mm., pubescent. Male flowers: bracteoles laciniate, plumose; stamens 3.3 mm. long. Female flower: perianth distinctly 3-lobed, with lobes rounded ± 1 mm. long; styles 2 mm. long, pubescent below, joined to halfway, with distinctly bifid ± thickened apices. Capsule exserted on a reflexed pubescent pedicel to 6 mm. long, ± 8 × 8 mm., pubescent. Seeds ovoid, 2.8 × 2.2 mm., greyish brown, minutely tuberculate; caruncle rudimentary. Fig. 101/8, 9, p. 536.

TANZANIA. Lushoto District: Makuyuni, 16 Dec. 1935, *Koritschoner* 1250!; Handeni District: Kideliko, 4 July 1954, *Faulkner* 1446!; Morogoro District: Uluguru Mts., Bunduki, 17 Mar. 1953, *Drummond & Hemsley* 1665!
DISTR. T 3, 5, 6; not known elsewhere
HAB. Sandy stony soils and rocky slopes with dry deciduous woodland; 300–1800 m.
SYN. *S. glaucescens* Pax var. *brevipes* N.E.Br. in F.T.A. 6(1): 467 (1911); T.T.C.L.: 227 (1949). Types: Tanzania, Lushoto District, Kwamkuyu, *Engler* 3399 (B, syn.†) & Bomole, *Braun* 733 (B, syn.†, K, drawing of syn.!, EA, isosyn.!)
VAR. Apart from habit there appears to be little to distinguish this species from *S. pereskiifolium* and the two may eventually prove to be more closely related. An apparently distinct form from the northern end of the South Pares (*Greenway* 6494 and *Wingfield* 2886), with long sparse hairs on the margins and midrib of the lower surface, may be the result of hybridisation with *S. molle*, which also occurs there.

8. **S. calycinum** *S. Carter* in K.B. 42: 670 (1987). Type: Tanzania, Ulanga District, Taweta, *Haerdi* 580/0 (K, holo.!)

Tree to 8 m. high. Leaves sessile, obovate, to 20 × 7 cm., base tapering, apex obtuse, margin entire or minutely toothed towards the apex, midrib keeled beneath, glabrous or pubescent on the margin and midrib beneath when young. Inflorescence lax, in pseudumbels of 2–5 cymes on glabrous peduncles to 5.5 cm. long; cymes 2–3-forked, with pubescent branches to 1.5 cm. long; bracts subquadrate, ± 2.5 × 3 mm., dentate, pubescent. Cyathia ± 2 × 6 cm., with broadly funnel-shaped involucres, densely pubescent below; glandular rim ± 1.5 mm. wide, deeply furrowed; lobes subquadrate, ± 1.5 × 1.5 mm., pubescent below and on the margins. Male flowers: bracteoles ± 2.5 mm. long, laciniate, plumose; stamens 3.3 mm. long. Female flower: perianth obvious, 3-lobed, pubescent, with lobes elongating to 3–4 mm. long in fruit, often divided at the apex; styles 1.75 mm. long, pubescent below, joined to halfway with deeply bifid, scarcely thickened apices. Capsule exserted on a pubescent reflexed pedicel to 6 mm. long, ± 8 × 9 mm., pubescent. Seeds ovoid, 3.3 × 2.5 mm., buff, shallowly and sparsely tuberculate; caruncle minute.

TANZANIA. Ulanga District: Taweta, Aug. 1960, *Haerdi* 580/0!; Rungwe District: near Kyimbila, June 1913, *Stolz* 2008!
DISTR. T 6, 7; known only from these 2 specimens
HAB. Deciduous woodland; 600 m.

9. **S. glabratum** *S. Carter* in K.B. 42: 667 (1987). Type: Zambia, Mbala District, Crocodile I., *Richards* 11209 (K, holo.!)

Shrub to ± 1.5 m. high. Leaves linear-lanceolate to oblanceolate, to ± 15 × 5 cm., base tapering to a petiole to 1 cm. long, apex obtuse, margin entire, midrib acutely prominent beneath, lamina glabrous. Inflorescence in pseudumbels of 3–5 cymes on glabrous peduncles to 4 cm. long; cymes 2-forked, with branches glabrous or rarely the uppermost shortly pubescent, 5–15 mm. long; bracts subquadrate, ± 3 × 3 mm., dentate, glabrous or the margin sometimes sparsely and minutely pilose. Cyathia ± 2 × 4 mm., with shallowly cup-shaped involucres glabrous or shortly puberulous at the base; glandular rim ± 0.8 mm. wide, yellow, darkening (? reddening) with age; lobes rounded, ± 1.5 × 1.5 mm., glabrous. Male flowers: bracteoles fan-shaped, laciniate, ± plumose; stamens 2.8 mm. long. Female flower: ovary and styles not seen. Capsule subsessile on a pubescent pedicel to 1 mm. long, ± 4.3 × 4.5 mm., minutely pubescent, reddish. Seeds ovoid, obtusely 4-angled, 2.5 × 2 mm., pale brownish grey, minutely tuberculate; caruncle sessile 0.3 mm. in diameter, yellow.

TANZANIA.   Ufipa District: Lake Tanganyika, Namkala I., Kasola village, 29 July 1964, *Richards* 19104!
DISTR.   T 4; Zambia (Mbala District); not known beyond a very restricted area at the southern end of Lake Tanganyika
HAB.   Amongst rocks on the shoreline and wooded slopes of islands; 780 m.

NOTE.   Most closely related to *S. cameronii* N.E. Br. from Malawi, this species is a much smaller, almost completely glabrous shrub, with similar but smaller inflorescences.

71. MONADENIUM

Pax in E.J. 19: 126 (1894); N.E. Br. in F.T.A. 6(1): 450 (1911); Bally, Genus Monadenium: 14 (1961)

Lortia Rendle in J.B. 1898: 30 (1898)

Stenadenium Pax in E.J. 30: 343 (1901); N.E. Br. in F.T.A. 6(1): 448 (1911); T.T.C.L.: 227 (1949)

Small trees, shrubs or perennial herbs, sometimes geophytic, with ± fleshy or succulent stems and branches and with caustic milky latex, monoecious; roots thick and fleshy, sometimes tuberous. Leaves fleshy, with stipules apparently absent, or modified as glands or spines. Inflorescences axillary, with sessile cyathia in dichotomously branching cymes, rarely solitary; bracts paired, persistent, free, or partly united along one edge to form a bract-cup behind and enveloping the involucre. Involucres with glands forming an entire rim notched on the lower side, surrounding 5 fringed lobes. Male flowers in 5 groups, bracteolate, scarcely exserted. Female flower pedicellate, with the perianth reduced to a rim below the ovary, rarely shortly 3-lobed; ovary 3-locular, with 1 pendulous ovule in each locule; styles 3, joined at the base and with bifid stigmas. Fruit a 3-lobed capsule, exserted through the notch in the glandular rim, dehiscent (a regma). Seeds oblong, with a cap-like pale yellow caruncle, rarely ecarunculate.

A well-defined genus of over 50 species of which 39 occur in the Flora area. Distribution extends throughout the eastern tropical regions of Africa, from northern Somalia southwards to the Transvaal. The genus is distinguished primarily by the entire horseshoe-shaped involucral gland, which has a wide rim extended to protect the ovary exserted through the notch. The persistent bracts, which envelope the involucre, are usually united behind the glandular rim and are sometimes large and showy. The seeds, with few exceptions, are oblong, with a relatively large mushroom-shaped caruncle capping the apex. Habit ranges from small geophytes and herbs with ± fleshy stems, to shrubs and small trees with succulent, sometimes spiny stems and branches, or fleshy-rooted herbs with thickly succulent and often prominently tubercled stems.

1. Cyathial bracts completely free to the base 2
   Cyathial bracts united to halfway or more, rarely less 11
2. Geophytes, with annual stems 2–20 cm. high; cymes on
      peduncles 8–60 mm. long 3
   Tuberous-rooted herbs with stems 25–90 cm. high; cymes
      subsessile 9
3. Leaves broadly obovate 4
   Leaves linear to lanceolate 7
4. Flowers appearing before the leaves . . . .   2. *orobanchoides*
   Flowers appearing with the leaves 5

5. Cyathia congested into a globose head of 1-forked cymes
with peduncles to 8 mm. long; seeds carunculate   *7. M. globosum*
Cyathia solitary on peduncles to 6 cm. long; seeds without
a caruncle 6
6. Leaves to 4 × 2 cm. on an elongating stem, with the cyathia
arising from their axils . . . . *1. M. pedunculatum*
Leaves to 9 × 4.5 cm., forming a rosette; cyathia few from
the stem apex . . . . *3. M. nervosum*
7. Cyathial bracts lanceolate, much longer than the involucre   *5. M. gracile*
Cyathial bracts oblong, shorter than the involucre 8
8. Leaves with entire margin; roots napoid-tuberous, 2–6 cm.
in diameter . . . . *1. M. pedunculatum*
Leaves with denticulate margin; roots forming chains of
cylindrical or subspherical tubers to 2 cm. in diameter   *4. M. catenatum*
9. Leaves with entire margin; stems smooth . . . . . *9. M. petiolatum*
Leaves with denticulate or crisped margin; stems
minutely papillose at least towards the apex 10
10. Marginal teeth of the leaves acute . . . . *8. M. capitatum*
Marginal teeth of the leaves rounded, or margin crisped   *10. M. crispum*
11. Geophytic herbs to 8 cm. high 12
Perennial herbs, shrubs or trees, often succulent 13
12. Seeds subglobose, black, speckled . . . . *6. M. pseudoracemosum*
Seeds oblong, 4-angled, grey, tuberculate . . . . . *11. M. nudicaule*
13. Herbs 30–100 cm. high, shrubs or small trees; cymes 2–8-
forked on peduncles (1–)3–15 cm. long 14
Herbs to 1 m. high, often succulents with tuberculate
stems; cymes 1–2(–3)-forked on peduncles 5–20 mm.
long, if longer then herbs less than 15 cm. high 24
14. Herbs with smooth, papillose or bristly stems 15
Shrubs or small trees, with usually spiny stems and
branches 19
15. Entire inflorescence bright red . . . . *19. M. coccineum*
Inflorescence yellowish green, often with purplish veining
on the bract-cup 16
16. Cyme-peduncles to 3 cm. long; bract-cup enveloping but
not much exceeding the involucre . . . . . . *12. M. depauperatum*
Cyme-peduncles 4–13 cm. long; bract-cup large and showy 17
17. Leaves glabrous or sparsely pilose, margin entire; bract-
cup glabrous . . . . *13. M. laeve*
Leaves with stiff bristles, at least on the margin; bract-cup
with stiff bristles at least on the midribs 18
18. Leaf-margins and midribs of the bract-cup with bristly
teeth; bract-cup green with purplish veining . . . *14. M. goetzei*
Leaf and bract-cup surfaces with stiff bristles; bract-cup
green with darker veining . . . . *15. M. echinulatum*
19. 1 or 3 stout spines 5–18 mm. long present below the
leaf-scars 20
Clusters of small prickles present below the leaf-scars and
along the stem-angles, or spines and prickles absent 22
20. Spines 1 . . . . *16. M. torrei*
Spines 3 21
21. Leaves spathulate, to 4 × 3 cm. . . . . *17. M. elegans*
Leaves obovate, to 17 × 6 cm. . . . . *18. M. spinescens*
22. Stems few to 1.5 m. high, 4–5-angled, with ridges of spiny
teeth along the angles . . . . *20. M. magnificum*
Stems solitary to 3–4 m. high, obscurely 5-angled, with or
without scattered clusters of small spines 23
23. Inflorescence brilliant red . . . . *21. M. spectabile*
Inflorescence pale green tinged with pink . . . . . *22. M. arborescens*
24. Stems less than 1 cm. thick, smooth or with only slightly
projecting tubercles 25
Stems at least 1 cm. thick, conspicuously grooved or
prominently tubercled 31

25. Stems erect or subscandent to 40–180 cm. long 26
    Stems decumbent to 25 cm. long, or erect to 15 cm. high 27
26. Stems smooth; leaves ovate to lanceolate, to 8 × 6.5 cm.;
    bract-cup shallowly notched, green with darker veining;
    rim of involucral gland spreading . . . .   23. M. invenustum
    Stems longitudinally ridged; leaves obovate, to 3 × 1.3 cm.;
    bracts almost free, greenish white; rim of involucral
    gland forming a tube . . . .   39. M. virgatum
27. Tubers clustered, rounded, 1.5–3 cm. in diameter; stems
    decumbent to 25 cm. long or erect to 5 cm. high 28
    Tubers 1–5 cm. thick but not rounded or in clusters; stems
    erect to 15 cm. high 29
28. Bract-cup bright pink . . . .   24. M. rubellum
    Bract-cup dark green striped with white . . . . . .   25. M. stoloniferum
29. Stems with slightly projecting tubercles, fleshy, smooth   28. M. rhizophorum
    Stems without tubercles, minutely papillose or scabrid 30
30. Stem scabridulous; leaves to 6 × 2.5 cm. . . . . . .   26. M. montanum
    Stem minutely papillose; leaves to 4.5 × 2.5 cm. . . . .   27. M. trinerve
31. Stem with 4 longitudinal grooves below each leaf-scar   38. M. ellenbeckii
    Stem tessellated, with leaf-scars at the apex of usually
    prominent tubercles 32
32. Prickles surrounding leaf-scar minute and evident only on
    young growth 33
    Prickles or scales surrounding leaf-scar 1–3 mm. long,
    persistent 35
33. Stem-tubercles projecting only slightly; bract-cup dark
    green with white midribs; rim of involucral gland
    yellow . . . .   29. M. yattanum
    Stem-tubercles prominent; bract-cup greenish-white
    flushed with pink; rim of involucral gland red 34
34. Stem tubercles very prominent, curving outwards; midribs
    of bract-cup crested . . . .   30. M. gladiatum
    Stem tubercles not curving outwards; midribs of bract-cup
    not crested . . . .   31. M. stapelioides
35. Stem-tubercles reflexed, tapering, 5–20 mm. long; leaf-scar
    surrounded by 2–5 toothed scales . . . .   37. M. reflexum
    Stem-tubercles conical or cylindrical, 3–10 mm. high, not
    reflexed; leaf-scar surrounded by 3–5 prickles 36
36. Bract-cup greenish white . . . .   33. M. heteropodum
    Bract-cup variously pink, pink-flushed or greyish green,
    but not white 37
37. Prickles surrounding the leaf-scar 3–5, very slender, often
    breaking off; cyme 1-forked but usually only one
    branch developing . . . .   32. M. ritchiei
    Prickles surrounding the leaf-scar stout, 3 or 5, persistent;
    cyme 1–3-forked, with both branches developing 38
38. Stems subscandent, with branches readily breaking off and
    rooting; tubercles in 5 loosely spiralled series   35. M. renneyi
    Stems erect and decumbent, branches persistent; tubercles
    in 8 tightly spiralled series 39
39. Stems 1.5–2 cm. thick, erect to 15 cm. high or decumbent to
    90 cm. long . . . .   34. M. guentheri
    Stems 3–5 cm. thick, erect to 90 cm. high or decumbent to
    125 cm. long . . . .   36. M. schubei

1. M. pedunculatum S. *Carter* in K.B. 42: 903 (1987). Type: Tanzania, Mpanda District, Milumba Plain, *Richards* 7053 (K, holo.!)

Glabrous perennial herb, with a napoid tuberous root to 6 cm. in diameter, producing 1–2 subterranean woody stems to 6 cm. long; annual stems 1–4, rarely branching, 2–15(–20) cm. high, longitudinally ridged, often minutely papillose. Leaves sessile, linear, to 9 × 0.5 cm., or lanceolate, to 6 × 1.5 cm., rarely obovate, to 4 × 2 cm., margin entire, midrib ± prominent beneath; stipules glandular, minute, evident on young growth only.

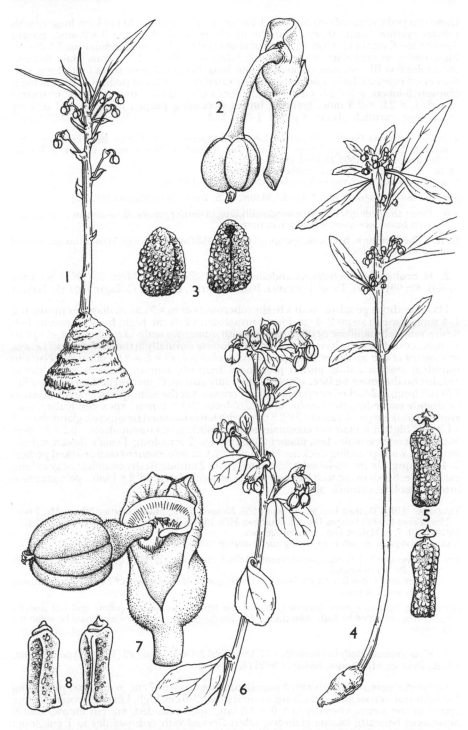

FIG. 102. *MONADENIUM PEDUNCULATUM* — **1**, habit, × ⅔; **2**,cyathium, with capsule, × 4; **3**, seeds, × 6. *M. CAPITATUM* — **4**, habit, × ⅔; **5**, seeds, × 8. *M. INVENUSTUM* var. *INVENUSTUM* — **6**, fruiting stem, × ⅔; **7**, cyathium, with capsule, × 4; **8**, seeds, × 6. 1, from *Robinson* 4121 (Zambia); 2, from *Richards* 20764; 3, from *Richards* 7053; 4, from *Polhill & Paulo* 1916; 5, from *Greenway & Kanuri* 14354; 6–8, from *Greenway & Kanuri* 13044. Drawn by Christine Grey-Wilson.

Cymes on peduncles to 5(-8) cm. long, 1-forked with cyme-branches to 1 cm. long, or with solitary cyathia; bracts shorter than the involucre, free, oblong, ± 3 × 2 mm., margin denticulate. Cyathia ± 5 × 3 mm., with barrel-shaped involucres; glandular rim 1.5–2 mm. high, entire to crenulate, white or pink; lobes rounded, ± 1 × 1 mm. Male flowers: bracteoles few, filamentous; stamens 3.2 mm. long. Female flower: styles 1 mm. long, with thickened rugulose bifid apices. Capsule exserted on a reflexed pedicel 6–12 mm. long, obtusely 3-lobed, ± 5 × 5.5 mm., smooth. Seeds conical with truncate base, obscurely 4-angled, ± 2.8 × 2.3 mm., greenish brown becoming purple, minutely and densely tuberculate; caruncle absent. Fig. 102/1–3, p. 543.

TANZANIA. Kigoma District: 40 km. from Uvinza on Mpanda road, 23 Nov. 1962, *Verdcourt* 3439!; Mpanda District: Kapapa Camp, 28 Oct. 1959, *Richards* 11628!; Chunya District: S. Rukwa, Mbangala, 14 Dec. 1963, *Richards* 18673!
DISTR. T 4, 7; Zaire, northern Zambia and Malawi
HAB. In wooded grassland with sandy or clay soils; 780–1200 m.

SYN. [*M. chevalieri* sensu Cribb & Leedal, Mount. Fl. S. Tanz.: 78 (1982), *non* N.E. Br.]

VAR. Stems and peduncles can be exceptionally long in denser grassland; specimens from around Mpanda have more obovate leaves than normal.

NOTE. The drawing in Bally's monograph of *Bullock* 1358 from Zambia (p. 31) is of this species and not of *M. chevalieri* N.E. Br.

2. M. orobanchoides *Bally* in Candollea 17: 30 (1959) & Gen. Mon.: 36 (1961); S. Carter in K.B. 42: 904(1987). Type: Tanzania, Iringa District, Dabaga, *Geilinger* 1861 (K, holo.!)

Hysteranthous geophyte, with a fleshy tuberous root to ± 5 cm. in diameter producing 1–4 subterranean stems 2–4 cm. long; annual stems 2–6 cm. high; flowering stems 1–4, fleshy, minutely papillose or often smooth, with numerous scale-like oblong leaves to ± 10 × 5 mm., crowded at the base, sometimes developing normally at the stem apices. Leaves in rosettes at the apices of the stems, sessile, obovate, to ± 6 × 2.5 cm., apex obtuse to rounded, margin entire, midrib prominent beneath, lamina glabrous, often tinted reddish on the lower surface; stipules apparently absent. Cymes on peduncles 0.5–2.5(-3.5) cm. long, 1–2-forked or cyathia solitary, crowded at the stem apices; bracts ± equal to or slightly exceeding the involucre, free, oblong, 2–4 × 2 mm., apex denticulate, pale green or pink tinged. Cyathia ± 3 × 2.5 mm., with barrel-shaped involucres; glandular rim ± 1.2 mm. high, undulate to ± crenulate, white or pink; lobes subquadrate, ± 0.5 × 0.5 mm. Male flowers: bracteoles few, filamentous; stamens 2 mm. long. Female flower: styles 1 mm. long, with spreading thickened bifid apices. Capsule exserted on a reflexed pedicel 3–4 mm. long, obtusely 3-lobed, ± 3.5 × 4 mm., with 2 minute fleshy crenulate ridges along each angle. Seeds ovoid with truncate base, obscurely 4-angled, 2 × 1 mm., pale greenish brown, areolate; caruncle absent.

TANZANIA. Ufipa District: Kito Mt., 13 Sept. 1956, *Richards* 6189!; Mbeya District: Mbogo Mt., 7 Nov. 1966, *Gillett* 17640!; Songea District: Unangwa Hill, 15 Jan. 1956, *Milne-Redhead & Taylor* 8341!
DISTR. T 4, 7, 8; Malawi and northern Zambia
HAB. Amongst grass, often appearing after burning; 1020–2400 m.

SYN. [*M. chevalieri* N.E. Br. var. *chevalieri* sensu Bally, Gen. Mon.: 32 (1961), quoad *Richards* 3480, *non* N.E. Br. sensu stricto]
      M. nervosum sensu Bally, Gen. Mon.: 23 (1961) quoad *Milne-Redhead & Taylor* 8341 & t. II, *non* Bally sensu stricto]

NOTE. The flowering stem, including that of the type, is minutely papillose and not densely puberulous as stated by Bally. Also the seed in his description and drawing should be without a caruncle.

3. M. nervosum *Bally* in Candollea 17: 29 (1959) & Gen. Mon.: 22 (1961). Type: Zambia, Mbala District, Mpulungu, *Richards* 3656 (K, holo.!)

Geophytic perennial herb with a napoid tuberous root to 7 cm. in diameter producing 1–3 subterranean stems 1–3 cm. long; annual stems 2–6 cm. high. Leaves in rosettes at the apices of the stems, obovate, to ± 9 × 4.5 cm., apex rounded, margin entire, midrib prominent beneath, lamina glabrous, often flecked with red; petioles to 1 cm. long; stipules apparently absent; a few scale-like leaves also present, crowded at the stem base, ovate, to ± 4 × 2 mm. Cymes with solitary cyathia on minutely papillose peduncles to 6 cm. long, occasionally hysteranthous; bracts shorter than the involucre, free, subquadrate, ± 2.5 × 2.5 mm., margin denticulate. Cyathia ± 4.5 × 3 mm., with barrel-shaped involucres;

glandular rim ± 1.5 mm. high, undulate to ± crenulate, pink or red; lobes rounded, ± 1 × 1 mm. Male flowers: bracteoles few, filamentous; stamens 2 mm. long. Female flower: styles 1 mm. long, with thickened deeply bifid apices. Capsule exserted on a reflexed pedicel 5–7 mm. long, obtusely 3-lobed, ± 4 × 4.5 mm., smooth. Seeds ovoid with truncate base, obscurely 4-angled, 2.3 × 1.3 mm., greenish brown, minutely and densely tuberculate; caruncle absent.

TANZANIA. Ufipa District: Kawa R., at top of gorge, 30 Dec. 1956, *Richards* 7396!
DISTR. T 4; known in East Africa from the one collection only; northern Zambia and possibly southeast Zaire
HAB. Sandy soil amongst rocks in deciduous woodland; 1500 m.

NOTE. *Milne-Redhead & Taylor* 8341 cited by Bally in Gen. Mon.: 23 & t. II (1961) is *M. orobanchoides*.

4. **M. catenatum** S. *Carter* in K.B. 42: 905(1987). Type: Tanzania, Songea District, 1.5 km. S. of Gumbiro, *Milne-Redhead & Taylor* 8523 (K, holo.!)

Glabrous perennial herb, with tuberous cylindrical or subspherical roots to 2 cm. in diameter, produced in chains; annual stems 1–2, to 20 cm. high, produced from ± 4 cm. below ground, purplish at ground-level, white below. Leaves subsessile, lanceolate, to 5 × 1.5 cm., apex acute, margin minutely toothed; stipules glandular, minute. Cymes of solitary cyathia on peduncles to 12 mm. long, usually much less; bracts shorter than the involucre, free, oblong, ± 2.3 × 1.8 mm., margin denticulate. Cyathia ± 3.5 × 3 mm., with barrel-shaped involucres; glandular rim entire, 1.5 mm. high, pink; lobes rounded, ± 1 × 1 mm. Male flowers: bracteoles few, filamentous, plumose; stamens 2.5 mm. long. Female flower: styles 1 mm. long, with thickened rugulose distinctly bifid apices. Capsule exserted on a reflexed pedicel 4–5 mm. long, obtusely 3-lobed, ± 3.3 × 4.2 mm., smooth. Seeds ovoid with truncate base, 2.3 × 1.8 mm., minutely and closely tuberculate; caruncle absent.

TANZANIA. Songea District: 1.5 km. S. of Gumbiro, 24 Jan. 1956, *Milne-Redhead & Taylor* 8523!; Tunduru District: 1.5 km. E. of Mawese bridge, 19 Dec. 1955, *Milne-Redhead & Taylor* 7833!
DISTR. T 8; known only from these two collections
HAB. *Brachystegia* woodland on sand; 450–830 m.

SYN. [*M. chevalieri* sensu Bally, Gen. Mon.: 30 (1961), quoad *Milne-Redhead & Taylor* 7833, *non* N.E. Br. sensu stricto]

NOTE. Bally's description of *M. chevalieri* is a composite one, and includes a drawing of *Milne-Redhead & Taylor* 7833, cited here, on p. 30.

5. **M. gracile** *Bally* in Candollea 17: 27 (1959) & Gen. Mon.: 40 (1961). Type: Tanzania, Dodoma District, Kazikazi, *B.D. Burtt* 3529 (K, holo.!)

Glabrous perennial herb, with an elongated tuberous root producing a woody subterranean stem to 2 cm. long; annual stems 1–3, to 5 cm. high, with up to 4 branches. Leaves sessile, linear, to 15 × 3 mm., margin entire; stipules apparently absent. Cymes of solitary cyathia on peduncles to 15 mm. long; bracts much longer than the involucre, free, lanceolate, ± 6.5 × 3 mm. Cyathia ± 4 × 3 mm., with barrel-shaped involucres; glandular rim ± 1.5 mm. high, white; lobes rounded, ± 1 × 1 mm. Male flowers: bracteoles few, filamentous, plumose; stamens 2.5 mm. long. Female flower: styles 1 mm. long, slender with slightly thickened bifid apices; ovary 3-lobed, slightly elongated, with 2 minute fleshy crested ridges along each angle. Capsule exserted on a reflexed pedicel to 1 cm. long, not seen entire but probably ± 4 mm. long. Seeds not seen.

TANZANIA. Dodoma District: Kazikazi, 4 Dec. 1931, *B.D. Burtt* 3529!
DISTR. T 5; known only from the type specimen
HAB. Grassland with hard pan soils; 1200 m.

6. **M. pseudoracemosum** *Bally* in Candollea 17: 30 (1959) & Gen. Mon.: 47 (1961). Type: Tanzania, Ufipa District, Kipili, *Bullock* 2374 (K, holo.!, EA, iso.)

Glabrous perennial herb, with a cylindrical tuberous root to 10 × 1.5 cm. producing a woody subterranean stem to 6 cm. long; annual stems to 7 cm. high. Leaves subsessile, obovate (in Tanzania) to 7.5 × 6 cm. (or oblanceolate to 11 × 2 cm. in Zambia), apex rounded, margin entire or crisped, midrib ± keeled beneath, lamina often tinged red on the undersurface; stipules glandular, minute. Cymes on peduncles to 2.5 cm. long, 1-forked with cyme-branches to 1.5 cm. long, or cyathia solitary, peduncles and cyme-branches with longitudinal winged ridges; bract-cup ± equal to the involucre, ± 6 × 10 mm., notched to ± halfway between acute apices and keeled midribs, green with darker veining. Cyathia ± 5 × 4 mm., with cup-shaped involucres; glandular rim ± 2 mm. high, entire, pink; lobes rounded, ± 1.3 × 1.3 mm. Male flowers: bracteoles few, filamentous; stamens 4 mm. long. Female flower: perianth an obvious ± lobed rim ± 0.5 mm. wide; styles 2 mm. long, joined to halfway with thickened rugulose distinctly bifid apices. Capsule exserted on a pedicel to 6 mm. long, obtusely 3-lobed, ± 5 × 6 mm., with a fleshy ridge along each angle. Seeds subglobose, 2.8 × 2.5 mm., black with shallow yellowish tubercles; caruncle sessile, 1 mm. in diameter.

var. **pseudoracemosum**

Leaves obovate.

TANZANIA. Ufipa District: Chapota, 6 Mar. 1957, *Richards* 8530! & Kalambo Falls, 26 Mar. 1960, *Richards* 12791!
DISTR. T 4; known only from near the shores of Lake Tanganyika
HAB. Stony ground amongst grass in open woodland; 1035–1650 m.

NOTE. Var. *lorifolium* Bally (Gen. Mon.: 49 (1961), from Zambia, Nmbulu I., very close to the Tanzanian border, is known only from the type collection, differing by its oblanceolate leaves with crisped margins.

7. **M. globosum** *Bally & S. Carter* in K.B. 42: 906 (1987). Type: Tanzania, Njombe District, Njombe–Milo road, *Richards* 14022 (K, holo.!, EA, iso.!)

Glabrous perennial herb, with a tuberous root ± 4 cm. in diameter, producing 1–2 woody subterranean stems to 2 cm. long; annual stems fleshy, to 8 cm. high, minutely papillose. Leaves subsessile, forming rosettes at the stem apices, obovate, to 6 × 3 cm., apex rounded, margin entire, lamina red tinged; stipules filamentous, to 3 mm. long, often much reduced. Cymes on peduncles 4–8 mm. long, 1-forked, much congested at the stem apex to form a globose head ± 3 cm. in diameter. Bracts ± equal to the involucre, free, oblong, ± 4 × 3 mm., denticulate and apiculate, flushed bright red. Cyathia ± 4 × 4 mm., with barrel-shaped involucres; glandular rim 2 mm. high, pink; lobes rounded, ± 1 × 1 mm. Male flowers: bracteoles filamentous, plumose; stamens 4 mm. long. Female flower: style 1 mm. long, joined to halfway, with thickened rugulose bifid apices. Capsule exserted on a reflexed pedicel to 4 mm. long, obtusely 3-lobed, ± 3.5 × 3.5 mm., with a small fleshy ridge along each angle. Seeds oblong, 4-angled, 2 × 1.3 mm., grey, minutely tuberculate; caruncle shortly stalked, 0.5 mm. in diameter.

TANZANIA. Njombe District: Lupembe Hill, 6 Nov. 1978, *Archbold* 2657! & Njombe–Milo road, 28 Jan. 1961, *Richards* 14022!
DISTR. T 7; known only from the type locality
HAB. Amongst flat rocks with gritty black soil; 1950–2500 m.

8. **M. capitatum** *Bally* in Candollea 17: 26 (1959) & Gen. Mon.: 34 (1961). Type: Tanzania, Rungwe District, N. of Tukuyu, *St. Clair-Thompson* 654 (K, holo.!)

Perennial herb to 90 cm. high, with a tuberous root to 16 × 6 cm.; stems unbranched, minutely and often densely papillose especially towards the apex. Leaves subsessile, oblanceolate to obovate, 9–13 × 2.5–4 cm., base tapering, apex acute, margin toothed with teeth near the base often larger and gland-tipped; stipules triangular, ± 1 mm. long, glandular. Cymes crowded towards the stem apex, subsessile, 1–2-forked; bracts subequal to a little shorter than the involucre, free, oblong, ± 5 × 4 mm., denticulate. Cyathia ± 5 × 3.5 mm., with barrel-shaped involucres; glandular rim 2 mm. high, margin crenulate, white or pink; lobes rounded, ± 1.2 × 1.2 mm. Male flowers: bracteoles few, filamentous, plumose; stems 3.5 mm long. Female flower: styles 0.8 mm. long, joined at the base, with slightly thickened bifid apices. Capsule exserted on a reflexed pedicel 4–5 mm. long, acutely 3-lobed, ± 4.5 × 4 mm., smooth to minutely papillose. Seeds oblong, 4-angled, 2.2 × 1.2 mm., pale grey, verrucose; caruncle pointed, 0.8 mm. in diameter Fig. 102/4, 5, p. 543.

TANZANIA. Mbeya District: Magangwe, 14 Apr. 1970, *Greenway & Kanuri* 14354!; Iringa District: Mtulingale, 28 Mar. 1962, *Polhill & Paulo* 1916!; Njombe District: Njombe–Igawa road, 4 Feb. 1961, *Richards* 14222!
DISTR. T 1, 4, 5, 7; NE. Zambia
HAB. In grassland and open deciduous woodland, with gritty sandy soil; 1200–2200 m.
SYN. [*M. crispum* sensu Bally, Gen. Mon.: 25 (1961), quoad *Bax* 106, *Hornby* 2051 & 2052, *non* N.E. Br.]

NOTE. Specimens from **T** 1, 4 and 5 are of poor quality and included in the distribution with reservation. *Azuma* 550 from Kigoma District, Kabogo Mts., is very tall (80 cm.), with very large leaves (15 × 4 cm.) and a single immature cyme. It may not belong here. *Bax* 106, *Hornby* 2051 & 2052, cited by Bally as *M. crispum*, and *Hornby* 714, are included here in the synonymy, but their leaves are large for *M. capitatum* and the marginal teeth are acutely pointed rather than rounded as in *M. crispum.*

9. M. petiolatum *Bally* in Candollea 17: 30 (1959) & Gen. Mon.: 23 (1961). Type: Tanzania, Dodoma District, Kazikazi, *B.D. Burtt* 3710 (K, holo.!)

Glabrous perennial unbranched herb ± 25 cm. high, with a tuberous root. Leaves obovate, to 8 × 3 cm., base tapering to a petiole 1–2 cm. long, apex acute, margin entire; stipules triangular, minute. Cymes subsessile, 1-forked, on minutely papillose peduncles, congested towards the stem apex; bracts slightly shorter than the involucre, free, subquadrate, ± 3 × 2.5 mm., margin minutely toothed. Cyathia ± 4 × 3.5 mm., with cup-shaped involucres; glandular rim 1.5 mm. high, crenulate; lobes rounded, ± 1 × 1.3 mm. Male flowers: bracteoles few, filamentous, plumose; stamens 3 mm. long. Female flower: styles 1 mm. long, joined at the base, with slightly thickened distinctly bifid apices. Capsule exserted on a reflexed pedicel ± 4 mm. long, sharply 3-lobed, oblong, ± 4 × 3.5 mm., smooth, with a pair of minute fleshy ridges along each angle. Seeds oblong, 4-angled, 2.8 × 1.5 mm., minutely tuberculate, pale brown; caruncle pointed, 0.8 mm. in diameter.

TANZANIA. Dodoma District: Kazikazi, 14 May 1932, *B.D. Burtt* 3710! & 15 km. S. of Itigi Station on Chunya road, 16 Apr. 1964, *Greenway & Polhill* 11579!
DISTR. **T** 5; known only from around the type locality
HAB. Closed bushland on sandy soil; 1280–1325 m.

NOTE. The differences between this species and *M. capitatum* appear to be very slight and the two may eventually prove to be conspecific. *M. petiolatum* differs by is non-papillose stems and petiolate leaves with entire margins, features which are possibly influenced by habitat conditions.

10. M. crispum *N.E.Br.* in F.T.A. 6(1): 1035 (1913); Bally, Gen. Mon.: 24 (1961); S. Carter in K.B. 42: 907 (1987). Type: Tanzania, Tanga District, Amboni Forest, *Sacleux* 2320 (P, holo., K, fragment & drawing of holo.!)

Glabrous perennial sparsely-branched herb to 75 cm. high, with a tuberous root ± 4 cm. in diameter; stems subfleshy, with obscure longitudinal ridges, sparsely and minutely papillose, often procumbent and rooting where they touch the ground. Leaves obovate to oblanceolate, to 8(–10) × 3(–4) cm., base tapering to a petiole to 1 cm. long, apex obtuse, margin crisped or denticulate with rounded teeth; stipules triangular, 1 mm. long, brown. Cymes subsessile, 1-forked; bracts ± equalling the involucre, free, ovate, ± 4 × 4 mm., apiculate. Cyathia ± 4.5 × 5 mm., with funnel-shaped involucres; glandular rim spreading, 1.5 mm. high, margin entire to crenulate, shallowly notched, pink; lobes rounded, ± 1.3 × 1.3 mm. Male flowers: bracteoles filamentous, plumose; stamens 3.5 mm. long. Female flower: perianth an obscurely 3-lobed rim below the ovary; styles 0.8 mm. long, with thickened rugulose scarcely bifid apices. Capsule exserted on a reflexed pedicel 4 mm. long, obtusely 3-lobed, ± 5.5 × 6 mm., smooth. Seeds oblong, 4-angled, 3.5 × 2 mm., minutely and shallowly tuberculate; caruncle subsessile, pointed, 1 mm. in diameter.

TANZANIA. Tanga District: 8 km. SE. of Ngomeni, 29 July 1953, *Drummond & Hemsley* 3538! & Kange Gorge, 30 Sept. 1955, *Faulkner* 1742!; Pangani District: Madanga, Mkuzi Katani, Kibubu, 25 July 1957, *Tanner* 3625!
DISTR. **T** 3; not known elsewhere
HAB. Rocky slopes in forest; 30–700 m.

SYN. *M. intermedium* Bally in Candollea 17: 28 (1959) & Gen. Mon.: 21 (1961). Type: Tanzania, Tanga District, Tongwe–Mlinga, *Greenway* 4666 (K, holo.!)

11. M. nudicaule *Bally* in Candollea 17: 29 (1959) & Gen. Mon.: 46 (1961). Type: Tanzania, Chunya District, Uleia, *Bally* E.133 (not E.135 as in original description) (EA, holo.!, K, iso.!)

A hysteranthous geophyte, with a tuberous root ± 4.5 cm. thick, producing a woody subterranean stem ± 2 cm. long; flowering stems 1–2, ± 8 cm. high, ± fleshy, with several scale-like glabrous ovate leaves ± 5 × 2 mm. and minute glandular stipules; fully developed leaves not seen. Cymes on longitudinally ridged peduncles to 2 cm. long, 2-forked; bract-cup ± equal to the involucre, ovate, ± 4 × 6 mm., notched to ± halfway between acute apices. Cyathia ± 4 × 3.5 mm., with barrel-shaped involucres; glandular rim 1.5 mm. high; lobes rounded, ± 1.3 × 1.3 mm. Male flowers: bracteoles filamentous, plumose; stamens 3 mm. long. Female flower: styles 1 mm. long, joined at the base, with spreading thickened rugulose distinctly bifid apices. Capsule exserted on a reflexed pedicel ± 5 mm. long, obtusely 3-lobed, ± 4.5 × 4.5 mm., with a pair of spreading lobed flanges along each angle. Seeds oblong, obtusely 4-angled, 2.5 × 1.5 mm., grey, minutely tuberculate; caruncle subsessile, 1 mm. in diameter.

TANZANIA.   Chunya District: Uleia, 2 Sept. 1942, *J. Adamson* cult. in *Bally* E.133!
DISTR.   **T** 7; known with certainty from the type collection only
HAB.   In open dry deciduous woodland on hard-pan soil; 1300 m.

NOTE.   The type specimen, description, illustration and photographs in Bally's book, were all made from a plant cultivated in Nairobi. In many respects these match specimens of *M. parviflorum* N.E.Br. collected in the region of the latter's type locality (Malawi, Chipita), which has, however, papillose peduncles and bracts a little longer than the involucre. More information is needed about the plants at Uleia before the exact status of *M. nudicaule* can be satisfactorily assessed. *Richards* 19819, from Igila Hill near Kibembawe (Chunya District), with leaves having a toothed margin, a slightly larger bract-cup and adnate flanges (not spreading) along each angle of the capsule, may be *M. nudicaule*, or may represent a distinct species.

12. M. depauperatum *(Bally)* S. *Carter* in K.B. 42: 908 (1987). Type: Tanzania, Rungwe District, Kiwira [Kibila], *Stolz* 1568 (K, holo.!, BM, iso.!)

Perennial herb to 1 m. high, with a thick woody rootstock; stems several, simple or few-branched, minutely papillose to sparsely hairy especially towards the apex. Leaves ovate, to 10 × 5 cm., base tapering to a winged petiole ± 1 cm. long, apex acute, midrib sharply keeled on the lower surface, young leaves, midribs and margins with long bristly hairs; stipules glandular. Cymes on peduncles 1–3 cm. long, 2-forked, with cyme-branches ± 1 cm. long, often distinctly papillose below the bracts; bract-cup distinctly longer than the involucre, ± 8 × 15 mm., notched ± ⅓ between obtuse apices. Cyathia ± 5 × 4 mm., with cup-shaped involucres; glandular rim 1.5 mm. high; lobes rounded, ± 1.3 × 1.3 mm. Male flowers: bracteoles laciniate; stamens 3 mm. long. Female flower: styles 1.5 mm. long, joined to halfway, with spreading, thickened rugulose deeply bifid apices. Capsule exserted on a reflexed pedicel to 5 mm. long, acutely 3-lobed, ± 4.5 × 4 mm., with a pair of small fleshy ridges along each angle. Seeds oblong, 2.7 × 1.3 mm., 4-angled and with 4 longitudinal grooves, grey becoming reddish-brown, minutely tuberculate; caruncle shortly stipitate, 1 mm. in diameter.

TANZANIA.   Ufipa District: road to Kasanga, 13 June 1957, *Richards* 10189B! & Kala Bay, 31 Dec. 1963, *Richards* 18722!
DISTR.   **T** 4, 7; northern Zambia, Malawi
HAB.   In grass in *Brachystegia* woodland on sandy soil and amongst rocks; 780–2100 m.

SYN.   [*M. laeve* sensu Bally, Gen. Mon.: 84 (1961), pro parte quoad *Bullock* 2682 & *Richards* 11019 & sensu Cribb & Leedal, Mount. Fl. S. Tanz.: 78 (1982), *non* Stapf]
        M. laeve Stapf forma *depauperata* Bally in Candollea 17: 35 (1959) & Gen. Mon.: 85 (1961)

NOTE.   The smaller size of the bracts and especially the shortly pedunculate inflorescence distinguishes this species from *M. laeve*. Specimens from near the type locality and from Malawi most nearly match the holotype, but all are immature. The two (mature) specimens cited here, and others from Ufipa District and near Mbala in Zambia, are morphologically very uniform, and with their more compact inflorescences may eventually prove to represent a distinct taxon.

13. M. laeve *Stapf* in Hook., Ic. Pl. 27, t. 2666, fig. 6 (1900); N.E.Br. in F.T.A. 6(1): 456 (1911); Bally, Gen. Mon.: 83 (1961). Type: Malawi, Kondowe to Karonga, *Whyte* (K, holo.!)

Perennial herb, with a thick woody rootstock; stems simple or sparsely branched to 1 m. high, or procumbent and rooting to 1.5 m. long, glabrous. Leaves obovate, to 15 × 5.5 cm., base tapering to a winged petiole to ± 1 cm. long, apex acute, margin entire, midrib sharply keeled on the lower surface, lamina glabrous to sparsely pilose with long spreading hairs on the upper or both surfaces; stipules glandular. Cymes on peduncles 4–10 cm. long, 2–4-forked, with cyme-branches ± 2 cm. long, occasionally minutely papillose below the bracts; bract-cup nodding, much longer and wider than the involucre, to 15 × 25 mm., notched to nearly halfway between rounded apices, green with darker veining. Cyathia ± 4 × 5 mm., with cup-shaped involucres; glandular rim ± 1.5 mm. high; lobes rounded, ± 1.5 × 1.5 mm. Male flowers: bracteoles few, laciniate; stamens 4.2 mm. long. Female flower: styles 2 mm. long, joined to halfway, with spreading slightly thickened rugulose deeply bifid apices. Capsule exserted on a reflexed pedicel to 7 mm. long, ± 5 × 5 mm., acutely 3-lobed, with a pair of minute fleshy ridges along each angle. Seeds oblong, 4-angled, truncate at both ends, 3 × 1.5 mm., pale brown, minutely tuberculate; caruncle shortly stipitate, 1 mm. in diameter.

TANZANIA. Mbeya District: Njerenje, 5 May 1975, *Hepper, Field & Mhoro* 5292!; Rungwe District: Bulambya, 16 Mar. 1913, *Stolz* 1955!
DISTR. T 7: northern Malawi
HAB. *Brachystegia* woodland; ± 1500 m.

14. **M. goetzei** *Pax* in E.J. 30: 342 (1902); N.E.Br. in F.T.A. 6(1): 456 (1911); Bally, Gen. Mon.: 83 (1961). Type: Tanzania, Njombe District, Mwigi Mt., *Goetze* 1017 (B, holo.†, K, drawing of holo.!)

Fleshy glabrous perennial herb, with a large woody rootstock; stems unbranched, to 75 cm. high, occasionally with a few small scattered spines. Leaves ± fleshy, broadly ovate, to 17 × 9 cm., base tapering sharply to a winged petiole to 1 cm. long, apex apiculate, margin with large bristly teeth to 2 mm. long especially towards the base, midrib keeled beneath; stipules glandular. Cymes on peduncles to 13 cm. long, 2–5-forked, with cyme-branches to 2(–3) cm. long, often with a few small fleshy spines especially below the bracts; bract-cup nodding, much longer and wider than the involucre, to 15 × 25 mm., notched to ± ⅓ between rounded apices and keeled crenulate midribs with a few fleshy teeth, green with purple veining. Cyathia ± 6 × 6.5 mm., with cup-shaped involucres; glandular rim 2 mm. high with a purplish edge; lobes transversely rectangular, ± 1.2 × 1.5 mm. Male flowers: bracteoles laciniate; stamens 3.5 mm. long. Female flower: styles 1.8 mm. long, joined for ⅓, with thickened rugulose bifid apices. Capsule on a reflexed pedicel to 6 mm. long, not seen complete, probably ± 5 mm. long, with 2 narrow fleshy ridges along each angle. Seeds oblong, truncate at each end, 2.8 × 1.3 mm., greyish brown, minutely tuberculate; caruncle stipitate, 0.8 mm. in diameter.

TANZANIA. Ufipa District: 48 km. from Sitalike on Sumbawanga road, 12 Feb. 1962, *Richards* 16089!; Mbeya District: Songwe R., 27 Jan. 1963, *Richards* 17546! & Kimani R Wayside Park, 26 May 1967, *Robertson* 660!
DISTR. T 4, 7; not known elsewhere
HAB. In *Brachystegia* woodland, usually amongst rocks in sandy soil; 1175–1500 m.

15. **M. echinulatum** *Stapf* in Hook., Ic. Pl. 27, t. 2666. figs 1–5 (1900); N.E.Br. in F.T.A. 6(1): 454 (1911); Bally, Gen. Mon.: 89 (1961); S. Carter in K.B. 42: 908 (1987). Type: Tanzania, coast ['Africa, W. coast' in error], cult. *Sander & Co.* St. Albans, England (K, holo.!)

Fleshy perennial herb, with a large woody tuberous rootstock; stems 1–3, unbranched, to 30(–50) cm. high, glabrous or with bristly hairs to 1.5 mm. long especially towards the apex. Leaves obovate, to 12 × 6.5 cm., base tapering to a winged petiole to 5 mm. long, apex apiculate, midrib keeled beneath, lamina glabrous or with a few fleshy bristles towards the base, to sparsely pilose above and densely bristly with fleshy sometimes branched hairs beneath; stipules glandular, minute. Cymes on peduncles to 9 cm. long, 1–4-forked, with cyme-branches to 2 cm. long, sparsely to densely covered with fleshy bristles 1–2 mm. long; bract-cup nodding, much longer and wider than the involucre, to 10 × 25 mm., notched to ± ⅓ between rounded apices and keeled toothed midribs, sparsely to densely aculeolate and scabrid, green with darker veining. Cyathia ± 6 × 6 mm., with barrel-shaped involucres, glabrous to minutely scabrid; glandular rim 1.5 mm. high, ± incurved; lobes subquadrate, ± 1.3 × 1.3 mm. Male flowers: bracteoles laciniate; stamens 3.8 mm. long. Female flower: styles 2.8 mm. long, scabrid, joined for ± ⅓, with ± thickened rugulose bifid apices. Capsule exserted on a reflexed pedicel to 6 mm. long, ± 3-lobed, 5 × 5 mm., scabrid or with fleshy hairs to 1 mm. long especially when young. Seeds oblong, truncate at both ends, 3.2 × 1.8 mm., greyish brown, shallowly and densely tuberculate; caruncle pointed, stipitate, 1 mm. in diameter. Fig. 103/1–3, p. 550.

TANZANIA. Pare District: 6.5 km. S. of Lembeni, 30 July 1957, *Bally* 11618!; Dodoma District: W. of Itigi station on Tabora track, 14 Apr. 1964, *Greenway & Polhill* 11566!; Iringa District: Ruaha National Park, by track between Ferry to Maganga Lookout, 4 May 1970, *Greenway & Kanuri* 14455!

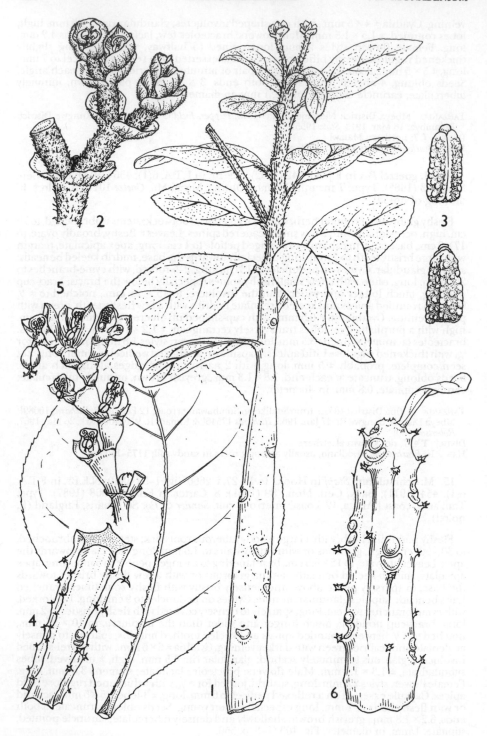

FIG. 103. *MONADENIUM ECHINULATUM* — 1, flowering stem, × ⅔; 2, part of inflorescence, × 1; 3, seeds, × 6. *M. MAGNIFICUM* — 4, leaf and portion of stem, showing spinescence, × ⅔; 5, part of inflorescence, × 2. *M. SPECTABILE* — 6, stem-apex, showing spinescence, × ⅔. 1, 2, from *Bally* E.44; 3, from *Richards* 24270; 4, 5, from *Bally* E.163; 6, from *Polhill & Paulo* 1336. Drawn by Christine Grey-Wilson.

DISTR. T 2, 3, 5–7; Zambia
HAB. In sandy soils amongst grass in open deciduous bushland or *Brachystegia* woodland; 300–1825 m.
SYN. *M. aculeolatum* Pax in E.J. 43: 89 (1909); N.E.Br. in F.T.A. 6(1): 455 (1911). Type: Tanzania, Pare District, Lembeni, *Uhlig* 92 (B, holo.†, K, drawing of holo.!)
    M. asperrimium Pax in E.J. 43: 90 (1909); N.E.Br. in F.T.A. 6(1): 455 (1911). Type: Tanzania, Masai District, Engaruka, *Merker* 577 (B, holo.†, K, drawing of holo.!)
    M. echinulatum Stapf forma *glabrescens* Bally in Candollea 17: 35 (1959) & Gen. Mon.: 93 (1961). Type: Tanzania, Ulanga District, junction of Kilombero R. and Luwegu R., *Schlieben* 2381 (B, holo.!, BM, iso.!)
    [*M. friesii* sensu Bally, Gen. Mon.: 93 (1961), *non* N.E. Br.]
    M. sp. aff. montanum Bally sensu Bally, Gen. Mon.: 101 (1961)

VAR. One of the most widespread and variable species, in height, leaf-shape, density of the fleshy bristles on all parts, size of the bract-cup and the extent to which it envelopes the cyathium, and the sculpturing of the seed-coat. Throughout its range some gatherings are almost glabrous whilst others from nearby are densely aculeate. Specimens from Chunya District, Lupa Forest Reserve, possess leaves which are glabrous when mature, and almost smooth seeds. Specimens from around the type-localities in the Pare Hills are particularly robust, densely aculeate and with large bract-cups. From the southwest of its distribution, in Iringa and Mbeya Districts, specimens of stunted growth to 5 cm. or less, have all been collected from dry sandy soils and may represent a varietal form.

NOTE. *Richards* 25798 from Mbeya District, Chunya Escarpment, on burnt ground, shows stunted distorted growth. Similarly, *Bally* E.125, which Bally used to describe and illustrate *M. friesii* (type locality Zambia, Kabwe [Broken Hill] was of cultivated material and appears to have produced atypical flowering leafless growth. The type of *M. friesii* is fragmentary, with shorter stems and much smaller flowering parts.

16. M. torrei *Leach* in Garcia de Orta, Sér. Bot. 1: 37 (1973). Type: Mozambique, Cabo Delgado, Montepuez, *Torre & Paiva* 11790 (LISC, holo., K, iso.!)

Sturdy succulent shrub erect to 3 m., with sparsely branched stem and smooth brown bark; branches ± 1.5 cm. thick, with shallow tubercles immediately beneath the leaf-scars in ± 5 longitudinal series, prominent on young growth, less so with age, each crowned by a stout curved spine to 8 mm. long. Leaves sessile, obovate to oblanceolate, to 12 × 3.5 cm., apex obtuse, margin prominently and irregularly toothed, midrib keeled on the lower surface, lamina bright green; stipules evident as acutely triangular weak spines to 1 mm. long. Cymes on peduncles to 10 cm. long, longitudinally ridged, pubescent towards the apex, 3–6-forked, with cyme-branches to 5 cm. long, progressively shorter and more densely pubescent above; bract-cup ± 1 × 1.5 cm., notched almost to the base between rounded apices and prominent midribs, shortly pubescent towards the base, yellowish green. Cyathia ± 10 × 5 mm., with barrel-shaped involucres narrowed at the top, pubescent; glandular rim 5–6 mm. high, glabrous, yellow; lobes oblong, ± 5 × 2 mm., divided to halfway into 3–4 segments, pubescent. Male flowers: bracteoles reduced to 2–3 plumose filaments; stamens 7 mm. long. Female flower: styles very slender, 3 mm. long, joined at the base, with scarcely thickened bifid apices. Capsule exserted on a pedicel to 4 mm. long, obtusely 3-lobed, ± 5.5 × 4.5 mm., pubescent at the base. Seeds oblong, 4-angled, truncate at both ends, 3 × 1.5 mm., pale grey, densely tuberculate; caruncle pointed, stipitate, 1 mm. in diameter.

TANZANIA. Masasi District: Lion Hill near Lupaso, 13 Sept. 1975, *Bally* 16936! & *Pfennig* 1199! & Lupaso, *Gerstner* 7179!
DISTR. T 8; northern Mozambique; known in Tanzania from this one area only
HAB. On granite, with deciduous woodland; 300–400 m.

SYN. *M. sp. aff. spinescens* Pax (Bally) sensu Bally, Gen. Mon.: 101 (1961)

17. M. elegans *S. Carter* in K.B. 42: 909 (1987). Type: Tanzania, Iringa District, 8 km. S. of Ruaha R. on Great North Road, *Polhill & Paulo* 1319 (K, holo.!)

Small tree or few-stemmed shrub to 3.5 m. high, sparsely branched, with flaking shiny yellowish brown papery bark; branches eventually pendulous, 7–15 mm. thick, with groups of 3 slender spines below the leaf-scars in 3–5 longitudinal series, the central spine to 18 mm. long, the lateral ones (modified stipules) to 10 mm. long. Leaves glabrous ± fleshy, spathulate, to ± 4 × 3 cm., tapering abruptly below the middle, apex rounded, margin toothed and undulate, midrib sharply keeled and prominently toothed on the lower surface. Cymes on peduncles to 3.5 cm. long, longitudinally ridged, with a pair of slender spines ± 5 mm. long at the apex, 2–3-forked, with cyme-branches 5–10 mm. long; bract-cup a little longer than the involucre, ± 9 × 10 mm., notched almost to the base between obtuse apiculate apices and ± prominent midribs, minutely pubescent towards the base, greenish white. Cyathia ± 8 × 4.5 mm., with barrel-shaped involucres, minutely pubescent; glandular rim 5.5 mm. high, pubescent on the inner surface, glabrous on the outer, white edged with red; lobes ± 5 × 2 mm., deeply divided into 5–7 segments, pubescent below. Male flowers: bracteoles very few, laciniate, plumose; stamens 6.5 mm. long. Female flower: styles 5 mm. long, pubescent, joined at the base, erect with thickened rugulose deeply bifid spreading apices; ovary pubescent, exserted on a reflexed glabrous pedicel. Capsule and seeds not seen.

TANZANIA. Dodoma District: 8 km. N. of Great Ruaha R., 18 Feb. 1932, *St. Clair-Thompson* 364!; Kilosa District: 15 km. N. of Mbuyuni, 4 Feb. 1974, *Bally & Carter* 16429!
DISTR. T 5–7; known only from near the Great Ruaha R. Gorge
HAB. Sandy soils and rocky slopes with dry mixed deciduous woodland; 650–875 m.

SYN. [*M. spinescens* sensu Bally, Gen. Mon.: 73 (1961), quoad *St. Clair-Thompson* (probably 364, cited above), *non* (Pax) Bally sensu stricto]

18. M. spinescens (*Pax*) *Bally* in Candollea 17: 34 (1959) & Gen. Mon.: 72 (1961). Type: Tanzania, Chunya District, Ilomo [Ilonia], Mt., *Goetze* 1099 (B, holo.†, K, fragment & drawing of holo.!, BM, iso.!)

Small tree to 6 m. high; stem ± 15 cm. in diameter, sparsely-branched, with peeling yellowish brown papery bark; branches 12–25 mm. thick, with groups of 3 stout spines below the leaf-scars in 5 longitudinal series, the central spine to 14 mm. long, slightly recurved, the lateral ones (modified stipules) to 7 mm. long. Leaves glabrous, ± fleshy, obovate, to 17 × 6 cm., apex acute, margin toothed and undulate, midrib keeled and toothed on the lower surface. Cymes on peduncles to 11.5 cm. long and to 13 mm. thick, longitudinally ridged, with a pair of short spines at the apex, to 8-forked, with cyme-branches 1–4 cm. long, the upper ones minutely pubescent, the whole inflorescence forming a dense head, green flushed with red; bract-cup ± 8 × 15 mm., notched to halfway between rounded apices and keeled crenulate midribs, pubescent. Cyathia ± 8 × 5 mm., with barrel-shaped involucres, pubescent; glandular rim 5 × 4 mm., pubescent on both surfaces; lobes oblong, 5 × 2 mm., deeply divided into 5 segments, pubescent. Male flowers: bracteoles very few, laciniate, plumose; stamens 7 mm. long. Female flower: styles 3 mm. long, pubescent, joined at the base, erect with thickened rugulose bifid apices. Capsule exserted on a pedicel to 1 cm. long, acutely 3-lobed, ± 8 × 7 mm., pubescent. Seeds oblong, truncate at both ends, 3.5 × 2.3 mm., brown, minutely tuberculate; caruncle stipitate, pointed, 1.2 mm. in diameter.

TANZANIA. Chunya District: Lake Rukwa, Ilomo [Ilonia] Mt., 17 July 1899, *Goetze* 1099! & Ilomo [Iromo] Hill, 3 Sept. 1942, *Bally* E.134!
DISTR. T 7; known only from the type locality
HAB. Rocky hillside with light deciduous woodland; 1200–1500 m.

SYN. *Stenadenium spinescens* Pax in E.J. 30: 343 (1901); N.E.Br. in F.T.A. 6(1): 449 (1911); T.T.C.L.: 227 (1949)

19. M. coccineum *Pax* in E.J. 19: 127 (1894); N.E.Br. in F.T.A. 6(1): 458 (1911); Bally, Gen. Mon.: 74 (1961). Type: locality unknown, probably Tanzania, Maswa District, Usukuma, *Fischer* 521 (B holo.†, K, fragment & drawing of holo.!)

Glabrous succulent perennial herb, with a large fleshy tuberous root; stems 1–4, erect or subscandent to 1.3 m. high, rarely branched, 1–1.5 cm. thick, with 5 longitudinal ridges, pale green sometimes red-flushed especially towards the apex, leaf-scars becoming calloused and prominent. Leaves sessile, fleshy, obovate, to 8.5 × 3.5 cm., apex obtuse, margin with fleshy teeth to 1.5 mm. long towards the base, midrib keeled and toothed on the lower surface; stipules scale-like, to 1.5 mm. long, occasionally dentate, glandular at the base. Cymes on peduncles to 6 cm. long and longitudinally ridged, with scattered fleshy teeth on the ridges, 2–5-forked, with cyme-branches to 3 cm. long, the whole inflorescence bright red; bract-cup to 8 × 12 mm., notched to halfway between acute apices and sharply keeled crenulate midribs. Cyathia ± 6 × 4.5 mm., with barrel-shaped involucres narrowed at the top; glandular rim 1.5 mm. high, lobes subquadrate, ± 1.5 × 1.5 mm. Male flowers: bracteoles laciniate; stamens 3.5 mm. long. Female flower: styles 2 mm. long, spreading from the base, with thickened deeply bifid apices. Capsule exserted on a pedicel to 5 mm. long, acutely 3-lobed, to 7.5 × 6 mm., with a pair of narrow fleshy crenulate ridges along each angle. Seeds oblong, 4-angled, with truncate ends, 2.8 × 1.5 mm., brown, shallowly and minutely tuberculate; caruncle stipitate, pointed, 1 mm. in diameter.

TANZANIA. Masai District: 9.5 km. S. of Endulen, 26 July 1957, *Bally* 11598!; Tabora District: 56 km. N. of Kitunda on Mbeya–Tabora road, 4 June 1963, *Boaler* 988!; Dodoma District: Chenene, 73.5 km. N. of Dodoma, 23 Jan. 1962, *Polhill & Paulo* 1246!
DISTR. T 1, 2, 4, 5; not known elsewhere
HAB. Sandy rocky soil in *Brachystegia* woodland and deciduous bushland; 1165–1550 m.

20. M. magnificum *E.A. Bruce* in K.B. 1940: 51 (1940); Bally, Gen. Mon.: 75 (1961). Type: Tanzania, Mpwapwa District, Gulwe valley, *B.D. Burtt* 6597 (K, holo.!)

Succulent glabrous shrub, with a large tuberous root; stems 1 to several, spreading or weakly erect to 1.5 m. high, sparsely branched, to 4 cm. thick at the base, ± 2 cm. thick above, 4–5-angled, with irregularly grouped ridges of closely spaced reddish brown spiny often branching teeth to 5 mm. long along the angles. Leaves sessile, reflexed, fleshy, obovate, to 15 × 10 cm., apex obtuse, margin toothed, midrib sharply keeled and toothed on the lower surface; leaf-scars to ± 6 mm. wide flanked by branching stipular spines ± 1.5 mm. long. Cymes on peduncles to 14 cm. long, 3–5-angled, with scattered fleshy teeth along the angles, 4–8-forked, with cyme-branches to 5 cm. long, progressively shorter above, the whole inflorescence bright red; bract-cup ± 5 × 9 mm., notched to halfway between obtuse apices and midribs sharply keeled and toothed or crenulate. Cyathia ± 4.5 × 3.5 mm., with barrel-shaped involucres narrowed at the top; glandular rim 2.5 mm. high; lobes subquadrate, ± 2 × 1.5 mm. Male flowers: bracteoles laciniate, minutely plumose; stamens 3.5 mm. long. Female flower: styles 1.3 mm. long, joined at the base, spreading, with thickened deeply bifid apices. Capsule exserted on a pedicel to 8 mm. long, obtusely 3-lobed, ± 5 × 4 mm., with a pair of narrow fleshy crenulate ridges along each angle. Seeds oblong, 4-angled, with truncate ends, 2.5 × 1.3 mm., greyish brown, minutely tuberculate; caruncle stipitate, pointed, 1 mm. in diameter. Fig. 103/4, 5, p. 550.

TANZANIA. Mpwapwa District: Gulwe, 26 Apr. 1932, *B.D. Burtt* 3914! & 7 June 1937, *Hornby* 849! & S. Mpwapwa, 3 Nov. 1942, *Greenway* 6654!
DISTR. T 5; known only from around Gulwe
HAB. Red soil in open *Commiphora* bushland; 825–1005 m.

NOTE. *Ward* P. 16 cited by Bruce, together with *Pole-Evans* 811 cited by Bally, belong to the following species, *M. spectabile.*

21. M. spectabile *S. Carter* in K.B.42:911(1987). Type: Tanzania, Iringa District, Iringa–Dodoma road above Great Ruaha R. bridge, *Greenway & Kanuri* 14522 (K, holo.!)

Sturdy succulent shrub, with an unbranched or few-branched stem, erect to 3 m. high; stem 3–5 cm. thick, obscurely 5-angled, with scattered spirally arranged projections or tubercles crowned by clusters of reddish brown usually branching spines 2–6 mm. long, surface pale green with a white waxy bloom. Leaves sessile, fleshy, obovate, to 33 × 23 cm., apex rounded, margin entire or with a few small teeth towards the base, midrib keeled with scattered spines on the lower surface; basal leaves and those of young plants pubescent on both surfaces, upper leaves completely glabrous; leaf-scars to 1 cm. wide, flanked by small triangular stipular spines to 0.8 mm. long. Cymes on peduncles to 15 cm. long, 3–5-angled, ± densely covered with reddish brown often branching spines to 4 mm. long, 4–6-forked (or more), with cyme-branches to 6 cm. long progressively shorter above, spiny, the whole inflorescence brilliant red; bract-cup ± 6 × 8 mm., notched to ± halfway between obtuse apices and sharply keeled rigidly toothed midribs. Cyathia ± 6 × 4 mm., with barrel-shaped involucres narrowed at the top; glandular rim 2 mm. high; lobes subquadrate, ± 2.3 × 1.8 mm. Male flowers: bracteoles laciniate, minutely plumose; stamens 4 mm. long. Female flower: perianth 3-lobed with lobes 1 mm. long; styles 1.8 mm. long, joined at the base, spreading, with thickened rugulose bifid apices. Capsule exserted on a pedicel to 7 mm. long, obtusely 3-lobed, ± 5 × 5.5 mm., with a pair of crenulate ridges along each angle. Seeds oblong, 4-angled, with truncate ends, 2.8 × 1.5 mm., grey, tuberculate; caruncle stipitate, pointed, 1 mm. in diameter. Fig. 103/6, p. 550.

TANZANIA. Iringa District: Pawaga, May–June 1936, *Ward* P.16!; & 56 km. N. of Iringa, 17 Apr. 1962, *Polhill & Paulo* 1336! & 1336A!
DISTR. T7; known only from the Iringa Escarpment

HAB.   Sandy soil on rocky slopes in dense deciduous mixed woodland; 800–1065 m.

SYN.   [*M. magnificum* sensu E.A. Bruce in K.B. 1940: 51 (1940) quoad *Ward* P.16 & sensu Bally, Gen.
Mon.: 76 (1961) quoad *Ward* P.16 & *Pole Evans* 811, *non* E.A. Bruce sensu stricto]

22. M. arborescens *Bally* in Candollea 17: 25 (1959) & Gen. Mon.: 76 (1961). Type:
Tanzania, Kilosa District, Malolo, *Harris* in *Bally* E.236 (K, holo.!)

Sturdy succulent sparsely branched shrub, erect to 4.25 m. high; stem to 10 cm. thick,
obtusely 5-angled, with scattered shallow projections along the angles occasionally
crowned by a minute reddish brown spine, surface pale green, angles more pronounced
and spines larger on young plants. Leaves sessile, fleshy, obovate, to ± 25 × 15 cm., apex
obtuse, margin entire or with a few minute teeth at the base, midrib ± prominent on the
lower surface, lamina glabrous but leaves of young plants pubescent on the lower surface;
leaf-scars to 1 cm. wide flanked by minute triangular stipular spines. Cymes on peduncles
to 6.5 cm. long, 3–7-forked, with cyme-branches to 2 cm. long, progressively shorter
above; bract-cup ± 6 × 5 mm., notched almost to the base between obscurely toothed
apices and ± prominent midribs, pale green tinged with pink. Cyathia ± 5.5 × 4.5 mm., with
barrel-shaped involucres; glandular rim 3 mm. high; lobes subquadrate,± 1.5 × 1.5 mm.
Male flowers: bracteoles spathulate, laciniate, minutely plumose; stamens 4.5 mm. long.
Female flower: styles 2.5 mm. long, joined at the base, spreading, with thickened deeply
bifid apices. Capsule exserted on a pedicel to ± 5 mm. long, subglobose, ± 6.5 × 7 mm.,
surface rugose, shallowly grooved along each suture and angle. Seeds oblong, 4-angled,
truncate at each end, 2.6 × 1.6 mm., brown, tuberculate; caruncle pointed, stipitate, 1 mm.
in diameter.

TANZANIA.   Kilosa District: Malolo, E. side of gorge, 11 Sept. 1943, *Harris* cult. in *Bally* E.236!
DISTR.   T 6; known only from the one collection, but reported also from near Morogoro
HAB.   In *Euphorbia* forest with other succulents and coarse grass; 900 m.

23. M. invenustum *N.E.Br.* in K.B. 1909: 329 (1909) & in F.T.A. 6(1): 459 (1911); W.F.K.:
36 (1948); Bally, Gen. Mon.: 50 (1961); U.K.W.F.: 225 (1974). Types: Kenya, Machakos
District, Kibwezi, *Kassner* 717 (BM, syn.!, K, drawing!) & Simba, *Kassner* 729 (BM, syn.!, K,
fragment!)

Very sparsely branched perennial herb, with a fleshy tuberous rootstock, minutely
puberulous in all parts; stems 1–2, fleshy, to 8 mm. thick, erect or subscandent to 80 cm.
Leaves ovate to subcircular or lanceolate, fleshy, apex apiculate, base tapering, margin
entire, often minutely sinuate, midrib prominent on the lower surface; stipules scale-like,
sharply 2–3-toothed, ± 1 mm. wide, brown. Cymes on peduncles to 8 mm. long often
flanked by a small toothed scale similar to the stipules, 1–2-forked; bract-cup ± 7 × 8 mm.,
shallowly notched between obtuse apices, green with darker reticulate veining, often
tinged pink. Cyathia ± 6 × 6 mm., with funnel-shaped involucres; glandular rim 2 mm.
high, spreading; lobes transversely oblong, ± 1 × 1.5 mm. Male flowers: bracteoles ligulate,
laciniate, pubescent; stamens 3 mm. long. Female flower: styles 1.2 mm. long, joined at the
base, with thickened rugulose bifid apices. Capsule shortly exserted on a reflexed
thickened pedicel ± 3.5 mm. long, oblong, obtusely 3-lobed, ± 5 × 5 mm., minutely
puberulous, with a pair of narrow ridges along each angle. Seeds oblong, base truncate,
3 × 2 mm., minutely tuberculate; caruncle 1 mm. in diameter.

var. **invenustum**

Stems green, to 80 cm. high, ± 8 mm. thick; leaves broadly ovate to subcircular, to 8 × 6.5 cm., base
tapering abruptly to a winged petiole ± 5 mm. long; bract-cup green. Fig. 102/6–8, p. 543.

KENYA.   Machakos District: 3 km. N. of Hunter's Lodge near Kiboko, 27 Dec. 1960, *Archer* 203! &
Kibwezi, 15 May 1938, *Bally* 592 in *E.A.H.* 7715!; Teita District: Tsavo National Park, W. of Lugard's
Falls, 12 Jan. 1967, *Greenway & Kanuri* 13044!
DISTR.   K 4, 7; not known elsewhere
HAB.   In red sandy soil in deciduous woodland and thicket; 60–900 m.

var. **angustum** *Bally* in Candollea 17: 35 (1959) & Gen. Mon: 50 (1961). Type: Kenya, Kitui District,
Yatta Gap at foot of Mutomo Hill, *Bally* 1693 (E.179) (K, holo.!, EA, iso.!)

Stems to 60 cm. high, ± 6 mm. thick, longitudinally streaked with dark green; leaves lanceolate, to 7
× 2 cm., subsessile; bract-cup greyish green, dark veined, flushed with pink.

KENYA. Kitui District: Yatta Gap at foot of Mutomo Hill, 28 Jan. 1942, *Bally* 1693 (E.179)!; Masai District: Nyeri Desert, NE. of Kilimanjaro, Jan. 1942, *Bally* E.45!
DISTR. **K** 4, 6; not known elsewhere
HAB. On red sandy soil in open bushland; 900–1000 m.

24. M. rubellum *(Bally) S. Carter* in K.B.42:913 (1987). Type: Kenya, Machakos District, Mbwinzao Hill, *Glover* in *Bally* E.390 (EA, holo.!, K, iso.!)

Glabrous perennial herb, with a fleshy rootstock forming clusters of rounded tubers 1.5–3 cm. thick; stems fleshy, 1–3 from each tuber, rarely branching, erect to ± 5 cm. or decumbent to ± 25 cm. long (more in cultivation), 4–5 mm. thick, longitudinally ridged and striped with purplish green. Leaves sessile, lanceolate, to 4.5 × 1 cm., apex acute, margin entire, lamina flushed with purple; stipules scale-like, acutely 1–4-toothed, ± 0.5 mm. across, brown. Cymes on peduncles to 1.7 cm. long, 1-forked; bract-cup exceeding the involucre, ± 6 × 4 mm., notched for ⅓ between acute apices and prominent midribs, rose-pink. Cyathia ± 4.5 × 4 mm., with barrel-shaped involucres; glandular rim 1.5 mm. high, red; lobes rounded, ± 1 × 1.3 mm. Male flowers: bracteoles few, laciniate, puberulous; stamens 2.6 mm. long. Female flower: styles 1 mm. long, joined to halfway, with thickened rugulose bifid apices. Capsule exserted on a pedicel 5 mm. long, acutely 3-lobed with truncate base, ± 4.5 × 4 mm. Seeds oblong, with truncate base, 2.5 × 1.3 mm., pale brown, surface roughened; caruncle stipitate, 1 mm. in diameter.

KENYA. Machakos District: Mbwinzao Hill, Feb. 1960, *Rauh* in *Bally* 12111!
DISTR. **K** 4; known only from the type locality
HAB. Amongst rocks, with *Xerophyta;* 1520 m.

SYN. *M. montanum* Bally var. *rubellum* Bally in Candollea 17: 28 (1959) & Gen. Mon.: 52 (1961)

NOTE. The longer decumbent glabrous stems and distinctive rose-pink cyathia, with the bract-cup far exceeding the involucre are the most obvious features which distinguish this plant from *M. montanum.*

25. M. stoloniferum *(Bally) S. Carter* in K.B.42:913 (1987). Type: Kenya, Masai District, Kapiti Plains, 33.5 km. S. of Athi R. Station, *Rauh* Ke.179 (G, holo., K, iso.!)

Glabrous perennial herb, with fleshy rounded tuberous roots to 2 cm. thick; stems fleshy, 1–3 from each tuber, ± 4 mm. thick, rarely branching, erect to 4 cm. or decumbent to 15 cm. long, rooting and forming further tubers. Leaves ovate, to 5 × 2.5 cm., base tapering abruptly, apex acute, margin entire, lamina usually purple tinged, with green veining; stipules filamentous to 1 mm. long, brown, usually deciduous. Cymes on peduncles to 1.5 cm. long, 1–2-forked; bract-cup ± equalling the involucre, ± 6 × 4 mm., notched for ± ⅓ between acute converging apices, dark green with prominent white midribs. Cyathia ± 5.5 × 3.5 mm., with barrel-shaped involucres; glandular rim 2 mm. high, yellow; lobes rounded, 1 × 1.2 mm. Male flowers: bracteoles laciniate; stamens 3.3 mm. long. Female flower: perianth obvious in fruit, of 3 3-toothed lobes and 4 mm. in diameter; styles 1 mm. long, joined at the base, with thickened rugulose bifid apices. Capsule exserted on a pedicel to 5 mm. long, oblong, acutely 3-lobed, with truncate base, 3.5 × 2.5 mm., and with a pair of narrow fleshy ridges along each angle. Seeds not seen.

KENYA. Masai District: 33.5 km. S. of Athi R. Station, 26 Jan. 1960, *Rauh* Ke.179! & 33–40 km. from Stony Athi Cement Factory, 10 Mar. 1960, *Bally* 12116!
DISTR. **K** 6; known only from the area of the type locality
HAB. Grassland with scattered *Acacia drepanolobium;* 1900 m.

SYN. *M. rhizophorum* Bally var. *stoloniferum* Bally, Gen. Mon.: 55 (1961)

NOTE. The different rootstock, slender decumbent stems and different leaves distinguish this as a species distinct from *M. rhizophorum.*

26. M. montanum *Bally* in Candollea 17: 28 (1959) & Gen. Mon.: 51 (1961); U.K.W.F.: 225 (1974). Type: Tanzania, Masai District, W. slope of Mt. Longido, *Bally* E.62 (EA, holo.!, K, iso.!)

Minutely scabridulous or occasionally glabrous perennial herb, with a tuberous rootstock to 2 cm. thick; stems 1–3, fleshy, rarely branching, erect to 15 cm. high or occasionally decumbent, to 5 mm. thick. Leaves obovate to 4.5 × 2.5 cm., apex acute, margin entire, lamina purplish mottled; stipules scale-like, minute, 1–2-toothed, 0.5 mm. across, brown. Cymes on peduncles to 3.5 cm. long, 1–2-forked; bract-cup ± equalling the involucre ± 4.5 × 4 mm., notched for ⅓ between acute apices, whitish green or pink-tinged with prominent white midribs. Cyathia ± 4.5 × 3.5 mm., with barrel-shaped involucres; glandular rim 1.5 mm. high; lobes rounded, ± 1 × 1.3 mm. Male flowers: bracteoles laciniate, pubescent; stamens 2.6 mm. long. Female flower: perianth of 3 2-toothed lobes and 2.5 mm. in diameter; styles 1 mm. long, joined at the base, with thickened rugulose bifid apices. Capsule exserted on a pedicel to 5 mm. long, oblong, acutely 3-lobed, with truncate base, 5 × 4.2 mm. Seeds oblong, obscurely 4-angled, 3 × 1.5 mm., pale brown, smooth with shallow depressions; caruncle 1 mm. in diameter.

KENYA. Masai District: 21 km. from Ol Tukai to Namanga, 14 Dec. 1959, *Verdcourt* 2578!
TANZANIA. Masai District, 48 km. N. of Arusha, Jan. 1966, *Beesley* 206! & Apr. 1967, *Beesley* CH.2!
DISTR. **K** 6; **T** 2; not known elsewhere
HAB. Grassland with scattered *Acacia* bushland; 1000–1500 m.

VAR. The two *Beesley* collections appear to be glabrous, but differ in no other way from other specimens.

27. M. trinerve *Bally* in Candollea 17: 33 (1959) & Gen. Mon.: 52 (1961); U.K.W.F.: 225 (1974). Type: Kenya, Machakos District, Emali Station, *MacArthur in Bally* E.109 (EA, holo.!, K, iso.!)

Glabrous, minutely papillose perennial herb with a tuberous root ± 5 cm. thick; stems 1–2, fleshy, erect to 15 cm. high, ± 7 mm. thick. Leaves very fleshy, obovate, to 6 × 2.5 cm., apex acute, margin entire, often undulate; stipules scale-like to 1 mm. wide, bearing a filament 1–2 mm. long which occasionally hardens into a minute prickle. Cymes on peduncles to 2 cm. long, 1–2-forked; bract-cup longer than the involucre, ± 7 × 5 mm., notched for ¼ between acute apices, green, often tinged purplish, with prominent whitish midribs. Cyathia ± 5 × 3.2 mm., with barrel-shaped involucres; glandular rim 1.8 mm. high; lobes rounded, ± 1 × 1.3 mm. Male flowers: bracteoles few, laciniate, minutely puberulous; stamens 2.8 mm. long. Female flower: perianth of 3 2-toothed lobes and ± 4 mm. in diameter; styles 1.2 mm. long, joined at the base, with thickened rugulose deeply bifid apices. Capsule oblong, obtusely 3-lobed, with truncate base, ± 5.5 × 4.8 mm. Seeds oblong, obscurely 4-angled, base truncate, 3.5 × 2 mm., pale brown, shallowly and sparsely tuberculate; caruncle stipitate, pointed.

KENYA. Kitui District: Emberre, Itabua, 26 Sept. 1932, *M.D. Graham* in *A.D.* 2241! & junction of Ginia and Thika rivers, 29 May 1960, *Archer* 107!; Masai District: Emali Hill, 19 Dec. 1971, *Faden & Holland* 71/903!
DISTR. **K** 4, 6; not known elsewhere
HAB. Wooded grassland; 1150–1400 m.

SYN. [*M. rhizophorum* sensu Agnew, U.K.W.F.: 225 (1974) quoad *Graham* in *A.D.* 2241, *non* Bally]

NOTE. This species appears to be very similar to *M. montanum*, but herbarium specimens are rather larger and not obviously scabridulous, with *Faden & Holland* 71/903 intermediate between the two. Until more is known of both in habitat, it seems best to continue regarding them as distinct species.

28. M. rhizophorum *Bally* in Candollea 17: 32 (1959) & Gen. Mon.: 54 (1961). Type: Kenya, Machakos District, Athi River station, *McArthur in Bally* E.117 (EA, holo.!, K, iso.!)

Succulent glabrous perennial herb, with a fleshy rhizomatous rootstock ± 1 cm. thick; stems unbranched, erect to 10 cm. high, ± 7 mm. thick. Leaves sessile on slightly projecting spirally arranged tubercles 5–15 mm. apart, obovate, to 4 × 2.2 cm., apex obtuse, margin entire, crisped, lamina sometimes deeply flushed with purplish brown; stipules evident as minute prickles 0.5 mm. long. Cymes on peduncles to 1.5 cm. long, arising immediately above the leaf-scar, 1-forked, or cyathia solitary; bract-cup ± equalling the involucre, ± 6.5 × 5.5 mm., shallowly notched between acute apices, dark green with prominent white midribs. Cyathia ± 5.5 × 4.5 mm., with cup-shaped involucres; glandular rim 2.5 mm. high, yellow; lobes rounded, 1.2 × 1.5 mm. Male flowers: bracteoles laciniate, puberulous; stamens 3.2 mm. long. Female flower: styles 1.8 mm. long, joined at the base, with thickened rugose bifid apices. Capsule exserted on a pedicel 5 mm. long, acutely 3-lobed, with truncate base ± 5 × 5 mm. Seeds not seen.

KENYA. Machakos District, Athi River station by Liebig's factory, 26 Mar. 1940, *MacArthur in Bally* E.117!
DISTR. **K** 4; known only from the type collection
HAB. Grassland, with black rock-strewn soil; 1650 m.

NOTE. The type collection has been widely distributed in cultivation.

29. M. yattanum *Bally* in Candollea 17: 34 (1959) & Gen. Mon.: 59 (1961); U.K.W.F.: 225 (1974). Type: Kenya, Machakos District, Makipenzi, *Bally* 9750 (EA, holo., K, iso.!)

Glabrous succulent perennial herb, with a fleshy tuberous rootstock ± 5 cm. thick; stems numerous, clustered, rarely branched, erect to 10 cm. or decumbent to 15 cm. long, 1–1.5 cm. thick, tessellated with scarcely projecting elongated tubercles 5–25 mm. apart in 3 loosely spiralled series bearing the leaves at the apices. Leaves sessile, oblanceolate, to 8 × 2.5 cm., apex obtuse and apiculate, margin entire, crisped towards the apex; stipules evident as minute prickles 0.5 mm. long. Cymes on peduncles to 2.5 cm. long arising immediately above the leaf-scar, 1-forked; bract-cup ± equalling the involucre, ± 6 × 5 mm., shallowly notched between acute converging apices, dark green with white sharply prominent midribs. Cyathia ± 5 × 4.5 mm., with cup-shaped involucres; glandular rim 2 mm. high, yellow; lobes transversely elliptic, ± 1 × 1.5 mm. Male flowers: bracteoles laciniate, puberulous; stamens 3.3 mm. long. Female flower: styles 1.5 mm. long, joined at the base, with thickened rugulose bifid apices. Capsule exserted on a pedicel to 8 mm. long, acutely 3-lobed, with truncate base, 7 × 6 mm., and with a pair of narrow fleshy ridges along each angle. Seeds oblong, 4-angled, with truncate base, 4 × 2 mm., brown, shallowly tuberculate; caruncle stipitate, pointed, 1 mm. in diameter.

KENYA. Machakos District: Makipenzi, 9 June 1954, *Bally* 9750! & Kapiti Plains at junction of Machakos road with Nairobi–Mombasa road, 4 Mar. 1960, *Bally* 12110!
DISTR. K 4; known only from Machakos District
HAB. Lightly wooded grassland among rocks in black cotton soil; 1175–1200 m.

30. M. gladiatum *(Bally) S. Carter* in K.B.42:913 (1987). Type: Tanzania, Masai District, 24 km. from Loliondo on road towards Ikoma, *Wakeford* in *Bally* E.329 (EA, holo., K, iso.!)

Succulent perennial herb, with a large fleshy branching rootstock; stems numerous, clustered, branching from the base, erect to 15 cm. or decumbent to 25 cm. long, to 1.5 cm. thick, tessellated, with elongated upward and outward curving tubercles (0.5–)1–4 cm. apart in 3–5 loosely spiralled series bearing the leaves at the apices. Leaves sessile, obovate, to 5 × 1.5 cm., base tapering gradually, apex obtuse, apiculate, margin entire, crenulate or irregularly toothed, midrib rounded, crenulate or occasionally toothed on the lower surface, lamina sometimes flushed with purplish red, young leaves often minutely puberulous; stipules evident as soft, non-persistent prickles to 1 mm. long, with occasionally a third prickle between them below the leaf-scar. Cymes on peduncles ± 2 cm. long arising 3–5 mm. above the leaf-scar, 1-forked, the whole cyme usually minutely puberulous; bract-cup exceeding the involucre, ± 7 × 5.5 mm., notched for ¼ between acute apices and sharply prominent to conspicuously crested midribs, pale greenish white or often flushed with purplish pink. Cyathia ± 5.5 × 5 mm., with cup-shaped involucres; glandular rim 2.5 mm. high, dull red; lobes rounded, ± 1.5 × 1.5 mm. Male flowers: bracteoles laciniate, puberulous; stamens 4 mm. long. Female flower: perianth obvious in fruit as 3 toothed lobes and to 4.5 mm. in diameter; styles slender, 2.8 mm. long, joined for ⅓, with thickened rugulose deeply bifid apices. Capsule exserted on a puberulous pedicel to 6 mm. long, obtusely 3-lobed with truncate base, ± 5 × 5.5 mm., minutely puberulous, and with a pair of narrow fleshy crested ridges along each angle. Seeds oblong, obtusely 4-angled, with truncate base, 3 × 2 mm., grey, areolate; caruncle stipitate, domed, 1 mm. in diameter.

TANZANIA. Musoma District: SW. Beacon Area, 27 Mar. 1962, *Greenway, Turner & Harvey* 10558!; Masai District: Lake Lgarya [Lagardia], 24 Apr. 1965, *Richards* 20251! & Soitaya, 16 Feb. 1968, *Greenway, Kanuri & S. Field* 13195!
DISTR. T 1, 2; not known elsewhere
HAB. In grassland on dark clay soil; 1500–2000 m.

SYN. *M. yattanum* Bally var. *gladiatum* Bally in Candollea 17: 34 (1959) & Gen. Mon.: 60 (1961)

VAR. Pubescence of young leaves and the whole of the inflorescence including the capsule, varies from plant to plant. The degree to which the leaf margin is toothed and the midrib prominent, can also vary on one individual; and the midribs of the bract-cup may be smooth or sharply keeled and crested on inflorescences of a single stem.

31. M. stapelioides *Pax* in E.J. 43: 89 (1909); N.E.Br. in F.T.A. 6(1): 454 (1911); W.F.K.: 36 (1948); Bally, Gen. Mon.: 57 (1961); U.K.W.F.: 225 (1974). Type: Tanzania, Masai District, Lamuniane Mt., *Jaeger* 359 (B, holo. †, K, drawing of holo.!)

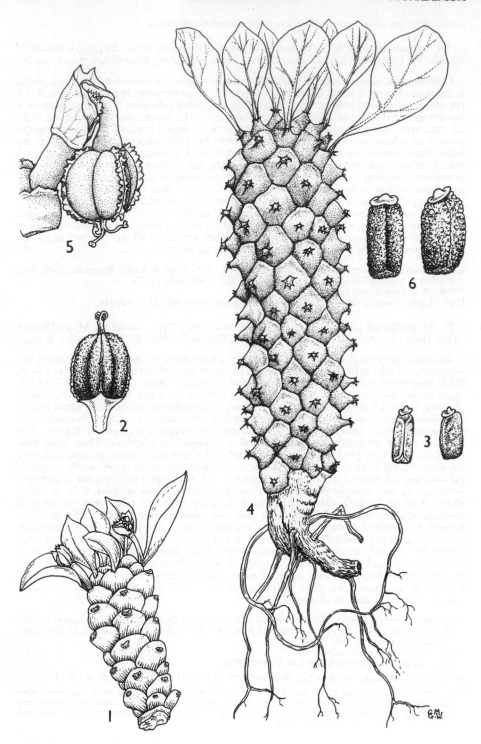

FIG. 104. *MONADENIUM STAPELIOIDES* var. *STAPELIOIDES* — 1, flowering branch, × ⅔; 2, capsule, × 4; 3, seeds, × 6. *M. SCHUBEI* — 4, rooted branch, × ⅔; 5, cyathium, with capsule, × 4; 6, seeds, × 6. 1–3, from *Bally* in C.M. 5605; 4–6, from *Harris* in *Bally* E.235.

Glabrous succulent perennial herb, with a large fleshy rootstock; stems numerous, dark green, branching mostly from the base to form a domed cushion to 60 cm. in diameter, erect or decumbent to 30 cm. long, 1–2 cm. thick, tessellated with ± prominent upward-pointing tubercles, 3–10 × 5–8 mm. at the base, bearing the leaves at the apices 1–4 mm. from the axils, in 5 or 8 tightly to ± loosely spiralled series. Leaves very fleshy, obovate to oblanceolate, to 5 × 1.5 cm., apex acute, apiculate, margin entire or minutely crenulate, lamina dark green, often marbled with purple; stipules rarely evident as minute prickles on young growth around the calloused leaf-scar. Cymes on peduncles to 1.5 cm. long, 1-forked, arising in the axils of the tubercles; bract-cup distinctly longer than the involucre, ± 7 × 6 mm., deeply notched between acute apices and scarcely prominent midribs, greenish white flushed with pink. Cyathia ± 5.5 × 4.5 mm., with cup-shaped involucres; glandular rim 2.5 mm. high, edged with dull red; lobes rounded, ± 1.2 × 1.2 mm. Male flowers: bracteoles laciniate, puberulous; stamens 4.3 mm. long. Female flower: perianth evident as 3 sometimes toothed lobes and ± 3.5 mm. in diameter; styles 2.5 mm. long, with thickened rugulose distinctly bifid apices. Capsule exserted on a stout pedicel ± 5 mm. long, sharply 3-lobed with truncate base, ± 5 × 5 mm., minutely papillose and with a pair of narrow fleshy toothed ridges along each angle. Seeds oblong, 4-angled, with truncate base, 3.5 × 2.8 mm., pale grey, minutely and shallowly tuberculate; caruncle domed, 1 mm. in diameter.

var. stapelioides

Stems erect or decumbent to 30 cm. long and 1.5 cm. thick, with tubercles to 1.5 × 1.3 cm. in usually 5 ± loosely spiralled series, tightly spiralled only on dwarfed growth in exposed situations; fleshy ridges on the capsule very narrow and shallowly toothed. Fig. 104/1–3.

KENYA. Laikipia District: S. bank of Ngobit R., 17 Jan. 1971, *Gillett* 19239!; Kiambu District: top of Kedong Escarpment, Aug. 1932, *Gardner* in *F.D.* 2900!; Masai District: Kajiado road, Nov. 1932, *V.G. van Someren* 2336!

TANZANIA. Arusha District: Songe Hill, 23 Feb. 1969, *Richards* 24193!

DISTR. **K** 1–4, 6; **T** 2; not known elsewhere

HAB. Amongst grass on rocky ground with open *Acacia* bushland; 1525–2125 m.

SYN. *M. succulentum* Schweik. in K.B. 1935: 274 (1935). Type: Kenya, Laikipia District, Ngobit R., *Gardner* in *F.D* 1478 (K, holo.!)
[*M. rhizophorum* sensu Agnew, U.K.W.F.: 225 (1974), quoad *Agnew* 10847, *non* Bally]

VAR. A variable species from small compact plants growing in exposed places, to plants with longer stems and larger cyathia growing in sheltered positions. *Joy Adamson* in *Bally* E.337 from Ol Asayeti, W. of Ngong Hills, has pubescent leaves and cyathia and is possibly distinct.

var. congestum *(Bally)* S. *Carter* in K.B.42:914 (1987). Type: Tanzania, Masai District, between Naidigidigo and Loliondo, *Bally* E.333 (EA, holo., K, iso.!)

Stems erect to 15 cm. high, rarely trailing to 30 cm. long, 2 cm. thick with crowded tubercles to 6 × 7 mm. in usually 8 tightly spiralled series; fleshy ridges on the capsule prominent and deeply toothed.

UGANDA. Karamoja District: Mt. Moroto, 23 May 1963, *J.G. Williams!*

KENYA. Masai District: Entasekera Escarpment, 12 July 1961, *Glover, Gwynne, Samuel & Tucker* 2271!

TANZANIA. Masai District: Loliondo Mt., 7 Nov. 1968, *Carmichael* 1592!

DISTR. **U** 1; **K** 6; **T** 2; not known elsewhere

HAB. Rocky ground of montane grassland with sparse deciduous bushland; 1425–2500 m.

SYN. *M. stapelioides* Pax forma *congestum* Bally in Candollea 17: 36 (1959) & Gen. Mon.: 59 (1961)

NOTE. *Wilson* 401 in *Bally* 11987 from Uganda, Karamoja District, Loro has trailing stems, but it is not clear from Bally's notes (unpublished) whether these were produced in habitat or only after some time in cultivation. In all other specimens of this variety, stems are usually ± 10 cm. long, and only occasionally to 15 cm.

32. M. ritchiei *Bally* in Candollea 17: 32 (1959) & Gen. Mon: 61 (1961); S. Carter in K.B. 42:914 (1987). Type: Kenya, Meru District, 12 km. SW. of Isiolo, *Ritchie* in *Bally* E.81 (EA, holo.!)

Succulent perennial herb, with a thick fleshy rootstock; stems erect, decumbent or rhizomatous to 40 cm. long, 1.5–3 cm. thick, tessellated, with prominent conical tubercles bearing the leaves in usually 5 spiralled series. Leaves sessile, broadly obovate, to 3 × 2.5 cm., apex obtuse, apiculate, margin entire and minutely crisped; stipules evident as 3–5 very slender prickles surrounding the leaf-scar, each to 2.5 mm. long but soon breaking off. Cymes on stout peduncles to 8 mm. long, 1-forked but often with only one branch developing, arising in the axils of the tubercles and occasionally flanked by a minute prickle; bract-cup slightly longer than the involucre, ± 7 × 6 mm., shallowly notched between acute apices and ± prominent midribs. Cyathia ± 6 × 5 mm., with cup-shaped involucres; glandular rim 2.5 mm. high; lobes rounded, ± 1.2 × 1.2 mm. Male flowers: bracteoles laciniate, puberulous; stamens 3.3 mm. long. Female flower: perianth evident in fruit as a 3-lobed rim, to 3 mm. in diameter; styles 1.5 mm. long, joined at the base, with thickened rugulose bifid apices. Capsule exserted on a stout pedicel to 6 mm. long, acutely 3-lobed with truncate base, ± 4.5 × 4.5 mm., and with a pair of narrow fleshy crenulate ridges along the angles. Seeds oblong, obtusely 4-angled, with truncate base, 3 × 1.75 mm., grey, minutely tuberculate; caruncle domed, 1 mm. in diameter.

KEY TO INFRASPECIFIC VARIANTS

1. Leaves glabrous; bract-cup entirely bright pink   . . .   b. subsp. **nyambense**
   Leaves puberulous; bract-cup greyish green, sometimes
   flushed with pink   2
2. Stems decumbent from the rootstock, to 3 cm. thick, pale
   green; tubercles in loosely spiralled series   . . . .   a. subsp. **ritchiei**
   Stems decumbent and rhizomatous, to 2 cm. thick, dark
   green; tubercles in tightly spiralled series   . . . .   c. subsp. **marsabitense**

a. subsp. **ritchiei**

Stems pale green, rarely branching, erect or decumbent to 40 cm. long and to 3 cm. thick; tubercles upward-pointing to 2 × 2 cm. at the base and to 8 mm. high, in fairly loosely spiralled series. Leaves puberulous. Cymes with usually only one branch developing. Bract-cup pale green flushed with pink, and glandular rim of the involucre bright red.

KENYA. Meru District: 12 km. SW. of Isiolo, Sept. 1939, *Ritchie* in *Bally* E.81! & 2 km. SW. of Isiolo, 22 July 1954, *Dalton* in *Bally* E.81a!
DISTR.   **K** 4; known only from SW. of Isiolo
HAB.   Steep stony slopes; ± 1300 m.

NOTE.   The *Archer* gathering from Marsabit, cited by Bally in his book, is distinct, see subsp. *marsabitense.*

b. subsp. **nyambense** *S. Carter* in K.B. 42:915 (1987). Type: Kenya, Meru District, Lagadema Hill, *Bally & Smith* 14719 (K, holo.!)

Stems dark green, numerous, branching above, erect to 15 cm. high and to 3 cm. thick; tubercles rounded to 1 × 1.5 cm. at the base and to ± 4 mm. high, in tightly spiralled series. Leaves glabrous. Cymes with usually both branches developing. Bract-cup and the whole cyathium bright pink, with a deep red margin to the glandular rim of the involucre.

KENYA. Meru District: NE. end of Nyambeni Hills, Lagadema Hill, 30 Dec. 1971, *Bally & Smith* 14719!
DISTR.   **K** 4; known only from the Nyambeni Hills on Lagadema Hill (only one herbarium specimen available) and Shaptiga
HAB.   In tussock grassland on rocky slopes; ± 1150 m.

c. subsp. **marsabitense** *S. Carter* in K.B. 42:915 (1987). Type: Kenya, Northern Frontier Province, Marsabit, *Lavranos & Bleck* 19627 (K, holo.!)

Stems darkish green, branching from the base, rhizomatous and decumbent to ± 25 cm. long and to 2 cm. thick; tubercles to 8 × 10 mm. at the base and to ± 3 mm. high, in tightly spiralled series. Leaves puberulous. Cymes with often only one branch developing. Bract-cup flushed with pink and glandular rim of the involucre red.

KENYA. Northern Frontier Province: Marsabit, Gof Redo, Mar. 1981, *Lavranos & Bleck* 19627!
DISTR.   **K** 1; known only from the type locality (only one herbarium specimen available)
HAB.   Slopes of volcanic crater amongst laval rock; c. 1200 m.

SYN. [*M. ritchiei* sensu Bally, Gen. Mon.: 62 (1961) quoad *Archer, non* Bally sensu stricto]

NOTE. The *Archer* specimen from Redo Gof, 6.5 km. from Marsabit on Moyale road was evidently in cultivation only and not preserved as a herbarium specimen.

33. M. heteropodum *(Pax) N.E.Br.* in F.T.A. 6(1): 453 (1911); T.T.C.L.: 220 (1949); Bally, Gen. Mon.: 62 (1961); S. Carter in K.B. 42: 915 (1987). Type: Tanzania, Lushoto District, Mbalu Mt., *Engler* 1472a (B, holo.†)

Succulent perennial herb, with a thick fleshy rootstock; stems numerous, bright green, branching from the base, erect or decumbent to ± 50 cm. long, 2–4.5 cm. thick, tessellated with prominent conical tubercles to 1.5 × 1.3 mm. at the base and to 7 mm. high, bearing the leaves in 5 or 8 tightly spiralled series. Leaves obovate to oblanceolate, to 6 × 2.5 cm., base tapering gradually, apex obtuse and apiculate, margin entire, crisped on young leaves; stipules evident as 3–5 prickles to 2 mm. long, from a horny pad surrounding the base of the leaf-scar. Cymes on stout peduncles ± 5 mm. long, 1–2-forked, arising in the axils of the tubercles; bract-cup slightly exceeding the involucre, ± 7 × 6 mm., shallowly notched between acute apices and prominent midribs, greenish white. Cyathia ± 6 × 5.5 mm., with cup-shaped involucres; glandular rim 3 mm. high edged with pink; lobes transversely elliptic, ± 1 × 1.5 mm. Male flowers: bracteoles few, laciniate, minutely puberulous; stamens 4.5 mm. long. Female flower: styles 2.5 mm. long, joined at the base, with thickened rugulose deeply bifid apices. Capsule exserted on a pedicel ± 5 mm. long, acutely 3-lobed, with truncate base, ± 5 × 6 mm., and with a pair of narrow fleshy toothed ridges along each angle. Seeds oblong, 4-angled, with truncate base, 3 × 2.3 mm., pale brown, minutely tuberculate; caruncle stipitate.

var. **heteropodum**

Stems to 35 cm. high when erect, otherwise decumbent, to 3 cm. thick, with upward pointing conical to depressed-conical tubercles in usually 5 series. Leaves glabrous or very minutely puberulous; stipular prickles usually 3, the apices soft, slender and eventually rubbing off the horny base. Bract-cup greenish white occasionally flushed with pink.

TANZANIA. Lushoto District: Lasa Mt., 6 Sept. 1935, *Greenway* 4047! & Mbalu Mt., 25 Jan. 1948, *Bally* 5766 (E.320)!
DISTR. **T** 3; known only from these 2 localities, with both the above collections distributed in cultivation
HAB. Amongst grass on lightly wooded rocky slopes; 975–1300 m.
SYN. *Euphorbia heteropodum* Pax in E.J. 34: 374 (1904)
      Monadenium guentheri Pax var. *mammillare* Bally in Candollea 17: 35 (1959) & Gen. Mon.: 63 (1961). Type: Tanzania, Lushoto District, Lasa Mt., *Greenway* 4047 (EA, holo., K, iso.!)

var. **formosum** *(Bally) S. Carter* in K.B. 42: 916 (1987). Type: Tanzania, Pare District, Makanya, *Bally* E.1 (EA, holo., K, iso.!)

Stems to 50 cm. high when erect, otherwise decumbent, to 4.5 cm. thick, with conical to cylindric-conical tubercles in usually 8 series. Leaves pubescent; stipular prickles 3–5, stout, persistent. Bract-cup white with faint green veining.

TANZANIA. Pare District: Makanya, Nov. 1934, *Bally* E.1!; Lushoto District: Mombo–Soni road, 24 June 1953, *Drummond & Hemsley* 2989!
DISTR. **T** 3; not known elsewhere
HAB. Steep rocky slopes with open scrub; 850–900 m.
SYN. *M. schubei* Pax var. *formosum* Bally in Candollea 17:36 (1959) & Gen. Mon.: 65 (1961)
      [*M. schubei* sensu Bally, Gen. Mon.: 65 (1961) quoad *Moore* in *Bally* E.128! & *Bally* E.406 (record only) & *Drummond & Hemsley* 2989!, *non* Pax]
NOTE. Var. *formosum* is more closely related to *M. heteropodum* from the same geographical region than to *M. schubei*, differing only in size and the number and prominence of tubercle spirals.

34. M. guentheri *Pax* in E.J. 43: 89 (1909); N.E. Br. in F.T.A. 6(1): 453 (1911); W.F.K.: 36 (1948); T.T.C.L.: 220 (1949); Bally, Gen. Mon.: 63 (1961). Type: Kenya, Teita District, Maktau near Bura Mt., *Uhlig* 48 (B, holo.†, K, drawing of holo.!, EA, iso.!)

Succulent perennial herb, with a thick fleshy rootstock; stems numerous, dark green, branching from the base, erect to ± 15 cm., or decumbent to 90 cm. long, 1.5–2 cm. thick, tessellated with prominent slightly upward pointing conical tubercles to 7 × 7 mm. at the base and ± 5 mm. high, bearing the leaves, in usually 8 tightly spiralled series. Leaves obovate, to 3 × 1.8 cm., apex obtuse, apiculate, margin entire ± crisped; stipules evident as 3 stout persistent prickles to 2 mm. long on a horny pad surrounding the leaf scar. Cymes on peduncles ± 4 mm. long, 1-forked, arising in the axils of the tubercles and usually flanked by a minute prickle; bract-cup slightly exceeding the involucre, ± 6 × 5 mm., shallowly notched between acute apices and prominent midribs, greenish white flushed with pink. Cyathia ± 5 × 4 mm., with cup-shaped involucres; glandular rim 2.5 mm. high, edged with red; lobes rounded, ± 1.25 × 1.5 mm. Male flowers: bracteoles laciniate, minutely puberulous; stamens 3.3 mm. long. Female flower: styles 2 mm. long with thickened rugulose deeply bifid apices. Capsule exserted on a pedicel 5 mm. long, acutely 3-lobed with truncate base, ± 5 × 5 mm., and with a pair of narrow fleshy toothed ridges along each angle. Seeds oblong, 4-angled, with truncate base, 2.8 × 1.5 mm., pale brown, minutely and shallowly tuberculate; caruncle stipitate.

KENYA. Teita District: Maktau, between station & Maktau Hill, Jan. 1940, *Bally* E.92! & foot of Maktau
    Hill, 17 Aug. 1969, *Bally* 13405!
DISTR. K 7; known only from the type locality; *Bally* E.92 is widespread in cultivation
HAB. Amongst grass in open scrubland; 900–1000 m.

35. M. renneyi *S. Carter* in K.B. 42: 916 (1987). Type: Kenya, Meru District, Meru Game Reserve, *Gillett* 16994 (K, holo.!, EA, iso.)

Succulent perennial herb, with fleshy roots; stems slender, subscandent to 50 cm. long, and ± 2 cm. thick, sparsely branching, with the branches easily breaking off and then quickly rooting, tessellated with very prominent cylindrical tubercles to 1.5 × 1.5 cm. at the rounded flattened base and to 7 mm. high in usually 5 fairly loosely spiralled series and bearing the leaves at the apices. Leaves broadly obovate, to 7 × 4.5 cm., base tapering to a winged petiole to 1 cm. long, apex obtuse-apiculate, margin entire, crisped, lamina often flushed and edged with red, puberulous; stipules evident as a cluster of 5 strong slender spines, each 2–3 mm. long from a horny pad, surrounding the leaf-scar. Cymes on stout peduncles to 1 cm. long, 1–3-forked, arising in the axils of the tubercles and often flanked by a minute prickle; bract-cup equalling the involucre, ± 6 × 7 mm., shallowly notched between acute apices and ± prominent midribs, yellowish green with dark veining. Cyathia ± 4.5 × 4 mm., with barrel-shaped involucres; glandular rim 2.5 mm. high edged with pale pink; lobes rounded, ± 1.2 × 1.2 mm. Male flowers: bracteoles laciniate; stamens 3.5 mm. long. Female flower: perianth evident in fruit as a 3-lobed rim 3 mm. in diameter; styles 1.5 mm. long, joined at the base, with thickened rugulose bifid apices. Capsule exserted on a stout pedicel to 1 cm. long, acutely 3-lobed, with truncate base, ± 4.5 × 4.5 mm., and with a pair of narrow fleshy many-toothed ridges along the angles. Seeds oblong, obtusely 4-angled, with truncate base, 3 × 1.7 mm., pale greyish brown, minutely tuberculate; caruncle domed.

KENYA. Meru District: Meru Game Reserve, 1–2 km. W. of Tana-Rojewero confluence, 3 Jan. 1966,
    Gillett 16994! & Tana-Rojewero confluence, 1975, *Renney!*
DISTR. K 1 (Kora Game Reserve), 4; not known elsewhere
HAB. Rocky slopes near river in *Acacia-Commiphora* bushland; 400 m.

36. M. schubei *(Pax) N.E.Br.* in F.T.A. 6(1): 453 (1911); T.T.C.L.: 220 (1949); Bally, Gen. Mon.: 64 (1961). Type: Tanzania, Iringa District, Geme by Lukose R., *Goetze* 485 (B, holo.†)

Succulent perennial herb, with fleshy roots; stems stout, dark bluish green, branching mostly from the base, erect to 90 cm. or decumbent to 1.25 m. long, 3–5 cm. thick, tessellated with very prominent cylindric-conical tubercles bearing the leaves, to 1.5 × 1.5 cm. at the base and to 1 cm. high, in usually 8 tightly spiralled series. Leaves obovate, to 8 × 5 cm., usually much less, base tapering to a winged petiole to 1.5 cm. long, apex obtuse and apiculate, margin entire, crisped on young leaves, lamina puberulous, sometimes edged and streaked with red; stipules evident as a cluster of usually 5 persistent prickles, each to 2 mm. long, from a horny pad surrounding the base of the leaf-scar. Cymes on stout peduncles to 6 mm. long, 1–2-forked, arising in the axils of the tubercles and occasionally flanked by a minute prickle; bract-cup slightly exceeding the involucre, ± 7 × 6 mm., shallowly notched between acute apices and sharply prominent midribs, greyish green often flushed with pink. Cyathia ± 5 × 4.5 mm., with cup-shaped involucres; glandular rim 2.2 mm. high, deep pink; lobes rounded, ± 1 × 1 mm. Male flowers: bracteoles laciniate, puberulous; stamens 2.8 mm. long. Female flower: perianth evident as a 3-lobed rim below the capsule, ± 3 mm. in diameter; styles 2 mm. long, joined at the base, with thickened rugulose bifid apices. Capsule exserted on a stout pedicel 4 mm. long, acutely

3-lobed with truncate base, ± 6 × 8 mm., and with a pair of narrow fleshy crenulate ridges along the angles. Seeds oblong, 4-angled, with truncate base, 3.2 × 2 mm., brown, minutely tuberculate; caruncle stipitate. Fig. 104/4–6, p. 558.

TANZANIA. Maswa District; Shanwa near the Simiyu R., June 1935, *B.D. Burtt* 5194!; Dodoma District: Chenene, 22 Jan. 1962, *Polhill & Paulo* 1244A!; Iringa District: Great Ruaha R., Nyamakuyu Rapids, 6 Apr. 1970, *Greenway & Kanuri* 14283!
DISTR. T 1, 4, 5, 7; not known elsewhere
HAB. Amongst rocks in grassland with open deciduous woodland; 800–1500 m.

SYN. *Euphorbia schubei* Pax in E.J. 34: 373 (1904)

NOTE. The specimens cited by Bally from the S. Pare and Usambara Mts. are of *M. heteropodum*. The apparently disjunct population from NE. Zimbabwe has larger yellowish cyathia with a yellow gland and a larger capsule. It is not this species but is more closely related to *M. lugardae* N.E.Br.

37. M. reflexum *Chiov.* in Webbia 8: 235 (1951); E.P.A.: 464 (1958); Bally, Gen. Mon.: 70 (1961). Type: Ethiopia, Sidamo, between Filtu and Neghelli, *Vatova* 845 (FT, holo., K, fragment of holo.!)

Succulent perennial herb, with a relatively small fleshy rootstock; stem simple or rarely few-branched, erect to 75 cm., cylindrical to 6 cm. thick, tessellated, with 10–15 spirally arranged series of closely set reflexed tapering elongated tubercles, each 5–20 mm. long, with a groove on the upper side and bearing a leaf at the tip. Leaves obovate, to 2.5 × 1.2 mm., apex obtuse, minutely pubescent, lamina often streaked with red; stipules evident as a pair of sharply 2–5-toothed flattened brown scales 1–3 mm. wide flanking the leaf-scar. Cymes arising from near the base of the tubercles in the groove, often flanked by a small irregularly placed toothed scale similar to the stipules; peduncles 1–2.5 cm. long, 1–2-forked, with the whole cyme yellowish green, sometimes tinged reddish; bract-cup ± 4 × 3.5 mm., shortly notched between the obtuse apiculate apices and ridged midribs, minutely puberulous. Cyathia ± 5 × 3.5 mm., with funnel-shaped involucres; glandular rim 2.5 mm. high; lobes rounded, ± 1 × 1 mm. Male flowers: bracteoles linear, puberulous; stamens 3.5 mm. long. Female flower: styles 1.5 mm. long, joined at the base, with thickened rugulose bifid apices. Capsule exserted on a reflexed pedicel to 6 mm. long, oblong, obtusely 3-lobed, ± 4 × 3 mm., minutely puberulous. Seeds oblong, base truncate, 4-angled, 3 × 1.2 mm.; caruncle stipitate, 1 mm. in diameter.

KENYA. Northern Frontier Province: Malka Murri near the Daua R., June 1951, *J.G. Williams* in Bally 8408 (E.383)! & *J. Adamson* in Bally 11631!
DISTR. K 1; known only from near the Ethiopian border
HAB. In sandy soil on rocky slopes near the river or dry stream beds with sparse *Acacia-Commiphora* bushland; 600–950 m.

38. M. ellenbeckii *N.E.Br.* in F.T.A. 6(1): 454 (1911); E.P.A.: 464 (1958); Bally, Gen. Mon.: 67 (1961). Type: Ethiopia, Sidamo, Tarre–Gumbi, *Ellenbeck* 2102 (B, holo. †, K, drawing of holo.! & iso.!)

Succulent perennial herb, with a thick fleshy shortly rhizomatous rootstock; stems several, erect to 80 cm., sparsely branched, cylindrical, to 15 mm. thick, pale green, with 4 longitudinal grooves below each of the numerous spirally arranged leaf-scars. Leaves sessile, broadly ovate, to 10 × 8 mm., thick and fleshy, minutely puberulous, quickly deciduous; stipules evident as a pair of 2–4-toothed flattened brown scales ± 1 mm. wide. Cymes arising immediately above the leaf-scars on peduncles to 5 mm. long, 1–2-forked; bract-cup ± 3.5 × 3 mm., notched for ¹⁄₃ between obtuse apices and keeled midribs, greenish yellow. Cyathia ± 3.5 × 2.8 mm., with barrel-shaped involucres; glandular rim 2 mm. high, yellow; lobes rounded, 1 × 1 mm. Male flowers: bracteoles linear, plumose; stamens 2.3 mm. long. Female flower: styles 1 mm. long, joined at the base, with thickened rugulose bifid apices. Capsule exserted on a thick fleshy recurved pedicel to 4 mm. long, obtusely 3-lobed, ± 5 × 5 mm., minutely and densely papillose, with a pair of narrow crested ridges along each angle. Seeds oblong, obtusely 4-angled, truncate at the base, 3 × 1.5 mm., pale brown, minutely tuberculate; caruncle stipitate, 1 mm. in diameter.

KENYA. Northern Frontier Province: Huri Hills, Gabr Bori, 27 Feb. 1963, *Bally* 12544! & 4 km. WNW. of Marsabit, Gar Jirimi, 24 Nov. 1977, *Carter & Stannard* 689! & Mt. Kulal, near Gatab airstrip, 20 Nov. 1978, *Hepper & Jaeger* 6944!
DISTR. K 1, Ethiopia and northern Somalia
HAB. Rocky slopes with low open scrub and succulent species; 1200–1735 m.

NOTE. The prostrate unbranched forma *caulopodium* Bally occurs in southern Ethiopia north of Mega. Plants occurring in northern Somalia south of Berbera are much sturdier and taller, with smooth capsules.

39. M. virgatum *Bally*, Gen. Mon.: 65 (1961). Type: Kenya, Lamu District, N. of Mlango-ya-Simba, *Greenway & Rawlins* 9457 (K, holo.!)

Glabrous succulent perennial herb, with a thick fleshy rhizomatous rootstock; stems numerous, erect to 40 cm. or scandent to 1.8 m., rarely branched, cylindrical, to 7 mm. thick, glaucous, with 3 longitudinal ridges below each leaf-scar. Leaves sessile, fleshy, obovate, to 3 × 1.3 cm., margin entire, midrib keeled on the lower surface; stipules evident as minute, denticulate, blackish brown scales. Cymes on peduncles 5–8 mm. long, 1–2-forked; bracts almost free, rounded, each ± 6 × 6 mm., midribs keeled, greenish white occasionally tinged with red. Cyathia ± 7 × 3 mm., with ± tubular involucres; glandular rim 3 mm. high; lobes oblong, 2 × 1 mm., deeply divided into 4–6 segments. Male flowers few: bracteoles linear, pubescent; stamens 5 mm. long. Female flower: styles 3 mm. long, joined at the base, slender, with thickened rugulose bifid apices. Capsule subsessile, conical, ± 5 × 4.5 mm., obtusely 3-lobed, with 6 fleshy shortly acute projections at the apex surrounding the base of the styles. Seeds oblong, truncate at both ends, obtusely 4-angled, 2.5 × 1.5 mm., grey, tuberculate; caruncle stipitate, pointed.

KENYA. Kwale District: 53.5 km. from Mombasa on Voi road, 27 Mar. 1960, *Bally* 12196!; Lamu District: N. of Mlango-ya-Simba, 6 Nov. 1957, *Greenway & Rawlins* 9457!
DISTR. **K** 4, 7; known from **K** 4 as living material from Kitui District, NE. of Tsavo National Park, Galana Ranch, coll. *Heath!*
HAB. In low fairly dense bushland, with other succulents; 50–200 m.

NOTE. Relationships of this species are difficult to ascertain. Its densely tufted habit, tubular cyathia and strangely ornamented capsule are unlike any other species. One, or possibly two, undescribed apparently related species occur in north-east Somalia.

72. PEDILANTHUS

A. Poit. in Ann. Mus. Hist. Nat. Paris 19: 388 (1812); Boiss. in DC., Prodr. 15(2): 4 (1862); Dressler in Contrib. Gray Herb. 182 (1957), nom. conserv.

Shrubs or small trees with woody or fleshy branches and a milky latex, monoecious. Leaves shortly petiolate, entire, with small stipules. Inflorescence with cyathia in dichotomous axillary or terminal cymes; bracts paired, persistent. Involucres with 5 unequal lobes and 2, 4 or 6 glands enclosed within an adaxial spur-like extension of the involucre formed from 4 gland appendages, often brightly coloured. Male flowers in 5 groups, usually bracteolate. Female flower pedicellate, with the perianth reduced to a rim below the ovary; ovary 3-locular, with 1-pendulous ovule in each locule; styles 3, connate, with bifid stigmas. Fruit 3-lobed, usually a dehiscent capsule, occasionally indehiscent. Seeds smooth or tuberculate, without a caruncle.

A genus of 14 species in Central America, the West Indies and northern S. America.

CULTIVATED SPECIES

P. tithymaloides (L.) A. Poit. subsp. *smallii* (Millsp.) Dressler in Contr. Gray Herb. 182: 152 (1957). A succulent stemmed shrub 1–2 m. high, grown as an ornamental and often used as a hedging plant. Its readily deciduous ovate fleshy leaves are slightly glaucous and sometimes variegated with yellowish green or pink. Its inconspicuous cyathia are enclosed by bright red beak-shaped 'bracts' ± 1.5 cm. long in terminal and axillary cymose clusters. *P. tithymaloides* is the most commonly cultivated species and includes a number of other varieties. Plants have occasionally become naturalized in India, but apparently not, so far as is known, in East Africa. Reference specimen: Kenya, Nairobi, East African Herbarium grounds, 5 Aug. 1969, *Gillett* 18785!

73. BISCHOFIA

Bl., Bijdr. 17: 1168 (1826/7); Muell. Arg. in DC., Prodr. 15(2): 478 (1866), as '*Bischoffia*'; G.P. 3(1): 280 (1880); Pax in E.P. IV. 147 (15): 312 (1922)

Dioecious or rarely monoecious trees with a simple (localized) indumentum. Leaves alternate, long-petiolate, stipulate, 3(–5)-foliolate, the leaflets petiolulate, crenate-serrate, penninerved; stipules readily caducous. Inflorescences axillary, paniculate, solitary, many-flowered; bracts 1-flowered, those subtending ♀ flowers readily caducous. Male flowers pedicellate; sepals 5, imbricate, strongly concave, cucullate, later reflexed; petals 0; disc 0; stamens 5, episepalous, free, filaments short, anthers basifixed, bilobate, longitudinally dehiscent; pistillode peltate, infundibuliform, pentagonal, shortly stipitate. Female flowers pedicellate; sepals 5, imbricate, flat, not cucullate, soon caducous; petals 0; disc 0; staminodes minute or 0; ovary 3(–4)-locular, with 2 ovules per locule; styles 3(–4), slightly connate at the base, undivided, filiform, spreading, the adaxial stigmatic surface running the length of the arms. Fruit a small depressed-globose indehiscent drupe; mesocarp fleshy; endocarp thin, horny. Seeds 3–6, oblong, ecarunculate; testa horny; albumen fleshy; cotyledons broad, flat.

An E. Asiatic and Indo-Pacific genus with 2 species, one of which has been introduced and has become ± naturalized in East Africa.

The genus was formerly placed as the sole member of a subtribe of *Phyllantheae*, but held by Airy Shaw, in Willis, Dict. Fl. Pl. & Ferns, ed. 7 (1966), to have only an illusory connection with Euphorbiaceae, but rather to be related to Staphyleaceae (cf. especially *Tapiscia* Oliv.). However, it differs from the latter in being apetalous, in having no disc, in the few ovules and in the long styles. It is provisionally retained as an anomalous genus of the Euphorbiaceae.

1. **B. javanica** *Bl.*, Bijdr. 17: 1168 (1826/7); Muell. Arg. in DC., Prodr. 15(2): 478 (1866); Pax in E.P. IV. 147 (15): 313 (1922). Type: W. Java, *Blume* 154 (L, holo.!, BO, iso.)

Tree 4–15 m. tall, with a narrow crown and drooping branches; bark brown, slightly scaly; wood very soft. Twigs sparingly lenticellate; shoots, petioles and petiolules glabrous. Petioles 9–17 cm. long, pulvinate at the base; median petiolules 3.5–5 cm. long, laterals 0.5–2 cm. long; leaflets 3(–5), elliptic-ovate to elliptic-obovate, (5–)7–14 cm. long, 3–9 cm. wide, the laterals slightly asymmetrical and smaller than the terminal, acutely or subacutely and often abruptly acuminate at apex, cuneate or rounded at base, crenate-serrate, firmly membranaceous to chartaceous, lateral nerves 7–9(–11), not or scarcely prominent above, slightly so beneath, weakly brochidodromous and reticulate towards the margin, tertiary nerves reticulate, glabrous above, sparingly puberulous at least along the midrib and otherwise ± glabrous beneath, dark green and somewhat shiny above, paler and duller beneath. Stipules lanceolate to triangular-lanceolate, 0.7–1(–2) cm. long, acutely acuminate, entire, subglabrous, brown. Male panicles (9–)11–14(–20) cm. long; peduncles ± 2–2.5 cm. long; axis and rhachides sparingly minutely puberulous; bracts ovate, 1–2 mm. long, acute, brown. Female panicles somewhat larger (up to 27 cm. long), with the peduncles up to 7 cm. long. Male flowers: pedicels 2 mm. long, slender, articulate ⅓ up from the base, pubescent below the articulation, glabrous above it; sepals broadly elliptic, 2 mm. long, rounded, ciliolate, yellowish green; stamens 1.1 mm. long, anthers 0.9 mm. long and wide, yellow; pistillode 0.5 mm. high, 0.7 mm. across. Female flowers: pedicels (1–)2–4 mm. long, articulate, pubescent throughout, extending to 2 cm. long in fruit; sepals ovate-lanceolate to oblong-lanceolate, 1.5–2 mm. long, acute, subentire, sparingly puberulous without at the base, otherwise glabrous, pale green with white margins; ovary subglobose, 1–1.5 mm. in diameter, ± smooth, glabrous; style-arms (4–)5–6 mm. long, stigmas slightly papillose. Fruit 5–6 mm. in diameter, ± smooth. Seeds 3 mm. long, 1.5 mm. wide, smooth, somewhat shiny, yellowish brown.

KENYA. Nairobi District: Nairobi Arboretum, 17 Mar. 1952, *G.R. Williams* 371!; N. Kavirondo District:Kakamega Forest, 25–26 Dec. 1969, *Faden & Rathbun* 69/2089! & 10 Apr. 1973, *Hansen* 885!
DISTR. K 4, 5; native from S. India eastwards to S. China and the Cook Is., occasionally cultivated elsewhere
HAB. Cultivated in gardens and arboreta, becoming naturalized as an understorey tree in evergreen forest; 1550–1770 m.

74. UAPACA

Baill., Étud. Gén. Euph.: 595 (1858); Muell. Arg. in DC., Prodr. 15(2): 489 (1866); G.P. 3 (1): 282 (1880); Hutch. in F.T.A. 6(1): 634 (1912); Pax & K. Hoffm. in E.P. IV. 147 (15): 298 (1922); De Wild., Contrib. Étud. Esp. *Uapaca* (1936)

Dioecious pachycaul vernicifluous trees or shrubs with a simple and/or lepidote indumentum; trunks often with buttress-roots; twigs often strongly marked with leaf-scars. Leaves alternate, usually crowded towards the ends of the branches, petiolate or subsessile, stipulate or not, simple, entire, penninerved. Inflorescences axillary, solitary or fasciculate, pedunculate, with a whorl of 5–10 imbricate involucrate tepaloid bracts surrounding the flowers; ♂ inflorescences many-flowered, the flowers in dense globose capitula; ♀ inflorescences 1-flowered. Male flowers sessile; calyx campanulate or turbinate, truncate, dentate, irregularly split or regularly 5–more-lobed, the lobes imbricate; petals 0; disc 0; stamens (4–)5(–6), free, episepalous, anthers erect, basifixed, introrse, thecae parallel, longitudinally dehiscent; pistillode cylindric-obconic, and often infundibuliform, hypocrateriform, or pileiform and sometimes lobate. Female flowers sessile; calyx minute, truncate or sinuate, disc-like; petals 0; disc 0; ovary (2–)3(–5)-locular, with 2 ovules per locule; styles (2–)3(–5), free, thick, recurved, enveloping the ovary, multipartite or laciniate. Fruit drupaceous, indehiscent; mesocarp spongy; pyrenes (2–)3(–4), dorsally bisulcate, indurate. Seeds mostly 1 per pyrene, compressed, ecarunculate; endosperm fleshy; cotyledons broad, flat.

A genus of 61 species, of which 49 are restricted to tropical Africa, the remainder in Madagascar.

Uapaca was formerly placed as the sole member of a subtribe of *Phyllantheae*, but held by Airy Shaw, in Willis, Dict. Fl. Pl. & Ferns, ed. 7 (1966) to be aberrant on account of anatomical, vegetative and inflorescence characters, and to show an affinity with the Anacardiaceae, Pistaciaceae and Picrodendraceae. It is provisionally retained here as an anomalous genus of the Euphorbiaceae. The grotesque name, reminiscent of clearing the throat, was derived by Baillon from the Malagassy vernacular name for certain species of the genus ('Uapac', 'Voa-paca', 'Voa-paka', 'Paka').

1. A shrub or small tree up to 3.5 m. tall; leaves sessile, or
    petiole not more than 7 mm. long; leaf-blades
    pubescent or pilose on both surfaces; pistillode
    hypocrateriform; fruit broader than long . . . .   *1. U. pilosa*
  Trees commonly up to 12–27 m. tall; leaves petiolate, the
    petioles up to 13 cm. long; leaf-blades usually glabrous
    above; pistillode infundibuliform, turbinate or
    cylindric; fruit as broad as long, or longer than
    broad 2
2. Lateral nerves up to 24 pairs; leaf-blades glabrous above,
    floccose-pubescent beneath; ovary densely tomentose   *2. U. kirkiana*
  Lateral nerves not more than 20 pairs; leaf-blades
    glabrous, sparingly minutely lepidote or pubescent
    above and beneath at first, later glabrescent, or else
    pubescent only along the midrib and main nerves
    beneath 3
3. Leaf-blades elliptic-oblong to elliptic-oblanceolate, not
    more than 13 cm. long and 6 cm. wide, shiny above   *3. U. nitida*
  Leaf-blades oblanceolate to broadly obovate, up to 45 cm.
    long and 25 cm. wide, shiny or not 4
4. Midrib commonly distally zigzag; stipules 0; fruit not more
    than 2 cm. long . . . .   *4. U. sansibarica*
  Midrib ± straight; stipules present; fruit up to 2.5–3 cm.
    long 5
5. Leaf-blades up to 25 cm. long and 17 cm. wide; stipules
    soon caducous; ovary glabrous . . . .   *5. U. guineensis*
  Leaf-blades up to 50 cm. long and 25 cm. wide; stipules
    foliaceous, persistent; ovary densely pubescent   *6. U. paludosa*

1. U. pilosa *Hutch.* in F.T.A. 6(1): 635 (1912), V.L. 3(2): 37 (1921); Pan in E.P. IV. 147 (15): 301 (1922); De Wild., Contrib. Étud. Esp. *Uapaca:* 163 (1936); F.F.N.R.: 206 (1962). Type: Malawi, Stevenson Road, *Scott-Elliot* 8272 (K, holo.!)

A shrub or short-trunked spreading tree with an open crown, 2–3.5 m. tall; bark smooth or rough, longitudinally, quadrangularly or reticulately-fissured, grey. Branches gnarled, brittle; branchlets stout, pubescent. Leaves sessile or shortly petiolate; leaf-blades broadly obovate, up to 40 cm. long and 25 cm. wide, rounded at the apex, cuneate-attenuate at the base, thinly coriaceous, lateral nerves 10–15 pairs, camptodromous or weakly brochidodromous, not prominent above, fairly prominent beneath, tertiary nerves not parallel, pubescent or pilose above and beneath (especially along the midrib and main nerves) when young, later ± glabrescent above, mid green and slightly shiny above, glaucous beneath. Stipules linear to subulate, 0.5–1 cm. long, evenly pubescent, soon caducous. Male peduncle up to 6 cm., evenly pubescent; bracts 10, oblong, 1–1.5 cm. long, (3–)5–7 mm. wide, pubescent without, glabrous within, yellowish; head of ♂ flowers 0.7–1 cm. across. Male flowers: calyx-lobes 5–6, oblong-linear to setaceous, 1.5 mm. long, pubescent at the apex; stamens 5, filaments very short, anthers 1 mm. long, yellow; pistillode hypocrateriform, pubescent at the apex. Female peduncle and bracts ± as in ♂. Female flower: calyx-lobes broadly triangular, 1 mm. long and wide, pubescent at the apex; ovary ellipsoid-subglobose, 4–5 mm. long, 3–4 mm. across, glabrous; styles 3–5, 7–8 mm. long, antler-like. Fruits depressed-subglobose, scarcely 3–5-lobed, up to 3 cm. long and 4.5 cm. across, smooth, glabrous, apple-green, often with brownish markings; mesocarp ± 3 mm. thick. Pyrenes 3–5, up to 2.3 cm. long and 1.4 cm. wide, strongly carinate.

TANZANIA. Mbeya District: Mbozi, Isalalo, Nov. 1966, *Knight* 65!
DISTR. T 7; ? Cameroun, Zaire (Shaba), Malawi, Zambia
HAB. *Brachystegia* woodland, fringing forest; 1000–1700 m.

NOTE. The fruits are edible.
A petiolate variant, var. *petiolata* Duvign., occurs in Zambia.

2. **U. kirkiana** *Muell. Arg.* in Flora 1864: 517 (1864) & in DC., Prodr. 15(2): 491 (1866); P.O.A. C: 237 (1895); Hutch. in F.T.A. 6(1): 636 (1912); V.E. 3(2): 37 (1921); Pax in E.P. IV. 147 (15): 302 (1922); De Wild., Contrib. Étud. Esp. *Uapaca:* 135 (1936); T.T.C.L.: 228 (1949); F.F.N.R.: 207 (1962). Type: Malawi, Southern Province, Soche Mts., 8 Mar. 1862, *Kirk* (K, holo.!)

A much-branched evergreen tree with a short trunk and spreading branches or with a longer trunk (to 9 m. clear) and a dense rounded crown, 3–12 m. tall; bole 25–45 cm. in diameter; bark dark grey or grey-brown, vertically fissured. Branchlets stout, pubescent. Leaves subsessile or shortly petiolate, or with a stout glabrescent petiole up to 2(–3.5) cm. long; leaf-blades obovate to broadly obovate or subcircular-obovate, (7–)12–36 cm. long, (5–)8–24 cm. wide, rounded to somewhat retuse at apex, attenuate-cuneate to rounded-cuneate at base, margin occasionally somewhat undulate, thinly coriaceous, lateral nerves often arising almost at right-angles from the midrib, (12–)16–20(–24) pairs, not prominent or slightly impressed above, prominent beneath, almost craspedodromous, tertiary nerves subparallel, glabrous above, floccose-pubescent beneath, dull glossy green above, paler beneath, with the midrib and lateral nerves often yellow above. Stipules filiform, 3–4 mm. long, evenly pubescent, soon caducous. Inflorescences borne among the leaves, or more often below them on the 2nd or 3rd season's wood. Male peduncles fasciculate, 0.7–1 cm. long in bud, sometimes extending to 2 cm., glabrous, often with 1–2 small empty bracts along them; main inflorescence-bracts 5(–7), oblong-elliptic or broadly elliptic, 6–7 mm. long, 3–5 mm. wide, sparingly pubescent or subglabrous without, glabrous within, pale yellow-green; head of ♂ flowers 5–7 mm. across. Male flowers: calyx-lobes 5–7, triangular and laciniate or linear, 2 mm. long, subglabrous, greenish white; stamens 5, filaments 2.5 mm. long, white, anthers 1 mm. long, cream; pistillode compressed-turbinate, 1 mm. high, truncate, pubescent in the upper part. Female bracts often tinged pink, otherwise ± as in ♂. Female flower: calyx shallowly cupular, with 5–8 small teeth or lobes, glabrous; ovary ovoid-subglobose, 4-locular, 3–4 mm. long and wide, densely fulvous-tomentose; styles 4, 4 mm. long, flabelliform, 7–8-partite, the lobes linear, subterete, yellow-green, completely occluding the ovary at first. Fruits subglobose or imperceptibly 4-lobed, 2.5–3.5 cm. long and across, smooth, patchily pubescent-puberulous at first, later glabrescent, soft, yellow, becoming brownish, 4-seeded; mesocarp 1.5 mm. thick when dry. Pyrenes 4, 1.6–2 cm. long, 1.2–1.4 cm. wide, carinate, apiculate, cordate.

TANZANIA. Biharamulo District: Lusahanga, 15 Nov. 1948, *Ford* 866!; Mpwapwa, 5 Mar. 1933, *Hornby* 448!; Songea District: near Kigonsera, 13 Apr. 1956, *Milne-Redhead & Taylor* 9631!

DISTR. T 1, 4, 5, 7, 8; Zaire (Shaba), Burundi, Mozambique, Malawi, Zambia, Zimbabwe
HAB. Common, often forming extensive stands in deciduous woodland, upland wooded grassland
and along streams, often on stony soils and rocky slopes; 720–1950 m.

SYN. *U. goetzei* Pax in E.J. 28: 418 (1900). Type: Tanzania, Iringa District, Uzungwa [Utschungwe] Mts.,
Goetze 598 (B, holo.†)

NOTE. The Angolan *U. gossweileri* Hutch., to which some short-petiolate or subsessile-leaved
Tanzanian gatherings have been referred from time to time, is very close to *U. kirkiana* and may be
conspecific with it. However, the Angolan examples tend to show much stouter and more densely
cicatricose branchlets than typical *U. kirkiana*.
See note under *U. sansibarica* for remarks on a probable hybrid.

3. U. nitida *Muell. Arg.* in Flora 1864: 517 (1864) & in DC., Prodr. 15(2): 491 (1866);
P.O.A. C: 237 (1895); Hutch. in F.T.A. 6(1): 639 (1912); V.E. 3(2): 37 (1921); Pax in E.P. IV.
147 (15): 307 (1922); De Wild., Contrib. Étud. Esp. *Uapaca:* 156 (1936); T.T.C.L.: 228 (1949);
F.F.N.R.: 206 (1962). Type: Zambia, Southern Province, Livingstone District, Batoka
Country, July–Oct. 1860, *Kirk* (K, holo.!)

A completely glabrous evergreen shrub or small tree up to 20 m. high, with a slender
trunk and rounded crown; bark dark brown or grey, smooth at first, later tessellated and
fissured. Branchlets fairly slender. Leaves long-petiolate, the petioles up to 5.7 cm. long,
slender, pale green; leaf-blades elliptic-oblong, oblong-oblanceolate or elliptic-
oblanceolate, (4–)7–10(–13) cm. long, 2–4(–6) cm. wide, rounded or obtuse, rarely
subacute at apex, rounded-cuneate, cuneate or attenuate at base, coriaceous, lateral
nerves 8–18 pairs, scarcely prominent above or beneath, brochidodromous or
camptodromous, tertiary nerves reticulate, pale green to deep glossy green above, duller
beneath, with the midrib rather paler green. Stipules triangular, ± 1 mm. long, very soon
caducous, or not developed. Inflorescences usually borne among the leaves, solitary.
Male peduncles 1–1.5 cm. long; inflorescence-bracts 7–10, elliptic-obovate, 5–10 mm.
long, 3–6 mm. wide, rounded, somewhat coriaceous, the outer reflexed, greenish, the
inner upright, yellow-green; head of ♂ flowers (5–)7–8 mm. across. Male flowers: calyx
turbinate, 1 mm. long, irregularly toothed, white; stamens 4–5; filaments 1.5 mm. long,
creamy-white, anthers 0.7 mm. long, sulphur-yellow; pistillode infundibuliform, 1 mm.
high, tripartite, the segments 2–3-lobed, sparingly pubescent without, glabrous within.
Female peduncle and bracts yellowish, otherwise ± as in ♂. Female flower: calyx shallowly
cupular, 6-lobed; ovary ovoid-subglobose, 3-locular, 4–5 mm. long, 4 mm. in diameter;
styles 3, 5 mm. long, flabelliform, irregularly 10-partite, the lobes somewhat flattened, pale
yellow-green, at first completely enveloping the ovary, later held horizontally. Fruits
somewhat compressed-ellipsoid or -ovoid, 1.6–2 cm. long, 1.4–1.5 cm. in diameter when
fresh (1.3–1.4×0.9–1 cm. when dry), ± smooth, longitudinally 9-ribbed, green at first, then
yellow-green, shiny, becoming dark purplish brown and beige flecked when dry. Pyrenes
3, 1–1.5 cm. long, 7–9 mm. wide, carinate, apiculate at both ends.

TANZANIA. Mwanza District: 24 km. S. of Geita, 13 Apr. 1937, *B.D. Burtt* 6477!; Mpanda District:
Pasagulu–Musenbantu, 10 Aug. 1959, *Harley* 9270!; Songea District: 3 km. NE. of Kigonsera, 13 Apr.
1956, *Milne-Redhead & Taylor* 9597!
DISTR. T 1, ?3, 4–8; Zaire (Kinshasa, Shaba), Burundi, Mozambique, Malawi, Zambia, Zimbabwe,
Angola
HAB. Very common in *Brachystegia* woodland and associated wooded grassland; 350–1600 m.

SYN. *U. microphylla* Pax in E.J. 23: 523 (1897). Type: Angola, Malange, *Teusz* 433 (B, holo.†)

NOTE. *U. similis* Pax & K. Hoffm. in E.P. IV. 147 (15): 305 (1922); De Wild., *op. cit.*: 177 (1936);
T.T.C.L.: 299 (1949), described from Dar es Salaam, T 6 (*Holtz* 3199), may have been based on a
large-leaved specimen of *U. nitida*.
The record from T 3 is based on a specimen received from the Conservator of Forests via the
Imperial Forestry Institute Herbarium, No. 477, labelled Lushoto, but probably not the site of
collection.

4. U. sansibarica *Pax* in E.J. 34: 370 (1904); Hutch. in F.T.A. 6(1): 636 (1912); V.E. 3(2):
37 (1921); Pax in E.P. IV. 147 (15): 304 (1922); De Wild., Contrib. Étud. Esp. *Uapaca:* 172
(1936); T.T.C.L.: 228 (1949); I.T.U., ed. 2: 143 (1952); F.P.S. 2: 100 (1952); F.F.N.R.: 206
(1962). Types: Tanzania, Uzaramo District, Dar es Salaam, *Stuhlmann* 32 (B, syn.†) &
Engler (B, syn.†) & Rungwe District, 'Massewe', *Goetze* 1325 (B, syn.†) & Mozambique,
Zambezia, Quelimane [Quilimane], *Stuhlmann* I. 577 (B, syn.†) & Malawi, *Buchanan* 221
(K, isosyn.!)

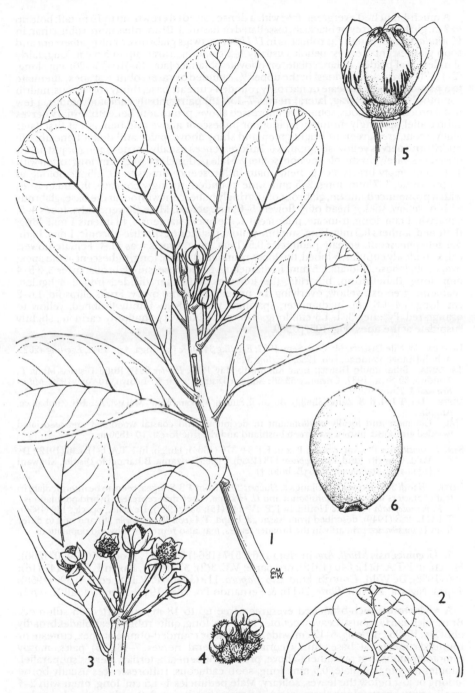

FIG. 105. *UAPACA SANSIBARICA* — **1**, flowering branch, × ⅔; **2**, tip of leaf, undersurface, × 2; **3**, ♂ inflorescence, × 1; **4**, ♂ flower, × 8; **5**, ♀ flower, with 2 petals removed, × 4; **6**, fruit, × 4. 1–4, from *Vesey-FitzGerald* 5165; 5, from *Davis* 176; 6, from *Procter* 904. Drawn by Christine Grey-Wilson.

A much-branched evergreen tree with a dense rounded crown, up to 15 m. tall; bole up to 2 m. girth; bark grey or blackish, tessellated or fissured. Branchlets more robust than in *U. nitida,* but not usually as robust as in *U. kirkiana;* twigs glabrous, or else pubescent and soon glabrescent. Leaves: petioles usually 1-2 cm. long, rarely up to 5.5 cm. long, fairly slender; leaf-blades oblanceolate or obovate-oblanceolate, (5-)10-15(-20) cm. long, (2-)4-7(-11) cm. wide, widest in the upper ⅓, rounded or rarely obtuse at apex, attenuate to a ± symmetrically cuneate or narrowly rounded-cuneate base, thinly coriaceous, midrib commonly distally zigzag, lateral nerves 7-13(-15) pairs, mostly camptodromous, a few distal brochidodromous, somewhat impressed above, prominent beneath, tertiary nerves subparallel, sparingly minutely lepidote or pubescent above and beneath at first, later glabrescent, grey-green or bright green and shiny above, paler and duller beneath, the midrib and nerves yellowish. Stipules 0. Inflorescences usually borne among or just below the leaves, rarely on the older wood, solitary. Male peduncles 0.5-2.5 cm. long, often with 1-3 small empty bracts along them; main inflorescence-bracts 9-10, elliptic-oblong to elliptic-ovate, 3-7 mm. long, 2-5 mm. wide, rounded, strongly concave, the outer ones with a pronounced midrib, glabrous or sparingly pubescent without at the base, glabrous within, yellow, sticky; head of ♂ flowers 5-6 mm. across. Male flowers: calyx-lobes 4-5, squarish, 1 mm. long, truncate, pubescent at the apex; stamens 4, filaments 1 mm. long, flattened; anthers 0.5 mm. long, sulphur-yellow; pistillode cylindric-obconic, 1 mm. high, densely pubescent. Female peduncles 0.5-1 cm. long; bracts ± as in ♂. Female flower: calyx shallowly cupular, 5-lobed, the lobes rounded, 0.5 mm. long, pubescent at the apex; ovary subglobose, 3-locular, 5 mm. in diameter, verruculose, subglabrous; styles 3, 3.5-4 mm. long, flabelliform, bipartite, the lobes each further divided into 2-3 ligulate segments, greenish yellow, overlying the upper half of the ovary. Fruits ellipsoid, 1.6-2 cm. long, 1.4-1.6 cm. in diameter, sparingly verruculose, faintly 9-ribbed, yellow to orange-red. Pyrenes 3, 1-1.3 cm. long, 8-9 mm. wide, very shallowly carinate, slightly cuspidate at the apex. Fig. 105, p. 569.

UGANDA. W. Nile District: Mt. Otzi, Sept. 1937, *Eggeling* 3420! & Ladonga, Apr. 1941, *Eggeling* 4262!; Acholi District: Paranga, Feb. 1938, *Eggeling* 3487!
TANZANIA. Biharamulo District: near Kimurani, May 1958, *Procter* 904!; Rufiji District: Mafia I., Tondwa, 30 Sept. 1937, *Greenway* 5339!; Songea District: near R. Likuyu, 10 Apr. 1956, *Milne-Redhead & Taylor* 9565!
DISTR. U 1; T 1, 4, 6-8; Zaire (Shaba), Burundi, Sudan, Mozambique, Malawi, Zambia, Zimbabwe, Angola
HAB. Common and locally co-dominant in deciduous and coastal woodland and associated wooded grassland, also in evergreen bushland and riverine forest; 10-1850 m.

SYN. *U. sansibarica* Pax var. *cuneata* Pax in E.J. 34: 370 (1904); Hutch. in F.T.A. 6(1): 636 (1912); De Wild., Contrib. Étud. Esp. *Uapaca:* 173 (1936). Type: Tanzania, Biharamulo District, Kimwani [Kimoani], *Stuhlmann* 3386 (B, holo. †)
NOTE. *Haerdi* 378/0 from T 6, Ulanga & *Hubbert* 2025 from T 8, Songea, seem to be intermediate in leaf characters between *U. sansibarica* and *U. kirkiana.* They may represent hybrid populations.
U. macrocephala Pax & K. Hoffm. in E.P. IV. 147 (15): 305 (1922); De Wild., op. cit.: 140 (1936); T.T.C.L.: 228 (1949), described from Kasanga, Ufipa, T 4 (*von Wangenheim* 36) and said to differ from *U. sansibarica* primarily in the broader leaves, may also represent a hybrid population.

5. **U. guineensis** *Muell. Arg.* in Flora 1864: 517 (1864) & in DC., Prodr. 15(2): 490 (1866); Hutch. in F.T.A. 6(1): 640 (1912), pro parte; V.E. 3(2): 37 (1921); Pax in E.P. IV. 147 (15): 306 (1922); De Wild., Contrib. Étud. Esp. *Uapaca:* 119 (1936); F.W.T.A., ed. 2, 1: 390 (1958). Types: Nigeria, R. Nun, *Barter* 2011b & Fernando Po, *Vogel* 194 & *Mann* 74 (all K, syn.!)

A ± glabrous much-branched evergreen tree up to 18 m., supported on stilt-roots. Branchlets fairly robust. Leaves: petioles 1.5-7 cm. long, quite robust; leaf-blades broadly obovate, 10-25 cm. long, 5-17 cm. wide, rounded or rounded-obtuse at apex, cuneate or rounded-cuneate at base, thinly coriaceous, lateral nerves 7-13(-15) pairs, mostly camptodromous, not prominent above, prominent beneath, tertiary nerves subparallel. Stipules minute, subulate, 0.5 mm. long, soon caducous. Inflorescences usually borne among or just below the leaves, solitary. Male peduncles 1-1.5 cm. long, often with 2-4 small empty bracts along them; main inflorescence-bracts 8-9, oblong to subcircular, 1-1.5 cm. long, 0.5-1 cm. wide, rounded, irregularly crenulate, somewhat ciliate, bright yellow; head of ♂ flowers 5-6 mm. across. Male flowers: calyx-lobes irregular, ± 9, with 5 ± oblong, 1 mm. long and 0.5 mm. wide, truncate and 4 ± subulate and shorter, all pubescent without and at the apex; stamens 5, filaments 1.5 mm. long, anthers 0.5 mm. long; pistillode cylindric, 1 mm. high, pubescent. Female peduncles 8-15 mm. long, stouter than in ♂; bracts ± as in ♂. Female flower: calyx shallowly cupular, 6-lobed, the lobes unequal, triangular or ± rounded, 1–1.5 mm. long, evenly pubescent; ovary subglobose, 3-locular, 2.5–3 mm. in diameter, smooth, glabrous; styles 3, 4 mm. long, twice bifid, the lobes flattened, somewhat undulate on the margins, pubescent at the base, otherwise glabrous. Fruits ovoid-subglobose, 1.8–2.5 cm. in diameter, smooth, glabrous, greenish. Pyrenes 3, ± 1.5–1.8 cm. long, 1.3–1.4 cm. wide, scarcely carinate, cuspidate at the apex, ± cordate at the base.

TANZANIA. Kigoma District: 16 km. S. of Uvinza, May 1956, *Procter* 468!; Pemba I., Pandani, 13 Dec. 1930, *Greenway* 2720! & Makongwe, 16 Dec. 1930, *Greenway* 2725! & Sengenya Dya, 9 Oct. 1951, *R.O. Williams* 105!

DISTR. T 4; P; from Sierra Leone eastwards to the Central African Republic and Zaire (Equateur), also in Malawi, Zambia and Zimbabwe

HAB. Common in bushland on steep slopes, in mixed evergreen forest and in riverine forest; 70–1060 m.

SYN. *U. sp. 1* sensu F.F.N.R.: 206 (1962)

NOTE. The disjunct distribution of this basically hygrophilous species would seem to be explicable in terms of the progressive desiccation of the African continent over the last 4000 years.

6. **U. paludosa** *Aubrév. & Leandri* in Bull. Soc. Bot. Fr. 82: 50, t. 1 (1935); F.W.T.A., ed. 2, 1: 390 (1958). Types: Ivory Coast, Abidjan, *Chevalier* 15317 (P, syn.!), Banco, *Aubréville* 361 (P, syn.!, K, isosyn.!) & several other *Aubréville* specimens from Ivory Coast (P, syn.!)

A much-branched deciduous tree with a spreading habit up to 40 m., supported on stilt-roots which may be produced from up to 4 m. above ground-level, often branching near the base; bark grey. Branchlets fairly robust; twigs ferruginous-pubescent. Leaves long-petiolate, the petioles (3–)5–9(–13) cm. long, quite robust; leaf-blades obovate to oblanceolate, 9–30(–50) cm. long, 4–17(–25) cm. wide, rounded or occasionally obtuse or shortly subacutely acuminate at apex, cuneate or rounded-cuneate at base, firmly chartaceous to thinly coriaceous, lateral nerves 7–15(–20) pairs, mostly camptodromous, somewhat impressed above, prominent beneath, tertiary nerves subparallel, sparingly minutely lepidote on both surfaces, and simply pubescent along the midrib and main nerves beneath at first, later ± glabrescent, mid-green and slightly glossy with a pale midrib above, pale green beneath. Stipules foliaceous, lanceolate to ovate-lanceolate, (0.5–)1–2 cm. long, (3–)5–7 mm. wide, entire, sparingly pubescent, persistent or subpersistent. Inflorescences usually borne among or just below the leaves, solitary. Male peduncles 2–3.5 cm. long, often with 2–3 small empty bracts along them; main inflorescence-bracts 7–10, elliptic-oblong to broadly elliptic-ovate, strongly concave, 5–9 mm. long, 2–7 mm. wide, rounded, the outer ones with a midrib, sparingly pubescent without, glabrous within, pale yellow or yellowish green; head of ♂ flowers 0.8–1 cm. across. Male flowers: calyx-lobes irregular, ± 8–9, with some oblong, 1 mm. long and 0.5 mm. wide, rounded, the others ± linear, all sparingly pubescent at the apex; stamens 5, filaments 2 mm. long, anthers 0.5 mm. long; pistillode infundibuliform, 1 mm. high, pubescent. Female peduncles 1–2 cm. long; bracts ± as in ♂. Female flower: calyx shallowly cupular, 6-lobed, the lobes somewhat unequal, ± triangular, 1 mm. long, densely pubescent; ovary ovoid, 3–4-locular, 4–5 mm. long, 3–4 mm. across, densely pubescent; styles 3–4.5 mm. long, subterete, 5–6-fid, the lobes themselves further divided, pubescent, the stigmas glabrous. Fruits ovoid-subglobose, (1.5–)2.5–3 cm. long, (1–)1.5–2 cm. across, conical at the apex, slightly verruculose, sparingly ferruginous-pubescent, greenish at first, later becoming brown. Pyrenes 3–4, (1–)1.5–2 cm. long, (0.8–)1–1.3 cm. wide, slightly carinate, ± acute.

UGANDA. Masaka District: S. Buddu, 1905, *Dawe* 329! & Bugala I., Nov. 1931, *Eggeling* 105!, & 20 Feb. 1933, *A.S. Thomas* 820!

TANZANIA. Morogoro District: Nguru Mts., Liwale R., Mahonda Sawmills, Turiani Falls, 4 Nov. 1947, *Brenan & Greenway* 8279! & Manyangu Forest, 1 Aug. 1951, *Greenway & Farquhar* 8629!; Iringa District: Mwanihana Forest Reserve, 7 June 1980, *Rodgers & Vollesen* 756!

DISTR. U 4; T 6, 7; from Liberia eastwards to Cameroun and south to Cabinda

HAB. Dominant in swamp forests, lakeside forests, fringing forest and also as an understorey in lowland rain-forest; 600–1400 m.

SYN. [*U. guineensis* sensu auct. plur. afr., *non* Muell. Arg.]

NOTE. *U. stipularis* Pax & K. Hoffm., from S. Cameroun and Equatorial Guinea, is very close to *U. paludosa*, but has smaller ♂ inflorescences.

75. ANTIDESMA

L., Sp. Pl.: 1027 (1753) & Gen. Pl., ed. 5, 451 (1754); Muell. Arg. in DC., Prodr. 15(2): 247 (1866); G.P. 3(1): 284 (1880); Hutch. in F.T.A. 6(1): 642 (1912); Pax in E.P. IV. 147 (15): 107 (1922)

Stilago L., Mant. 1: 16 (1767), *non Stylago* Salisb. (1866)

Dioecious shrubs or small trees with a simple indumentum. Leaves alternate, shortly petiolate, stipulate, simple, entire, penninerved, usually brochidodromous, provided with acarodomatia, eglandular. Inflorescences axillary, leaf-opposed, terminal or cauliflorous, shortly pedunculate, simple or compoundly spicate or racemose, solitary or subfasciculate, usually densely flowered; bracts small, 1-flowered. Male flowers: calyx ± cupular, 3–5(–8)-toothed or -lobed, the lobes imbricate; petals 0; disc-glands free or ± connate, extrastaminal or covering the receptacle; stamens (2–)3–5(–10), episepalous, filaments free, long-exserted, arising from within or from among the disc-glands, anthers inflexed in bud, apicifixed, bilobate, thecae ± divergent, almost free, basally dehiscent, connective thick; pistillode small, often ± cylindric. Female flowers: calyx ± as in ♂; disc hypogynous, annular or cupular; ovary 1(–2)-locular, with 2 apical pendulous anatropous ovules per locule; styles (2–)3(–5), very short, usually bilobed, terminal or lateral. Fruit a small, often oblique and compressed-ellipsoid red or black indehiscent drupe; endocarp hardened, laxly reticulate-foveolate. Seed solitary by abortion (very rarely 2 developed), ecarunculate; albumen fleshy; cotyledons broad, flat.

A genus of ± 170 species in the Old World tropics and subtropics, of which less than 10 are African.

The genus has been customarily included in the Euphorbiaceae as a member of the somewhat artificial subtribe *Antidesminae* of the tribe *Phyllantheae*, although in 1825, C.A. Agardh, in his 'Aphorismi Botanici' had established the Stilaginaceae for it, showing how it differed from the Euphorbiaceae. Airy Shaw, in the 8th edn. of 'Willis' (1973), upheld the family, and regarded it as to some extent intermediate between the Euphorbiaceae and the Icacinaceae, owing to the similarity of the fruits to those of several genera of the latter family, but for present purposes the genus is retained as an anomalous member of the Euphorbiaceae.

Two widespread Asiatic species are on record as having been cultivated in East Africa: *A. bunius* (L.) Sprengel (Kenya, C. Kavirondo District, Siriba, 'S.A.D. H 146/56'!; Zanzibar, Dec. 1930, *Greenway in Agr. Dept.* 3!), and *A. ghaesembilla* Gaertner (Tanzania, Lushoto District, Amani, 17 Feb. 1921, *Soleman* 6093! & 20 Jan. 1931, *Greenway* 2817! — cfr. T.T.C.L.: 199 (1949)).

1. Stipules divided into 6–9 filiform or branched segments . . . *1. A. laciniatum*
   Stipules simple, entire, narrowly lanceolate to linear-lanceolate 2
2. Leaves usually rounded, obtuse, subacute or shortly obtusely acuminate at the apex, sparingly pubescent to densely fulvous-ferruginous-tomentose beneath; dried fruits 5–7(–8) mm. . . . . *2. A. venosum*
   Leaves usally distinctly acutely acuminate 3
3. Dried fruits 3–5 mm. long; leaves membranaceous, often evenly to densely pubescent beneath . . . *3. A. membranaceum*
   Dried fruits (5–)7–9(–10) mm. long; leaves coriaceous, sparingly pubescent along the midrib above and beneath, usually otherwise ± glabrous . . . . . *4. A. vogelianum*

1. **A. laciniatum** *Muell. Arg.* in Flora 47: 520 (1864) & in DC., Prodr. 15(2): 260 (1866); Hutch. in F.T.A. 6(1): 643 (1912); V.E. 3(2): 20 (1921); Pax in E.P. IV. 147 (15): 145 (1922); F.W.T.A., ed. 2, 1: 374 (1958). Types: Fernando Po, *Mann* 201 & 256 (K, syn.!)

A small understorey tree up to 12 m., with spreading branches; wood hard, yellow. Twigs, petioles and stipules evenly to densely pubescent and/or pilose. Leaves: petioles 3–5(–8) mm. long; blades elliptic-oblanceolate or elliptic-oblong, 7.5–20 cm. long, 3–7 cm. wide, shortly acutely acuminate and mucronate at apex, rounded to shallowly and sometimes asymmetrically cordulate at base, firmly membranous to chartaceous, lateral nerves 9–14 pairs, somewhat impressed above, prominent beneath, tertiary nerves subparallel, evenly to densely pubescent and/or pilose along the midrib above and along the midrib and main nerves beneath, otherwise sparingly so at first and later glabrescent. Stipules laciniate, with 3–9 lanceolate, filiform or branched segments, (0.7–)1–1.5 cm. long, fairly persistent. Inflorescences simply spicate or with 1-2 lateral spikes at the base, commonly not exceeding 10 cm. in length, peduncles evenly pubescent; bracts linear, 1 mm. long, pubescent and/or pilose. Male flowers: calyx 0.5-1 mm. long, ± glabrous, 3(-4)-toothed, the teeth acute or subacute, ciliate-pubescent, reddish pink or reddish purple; disc extrastaminal, annular, 1 mm. across; stamens 3(-4), filaments 1.5-2 mm. long, anthers 0.3 mm. long, purplish; pistillode narrowly cylindric, minute, glabrous. Female flowers: pedicels 0.5 mm. long; calyx ± 1.5 mm. long, sparingly pubescent, brownish, otherwise ± as in ♂; disc saucer-shaped; ovary ellipsoid, 2 mm. long, glabrous or pubescent; styles terminal, 3, bifid, or 2, trifid, very small. Infructescences often up to 15-20 cm. long. Fruits (4-)5-6(-7) mm. long (dry), coarsely foveolate-rugose when dried, glabrous or sparingly pubescent, yellowish to orange when fresh.

var. **membranaceum** *Muell. Arg.* in Flora 47: 520 (1864) & in DC., Prodr. 15(2): 260 (1866); Hutch. in F.T.A. 6(1): 644 (1912); Pax in E.P. IV. 147 (15): 145 (1922); I.T.U., ed. 2: 116 (1952); F.W.T.A., ed. 2, 1: 375 (1958). Type: Sierra Leone, Bagru [Bagroo] R., 1861, *Mann* (K, holo.!)

Twigs, petioles, stipules, midrib on both surfaces and main nerves beneath densely pilose. Stipules with 6-9 filiform or branched segments.

UGANDA. Bunyoro District: Budongo Forest, June 1935, *Eggeling* 2067!; Mengo District: Kyagwe, Mau Forest, May 1932, *Eggeling* 429! & Namiryango, Feb. 1939, *Chandler* 2740!
DISTR. U 2, 4; ?T 7; from Guinée eastwards to Zaire (Oriental & Kivu)
HAB. As an understorey tree in dense forest, and at the edge of forest patches; 975-1200 m.

SYN. *A. pseudolaciniatum* Beille in Bull. Soc. Bot. Fr. 57, Mém. 8c: 122 (1910). Type: Ivory Coast, Aboisso, *Chevalier* 17875 (P, holo.!, K, photo.)

NOTE. The typical variety has a more restricted distribution than var. *membranaceum;* it occurs from Ivory Coast eastwards to Cameroun, Equatorial Guinea and Gabon.
Perdue & Kibuwa 11643 from **T** 7, Rungwe, coll. 1 Oct. 1971, may be this species, but the material is too imperfect to admit of precise determination.

2. **A. venosum** *Tul.* in Ann. Sci. Nat., sér. 3, 15: 232 (1851); Muell. Arg. in DC., Prodr. 15(2): 260 (1866); P.O.A. C: 237 (1895); Hutch. in F.T.A. 6(1): 646 (1912); V.E. 3(2): 19 (1921); Pax in E.P. IV. 147 (15): 139 (1922); T.S.K. 48 (1936); T.T.C.L.: 200 (1949); F.P.S. 2: 55 (1952); F.W.T.A., ed. 2, 1: 375 (1958); K.T.S.: 185 (1961); F.F.N.R.: 193 (1962). Types: South Africa, Natal, near Durban, 1839, *Drège* (P, syn.!, G, K, isosyn.!) & near the Umlaas R., *Krauss* 138 (P, syn.!, G, K, isosyn.!)

A spreading shrub or small evergreen tree, sometimes straggling, 1-9(-15) m. tall, many- or single-stemmed and sometimes with drooping branches; bark fibrous, fairly smooth to lightly fissured, flaking or roughened, grey-brown; wood hard, white or pale brown. Twigs sparingly lenticellate; buds ferruginous-tomentose; young shoots sparingly pubescent to densely fulvous- or ferruginous-tomentose. Leaves: petioles 3-7(-10) mm. long, evenly to densely pubescent or tomentose; blades elliptic-obovate to oblong-oblanceolate, (1.5-)3-10(-17) cm. long, (1-)2-5(-9) cm. wide, rounded, obtuse, subacute or shortly obtusely acuminate at apex, rounded or rounded-cuneate or sometimes cuneate at base, chartaceous to thinly coriaceous, lateral nerves 6-8(-9) pairs, camptodromous to weakly brochidodromous, impressed above, prominent beneath, evenly pubescent along the midrib but otherwise glabrous or sparingly pubescent and glabrescent above, sparingly pubescent to densely fulvous- or ferruginous-tomentose beneath, light to dark green and shiny above, paler and dull beneath. Stipules simple, narrowly lanceolate to linear-lanceolate, 4-8 mm. long, rarely longer, entire, evenly to densely pubescent or tomentose. Male inflorescences pendent, compoundly spicate, usually with 1-2 lateral spikes at the base, (4-)7-10(-12) cm. long, peduncles and axis sparingly to densely pubescent or tomentose; bracts ovate, 0.5 mm. long, evenly pubescent. Male flowers: calyx 0.5-1 mm. long, sparingly to evenly pubescent, (3-)4(-5)-lobed, the lobes somewhat unequal, subacute or obtuse, ciliate-pubescent, greenish white or yellowish, sometimes tinged reddish or brownish; disc (3-)4(-5)-lobed, the lobes enveloping the filaments, subglabrous; stamens (3-)4(-5), filaments 2 mm. long, anthers 0.5 mm. long, yellow; pistillode hemispherical-cylindric, 0.5 mm. high, sparingly pubescent. Female inflorescences usually erect or held laterally, sometimes pendent, compoundly spicate, usually with 1-4 lateral spikes in the lower part, but often galled due to the action of a mite and then densely and irregularly paniculate, especially in the upper half, otherwise ± as ♂. Female flowers: pedicels 0.5-0.8 mm. long, sparingly pubescent; calyx 1 mm. long, suburceolate, otherwise ± as ♂; disc shallowly cupular; ovary somewhat asymmetrically lenticular, 2 mm. long, smooth, glabrous; styles terminal, 2, bifid, 0.75 mm.

long, borne on a tumid stylopodium. Infructescences not elongated. Fruits 5–7(–8) mm. long (dry), irregularly and coarsely reticulate-rugose when dried, glabrous, dull greenish yellow or whitish at first, later becoming pinkish, reddish, purplish, brownish or almost black when ripe.

UGANDA. Karamoja District: Mt. Kadam, Apr. 1959, *J. Wilson* 773!; Kigezi District: Bwambara, Feb. 1950, *Purseglove* 3273!; Mbale District: near Apoli, 5 July 1971, *Katende* 1141!

KENYA. Machakos District: Nzaui Hill, 16 Feb. 1969, *Kokwaro* 1868!; C. Kavirondo District: near Kururuma, 19 Feb. 1964, *Brunt* 1490!; Kwale District: Shimba Hills, 25 Dec. 1968, *Mwangangi* 1313!

TANZANIA. Mwanza District: Rumara, 7 Dec. 1953, *Tanner* 1878!; Handeni District: Mgambo Forest Reserve, 5 Feb. 1971, *Shabani* 647!; Songea District: Maweso, 4 July 1956, *Milne-Redhead & Taylor* 10949!; Zanzibar I., Jozani forest, 22 Dec. 1960, *Faulkner* 2740! & Unguja Ukuu, 14 May 1963, *Faulkner* 3188!; Pemba I., Ngezi Forest, 18 Feb. 1929, *Greenway* 1487!

DISTR. U 1–4; K 4, 5, 7; T 1–8; Z; P; from the Gambia eastwards to Ethiopia and south to Namibia and South Africa (Natal, Transkei); ? Madagascar

HAB. Forest edges or riverine, dry evergreen forest and associated bushland, deciduous woodland and wooded grassland, sometimes on termite mounds or in thickets; 0–1830 m.

SYN. *A. bifrons* Tul. in Ann. Sci. Nat., sér. 3, 15: 229 (1851). Type: Sudan, Blue Nile Province, Fazughli [Fazokl], *Kotschy* 397 (G, K, P, iso.!)
  A. boivinianum Baill., Adansonia 2: 45 (1861–2). Type: Zanzibar, 1848, *Boivin* (P, holo.!)
  A. membranaceum Muell. Arg. var. *molle* Muell. Arg. in Linnaea 34: 68 (1865–6), & in DC., Prodr. 15(2): 261 (1866), pro parte excl. spec. *Welwitsch* 404. Types: Angola, Banza de Bango Aquitamba, *Welwitsch* 405 (BM, syn., G, K, isosyn.!) & Sange, *Welwitsch* 406 (BM, syn., G, isosyn.!)

VAR. This species varies somewhat in leaf-shape, size and indumentum. There also appears to be a certain degree of vegetative sexual dimorphism, the male plants often having smaller, more acuminate leaves than the female, and consequently approaching the male plants of *A. membranaceum*. Keay (in F.W.T.A., ed. 2) mentions the difficulty of naming male specimens satisfactorily, whilst Pax (in E.P.) draws attention to the similarity of the two species in the juvenile state.

NOTE. *Conrads* 5574 from T1 (? Ukerewe I.) presents some anomalous features in that the foliage has the leaf-blades irregularly and shallowly denticulate at the apex, and the male inflorescences are much-branched, although in this case it does not appear to be as the result of gall-formation. However, in other respects, it is clearly referable to *A. venosum*. Fruits forming on galled inflorescences may be abnormal in shape and size, e.g. fusiform and 9–12 mm. long in *Gitonga* 7 from K4, Meru.
  According to J. Léonard (personal communication and in press), *Tanner* 2564, 2638 & 3354 from T3, *Schlieben* 2333 & *Harris* 3575 from T6, and *Schlieben* 5641 from T8, constitute a species distinct from *A. venosum* and *A. vogelianum*; it has closely reticulate sparingly pubescent leaves, free or ± free ♂ disc-glands, lateral or terminal styles, and is nearly always found in or near water. It also occurs from Gambia to Central African Republic and south to Namibia (Caprivi), Zambia & Mozambique. It would appear that all this material should be referred to *A. rufescens* Tul. in Ann. Sci. Nat., sér. 3, 15: 231 (1851). Type: Gambia, near Albreda, June 1827, *Leprieur* (P, holo.!, K, photo.!) (Syn.: *A. membranaceum* Muell. Arg. var. *glabrescens* Muell. Arg. in Linnaea 34: 68 (1865–6) & in DC., Prodr. 15(2): 261 (1866). Type: N. Nigeria, Nupe, *Barter* 1557 (K, holo.!)) which has been considered by many authors hitherto as being synonymous with *A. venosum*.

3. A. membranaceum *Muell. Arg.* in Linnaea 34: 68 (1865–6) & in DC., Prodr. 15(2): 261 (1866), pro max. parte, excl. spec. *Welwitsch* 405 & 6; P.O.A. C: 237 (1895); Hutch. in F.T.A. 6(1): 645 (1912), pro parte; V.E. 3(2): 20 (1921); Pax in E.P. IV. 147 (15): 141 (1922); T.S.K., ed. 2: 48 (1936); T.T.C.L.: 199 (1949); I.T.U., ed. 2: 116 (1952); F.P.S. 2: 55 (1952); F.W.T.A., ed. 2, 1: 375 (1958); K.T.S.: 184 (1961); F.F.N.R.: 193 (1962); Troupin, Fl. Pl. Lign. Rwanda: 249, t. 85.2 (1982) & Fl. Rwanda 2: 206, t. 61, 2 (1983). Type: Angola, Malange, Pungo Andongo, *Welwitsch* 404 (G, lecto. — J. Léonard, ined.)

Very similar to *A. venosum*, but differing in having the leaves more markedly and acutely acuminate, with rather less dense indumentum, and in having smaller fruits (3–5 mm. long when dry).

UGANDA. Bunyoro District: near Busingiro, 4 Sept. 1968, *Katende* 16/68!; ? Busoga District: 'Nabukongolo', Aug. 1927, *African Staff* 120!; Masaka District: Bugabo, 1 Feb. 1969, *Lye et al.* 1891!

KENYA. S. Nyeri District: Iganjo, 9 Dec. 1963, *Kibui* 99!; Kwale District: Pengo Hill, 9 July 1968, *Gillett* 18669!; Mombasa District: 'Nyika Country', *Wakefield!*

TANZANIA. Tanga District: Potwe, Jun. 1960, *Semsei* 2999!; Buha District: Kakombe valley, 22 Dec. 1963, *Pirozynski* 61!; Lindi District: Rondo Plateau, Mchinjini, Jan. 1952, *Semsei* 618!; Zanzibar I., Masingini, 5 Oct. 1930, *Vaughan* 1612!; Pemba I., Ngezi Forest, 11 Dec. 1930, *Greenway* 2702!

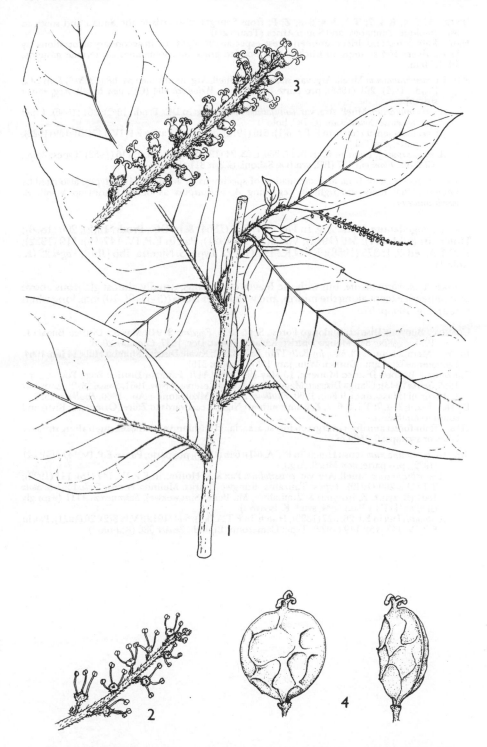

FIG. 106. *ANTIDESMA VOGELIANUM* — 1, flowering branch, × ⅔; 2, ♂ inflorescence, × 4; 3, ♀ inflorescence, × 4; 4, fruit, two views, × 4. 1, 3, from *Faulkner* 2528; 2, from *Faulkner* 3327; 4, from *Faulkner* 2496. Drawn by Christine Grey-Wilson.

DISTR. U 2–4; **K** 4, 7; **T** 1, 3, 4, 6–8; **Z**; **P**; from Senegal eastwards to the Sudan and south to Mozambique, Zimbabwe and South Africa (Transvaal)

HAB. Rain-forest and drier evergreen forest, especially at edges and in secondary associations, by lakes, rivers and swamps, in thickets and wooded grassland, sometimes on termite mounds; 10–1530 m.

SYN. [*A. membranaceum* Muell. Arg. var. *molle* sensu Muell. Arg. in Linnaea 34: 68 (1865–6) & in DC., Prodr. 15(2): 261 (1866), pro parte quoad spec. *Welwitsch* 404 (G!), *non* Muell. Arg. sensu stricto]

    A. membranaceum Muell. Arg. var. *tenuifolium* Muell. Arg. in DC., Prodr. 15(2): 261 (1866). Type: Sierra Leone, *Smeathman* (G-DC, holo.!)

    [*A. venosum* sensu Hutch. in F.T.A. 6(1): 646 (1912), pro parte; Pax in E.P. IV. 147(15): 139 (1922), pro parte, *non* Tul.]

    A. meiocarpum J. Léon. in B.J.B.B. 17: 260, t. 23, 24 (1945); I.T.U., ed. 2: 116 (1952). Types: Zaire, Kitobola, *Flamigni* 487 (BR, syn.!) & Saboni, *Gilbert* 311 (BR, syn.!)

NOTE. It seems clear, from an examination of specimens in addition to the types, also cited by Léonard, that Keay was right in regarding *A. meiocarpum* as being synonymous with *A. membranaceum*.

4. A. vogelianum *Muell. Arg.* in Flora 47: 529 (1864) & in DC., Prodr. 15(2): 260 (1866); Hutch. in F.T.A. 6(1): 645 (1912); V.E. 3(2): 20 (1921); Pax in E.P. IV. 147(15): 119 (1922); F.W.T.A., ed. 2, 1: 375 (1958); F.F.N.R.: 194 (1962). Type: S. Nigeria, Ibo [Ibu], *Vogel* 22 (K, holo.!)

Like *A. membranaceum*, but with the leaves coriaceous, usually almost glabrous above and beneath except along the midrib, and with larger fruits ((5–)7–9(–10) mm. long when dried). Fig. 106, p. 575.

UGANDA. Bunyoro District: Budongo Forest, Sept. 1935, *Eggeling* 2194!; Masaka District: Bugala I., Nov. 1931, *Eggeling* 84!; Mengo District: Kayansi Forest, Dec. 1937, *Chandler* 2048!

KENYA. Meru, 2 Sept. 1943, *Mrs. Joy Bally* 17 in *Bally* 3217!; Kwale District: Shimba Hills, 14 Jan. 1964, *Verdcourt* 3932!; Lamu District: Witu, Jan. 1957, *Rawlins* 315!

TANZANIA. Bukoba District: Munene, 10 May 1948, *Ford* 447!; Pangani District: Bago Forest, Sept. 1955, *Semsei* 2343!; Ulanga District: Magombera Forest Reserve, 1 Nov. 1961, *Semsei* 3372!; Zanzibar I., Mazizini [Massazine], 9 Feb. 1960, *Faulkner* 2496! & Mkokotoni, 1 Apr. 1960, *Faulkner* 2528!

DISTR. U 2, 4; **K** 4, 7; **T** 1, 3, 6, 7; **Z**; from southern Nigeria eastwards to Zaire (Oriental & Kivu) and south to Zimbabwe

HAB. Rain-forest and drier evergreen forest, associated bushland and thicket, often along streams, lakes or swamps; 0–1500 m.

SYN. [*A. membranaceum* sensu Hutch. in F.T.A. 6(1): 645 (1912), pro parte; Pax in E.P. IV. 147 (15): 141 (1922), pro parte, *non* Muell. Arg.]

    A. membranaceum Muell. Arg. var. *crassifolium* Pax & K. Hoffm., in E.P. IV. 147 (15): 141 (1922); T.T.C.L.: 200 (1949). Types: Tanzania, Rungwe District, Mulinda Forest, *Stolz* 1451 & *Stolz* 1804 (B, syn.†, K, isosyn.!) & Zimbabwe, Mt. Pene (Singwekwe), *Swynnerton* 1111 (wrongly cited as '111' by Pax) (BM, syn.!, K, isosyn.!)

    A. staudtii Pax in E.J. 26: 327 (1899); Hutch. in F.T.A. 6(1): 644 (1912); V.E. 3(2): 20 (1921); Pax in E.P. IV. 147 (15): 149 (1922). Type: Cameroun, Bipindi, *Zenker* 988 (K, holo.!)

HYMENOCARDIACEAE

A.R.-SMITH

Dioecious deciduous trees or shrubs. Leaves alternate, rarely opposite, and then only on sucker shoots, simple, entire, penninerved or occasionally subtriplinerved, shortly petiolate, stipulate, sometimes densely gland-dotted beneath. Male flowers precocious, in axillary spikes or subpaniculate clusters of spikes. Female flowers in few-flowered terminal racemes or axillary, solitary. Male flowers: calyx cupular, lobes 4–6, imbricate; petals 0; disc 0; stamens 4–6, opposite the sepals; filaments short, spreading, free or connate at the base; anthers large, extrorse, often with a dorsal gland, thecae opening right out on dehiscence; pistillode minute. Female flowers: calyx-lobes 4–6(–8), ± free to base, linear, caducous; petals 0; disc 0; ovary 2-locular, compressed at right-angles to the plane of the septum, with 2 apical pendulous anatropous ovules per locule, and 2 long, free, simple glabrous or papillose styles. Fruit with 2 broad flattened cocci, winged (in Africa) or not, separating from the persistent central axis; pericarp subcrustaceous; endocarp thin, membranous. Seeds usually 1 per coccus, flat, with sparse endosperm.

One genus only, *Hymenocardia.*

The family comes close to the Ulmaceae (see note under *Holoptelea grandis* (Hutch.) Mildbr. in F.T.E.A. *Ulmaceae* (1966)). The pollen resembles that of *Celtis* L. and *Chaetacme aristata* Planch. (Livingstone, Ecol. Monogr. 37: 41 (1967)). *Hymenocardia* was formerly included in the Euphorbiaceae, but the staminal characters, long simple styles and winged fruits are anomalous in that family, and are more suggestive of the Ulmaceae. See also J. Léonard & M. Mosango, F.A.C., Hymenocardiaceae (1985), also R. Dechamps et al. in B.J.B.B. 55: 473–485 (1985) for further discussion of the pollen and wood anatomy.

HYMENOCARDIA

Lindl., Nat. Syst. Bot., ed. 2: 441 (1836); Endl., Gen. 1: 288 (1837); Muell. Arg. in DC., Prodr. 15(2): 476 (1866); G.P. 3: 285 (1883); Hutch. in F.T.A. 6(1): 648 (1912); Pax & K. Hoffm. in E.P. IV. 147 (15): 72 (1922)

Characters of the family.

A genus of 7–8 species, one Asiatic, the rest African, of which two occur in the Flora area.

Leaves elliptic-ovate, usually somewhat acuminate; ♂
    inflorescences 1.5 cm. long, lax; fruit ± suborbicular,
    almost surrounded by the wing . . . . 1. *H. ulmoides*
Leaves ± oblong, obtuse and never acuminate; ♂ inflorescences
    3–7 cm. long, dense; fruit V-shaped, with 2 apical divergent
    wings . . . . 2. *H. acida*

1. **H. ulmoides** *Oliv.* in Hook., Ic. Pl. 12, t. 1131 (1873); P.O.A. C: 236 (1895); Hutch. in F.T.A. 6(1): 648 (1912); V.E. 3(2): 18, fig. 6 (1921); Pax & K. Hoffm. in E.P. IV. 147 (15): 73 (1922); T.T.C.L.: 216 (1949); F.P.S. 2: 80 (1952); K.T.S.: 206 (1961), excl. spec. cit.; F.F.N.R.: 200 (1962); J. Léon. & M. Mosango in F.A.C., Hymenocard.: 3, fig. 1/A–E (1985). Type: Tanzania, Uzaramo District, Dar es Salaam, *Kirk* 117 (K, holo.!)

A medium-sized or small tree or occasionally scandent shrub up to 12 m. in height, with smooth grey bark and very slender twigs pubescent at first, later glabrescent. Leaf-blades elliptic-ovate, 1.5–4(–5) cm. long, 0.7–2(–3.5) cm. wide, subacute or obtuse, usually acuminate, rounded or cuneate, entire, nerves inconspicuous, 4–5 pairs, sparsely puberulous along the midrib and main nerves above, otherwise glabrous, almost glabrous beneath except for the angles between the midrib and main nerves, where there are sparse to dense tufts of hairs in the domatia, sparsely gland-dotted beneath, thinly coriaceous; petiole 4–8 mm. long, pubescent. Stipules linear-lanceolate or lanceolate, 4 mm. long, puberulous, soon caducous. Male inflorescences 1.5 cm. long, many-flowered, laxly spicate or subpaniculate, pendent, leafy or not, often borne on lateral short shoots; axis very slender, puberulous; bracts spathulate, 2–3 mm. long, puberulous. Male flowers:

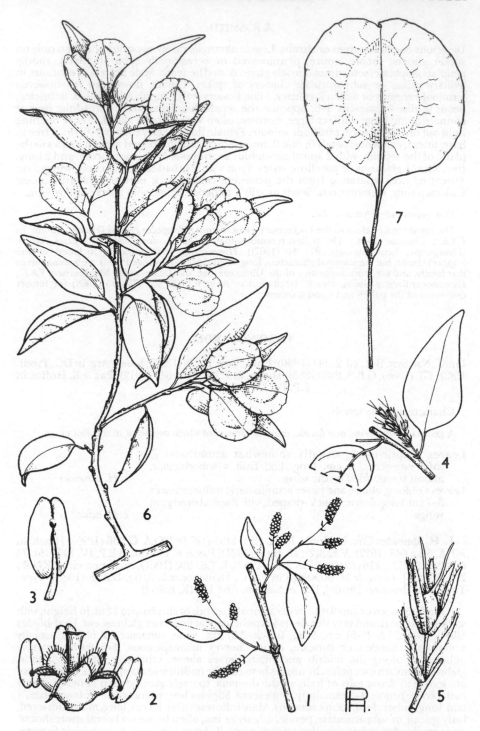

FIG. 107. *HYMENOCARDIA ULMOIDES* — 1, ♂ flowering branchlet, × 1; 2, ♂ flower, × 12; 3, stamen, rear view, showing gland, × 20; 4, ♀ inflorescence, × 1; 5, ♀ flower, × 6; 6, fruiting branch, × ⅔; 7, fruit, × 2. 1–3, from *Eggeling 6400*; 4, 5, from *Eggeling 6398*; 6, from *B.D. Burtt 4633*; 7, from *Semsei 548*. Drawn by Pat Halliday.

calyx 1.5 mm. long, lobes 5, rounded, 0.5 mm. long, ciliate, greenish white; anthers 1 mm. long, pale yellow or whitish; pistillode cylindric, 1 mm. long. Female inflorescences 1.5 cm. long, 4–5-flowered, leafy at the base, terminal on lateral shoots; axis slender, sparsely to evenly pubescent, gland-dotted; bracts elliptic-lanceolate, 2 mm. long, sparsely puberulous. Female flowers: pedicels 1.5 mm. long, sparingly puberulous and gland-dotted, extending to 1 cm. in fruit, very slender; sepals 5, linear, 2.5–3 mm. long, free almost to base, sparingly pubescent; ovary elliptic-oblong to obovate, compressed, 1 mm. long, glabrous, with a few scattered gland-dots; styles filiform, 2 mm. long at first, later extending to 5 mm., glabrous, reddish. Fruits almost surrounded by the wing, suborbicular-obcordate or obovate in outline, compressed, 1.8–2.3 cm. long, 1.3–2.3 cm. wide, borne on a stipe 1–3 mm. long, emarginate at apex, with the wing-extremities sometimes overlapping, rounded at the base with the wings slightly decurrent on the stipe, glabrous, membranous, pink or pale yellow. Seeds semicircular-oblong, compressed, 9 mm. long, 4 mm. wide, smooth, purplish brown. Fig. 107.

TANZANIA.   Handeni District: Handeni–Bagamoyo road, Oct. 1950, *Semsei* 574!; Bagamoyo District: Bana Forest Reserve, 28 Oct. 1965, *Mgaza* 758!; Tunduru District: Tunduru–Masasi road, 19 Nov. 1966, *Gillett* 17924!
DISTR.   T 3, 5, 6, 8; from Cameroun south to Angola, east to S. Sudan & Tanzania and southeast to South Africa (Natal)
HAB.   Dry evergreen forest and associated bushland or thicket, persisting in wooded grassland and disturbed places; 50–1550 m.

SYN.   *H. poggei* Pax in E.J. 15: 528 (1893). Types: Zaire, Kasai, Mukenge, *Pogge* 1361 & 1362 (B, syn.†)
        H. ulmoides Oliv. var. *capensis* Pax in E.J. 28: 22 (1899). Type: Mozambique, Maputo [Delagoa Bay], Matola, *Schlechter* 11725 (B, holo.†)
        H. capensis (Pax) Hutch. in K.B. 1920: 334 (1920)

NOTE.   The record in K.T.S. from K 7 is based on *Swynnerton* K55, which is not a *Hymenocardia*. It is probably a *Pteleopsis* (Combretaceae), but as it is a sterile specimen, it is not possible to be more definite about it.

2. **H. acida** *Tul.* in Ann. Sci. Nat., sér. 3, 15: 256 (1851); Muell. Arg. in DC., Prodr. 15(2): 477 (1866); P.O.A. C: 236 (1895); Hutch. in F.T.A. 6(1): 651 (1912); V.F. 3(2): 18, fig. 6 (1921); Pax and K. Hoffm. in E.P. IV. 147 (15): 76 (1922); T.S.K., ed. 2: 48 (1936); T.T.C.L.: 216 (1949); I.T.U., ed. 2: 130 (1952); F.P.S. 2: 80 (1952); F.W.T.A., ed. 2, 1: 377, fig. 132 (1958); K.T.S.: 204 (1961); F.F.N.R.: 200 (1962); Troupin, Fl. Pl. Lign. Rwanda: 264, t. 90.1 (1982) & Fl. Rwanda 2: 227, t. 62/3 (1983). Types: Guinée, Fouta Djallon, *Heudelot* 779 (P, syn.!, K, isosyn.!) & Gambia, Albreda, 3 May 1827, *Leprieur* (P, syn.!, K, isosyn.!)

Shrub or small tree up to 10 m. in height, with smooth or flaking yellowish grey bark; twigs densely, evenly or sparingly pubescent or almost glabrous, gland-dotted. Leaf-blades oblong, oblanceolate-oblong or obovate-oblong, rarely ± ovate, (3–)5–7(–9) cm. long, 1.5–4 cm. wide, obtuse or rarely subacute, rounded or cuneate, entire, nerves 7–12 pairs, sparingly puberulous or glabrous above, sparingly or evenly puberulous beneath, or glabrous except for the midrib and angles between the midrib and main nerves, where there may be dense tufts of hairs in the domatia, densely gland-dotted beneath, ± coriaceous when mature; petioles 3–10 mm. long, evenly or densely pubescent. Stipules linear-lanceolate to linear-filiform, 2–3 mm. long, minutely puberulous, soon caducous. Male inflorescences 3–7 cm. long, spicate, the spikes often fascicled on older wood, with the flowers densely packed spirally on the axis; axis densely pubescent; bracts spathulate, 1 mm. long, ciliate-fringed. Male flowers: calyx 1 mm. long, lobes 5, short, acute, ciliate, dark red; stamens 5, anthers 1.5 mm. long, thecae pubescent near the top, whitish, with the apical gland gold-coloured and prominent; pistillode cylindric, 1 mm. long, bifid and pubescent at apex. Female inflorescences 2–3 cm. long, 5–6-flowered, leafy at the base, terminal on lateral shoots, axis pubescent, gland-dotted; bracts elliptic-oblanceolate, 1.5 mm. long, readily caducous, or flowers axillary, solitary. Female flowers: pedicels 1 mm. long, sparsely pubescent, extending to 1(–2) cm. in fruit; sepals 5–8, linear, 2.5–3 mm. long, free, pubescent, gland-dotted, pinkish; ovary obovate-oblong to obcordate, compressed, 2-winged, 3 mm. long, glabrous or pubescent, with or without a few scattered gland-dots, crimson; styles filiform, 3 mm. long at first, later extending to 1.2 cm., rugose-papillose, purplish crimson. Fruits V-shaped, compressed, with 2 divergent rounded-rhomboid wings at the apex, 2–2.5 cm. long (base to style-base), 2.5–4.5 cm. across the wings, on a stipe 2 mm. long, rounded or cordate at the base, reticulate, glabrous or pubescent, sparingly gland-dotted or not, the wings firmly membranous, longitudinally striate; axis fusiform, remaining after dehiscence. Seeds ± semicircular, compressed, 1 cm. long, 0.5 cm. wide, smooth, shiny, dark purplish brown streaked with black.

var. **acida**

Fruits glabrous.

UGANDA. W. Nile District: Koboko, Apr. 1938, *Hazel* 511!; Teso District: Serere, Jan. 1932, *Chandler* 397!; Busoga District: Bunyantole [Buyamtole], Mar. 1931, *Harris* 6 in *F.H.* 73!
KENYA. N. Kavirondo District: Oct. 1930, *Gardner* in *F.D.* 2471!; & Nzoia R. 2 km. N. of Lusumu, 31 Mar. 1971, *Kokwaro* 2569!; S. Kavirondo District: Kisii, Mar. 1953, *Beecher* in *Bally* 8839!
TANZANIA. Ngara District: Busubi, 5 Oct. 1960, *Tanner* 5258!; Mpanda District: Mahali Mts., 1 Aug. 1958, *Jefford, Juniper & Newbould* 1288!; Ulanga District: Ifakara, July 1959, *Haerdi* 287/0!
DISTR. U 1–4; K 5; T 1, 4, 6, 8; ± throughout tropical Africa from Senegal eastwards to Ethiopia (Wolega) and southwards to Angola (Huila) and Mozambique (Manica e Sofala)
HAB. Wooded grassland and mixed deciduous woodland, sometimes riverine, relatively tolerant of fire; 600–1700 m.

SYN. *H. mollis* Pax var. *glabra* Pax in E.J. 15: 528 (1893). Type: Zaire, Lomami R., *Pogge* 1349 (B, holo.†)
      H. obovata Beille in Bull. Soc. Bot. Fr. 55, Mém. 8: 62 (1908). Types: Guinée, Guelila [Guélia], *Chevalier* 311 & Kouroussa, *Chevalier* 383 (P, syn.!)
      H. granulata Beille in Bull. Soc. Bot. Fr. 55, Mém. 8: 62 (1908). Type: Central African Republic, E. Dar Banda, *Chevalier* 7366 (P, holo.!)
      H. lanceolata Beille in Bull Soc. Bot. Fr., Mém. 8: 63 (1908). Type: Guinée, Kankan, *Chevalier* 573 (P, holo.!)

var. **mollis** (*Pax*) *A. R.-Sm.* in K.B. 28: 324 (1973). Type: Tanzania, Mwanza Distict, Kayenzi [Kagehi], *Fischer* 533 (B, holo.†, K, iso.!)

Fruits pubescent.

TANZANIA. Shinyanga/Kahama Districts: Shinyanga–Kahama road, 9 Jan. 1933, *B.D. Burtt* 4519!; Tabora District: Uyui-Kigwa-Rubuga Forest Reserve, 22 Dec. 1967, *Ruffo* 36!; Tunduru District, Muhuwesi [Mawese], 19 Dec. 1955, *Milne-Redhead & Taylor* 7713!
DISTR. T 1, 4, 6, 8; from Rwanda southwards to Zimbabwe (Mutari) & Mozambique (Zambezia)
HAB. Deciduous woodland and wooded grassland, also along streams and lakesides; 15–1700 m.

SYN. *H. mollis* Pax in E.J. 15: 528 (1893); P.O.A. C: 237 (1895); Hutch. in F.T.A. 6(1): 651 (1912); V.E. 3(2): 18 (1921); Pax & K. Hoffm. in E.P. IV. 147(15): 75 (1922); T.T.C.L.: 216 (1949)
      H. lasiophylla Pax in E.J. 19: 79 (1894); P.O.A. C: 237 (1895). Type: Tanzania, Tabora/Mpanda/Kigoma Districts, Ugalla R., *Böhm* 117a (B, holo.†, K, iso.!)
      H. mollis Pax var. *lasiophylla* (Pax) Pax in E.P. IV. 147(15): 76 (1922); T.T.C.L.: 216 (1949)

PANDACEAE

A. R.-SMITH

Dioecious trees or shrubs. Leaves alternate, distichous, simple, entire, crenate or serrate, penninerved, stipulate, borne horizontally on shoots which subtend 'axillary buds' and thus resemble pinnate leaves; stipules usually persistent, the adaxial one arising below the abaxial. Flowers small, regular, unisexual, in axillary or supraaxillary fascicles, abbreviated racemes or cymes, or in terminal or cauliflorous racemiform thyrses. Calyx cupular, truncate or 5-lobed, the lobes imbricate or open. Petals 5, contorted, imbricate or valvate, flattened or galeate-cucullate. Disc small, rarely large, or absent. Stamens 5, 10 or 15, uniseriate or biseriate, sometimes unequal; anthers introrse, 2-thecous, opening lengthwise; staminodes sometimes present in ♀ flowers; pistillode of ♂ flowers linear-subulate, cylindric or narrowly conical, sometimes pentagonal at base. Ovary superior, 2–5-locular, with 1 (–2) apical pendulous orthotropous or anatropous ovules per locule, without obturator; styles 1–5, simple or bipartite. Fruit a drupe, sometimes flattened, more rarely a capsule; mesocarp fleshy; endocarp woody, bony or stony, thin to thick and massive, usually tubercled, muricate, pitted or ridged, sometimes dehiscent by valves. Seeds usually flattened-concave, more rarely ovoid, ecarunculate, with endosperm. Cotyledons broad, sometimes cordate.

An Old World family of tropical rain-forests, being confined to W. and C. Africa, SE. Asia, Malaysia and Indonesia, comprising 3 genera (*Panda* Pierre, *Galearia* Zoll. & Mor., *Microdesmis* Planch.) and 27 species following Forman in K. B. 20: 309 (1966); Airy Shaw (Willis, Dict. Fl. Pl. & Ferns, ed. 7 (1966)) adds also the monotypic genus *Centroplacus* Pierre. *Microdesmis* just gets into W. Uganda. *Galearia, Microdesmis* and *Centroplacus* were formerly included in the Euphorbiaceae, but in the structure of the endocarp, and also anatomically, they show more of an affinity with *Panda*.

MICRODESMIS

Planch. in Hook., Ic. Pl. 8, t. 758 (1848); Muell. Arg. in DC., Prodr. 15(2): 1041 (1866); G.P. 3: 287 (1880); Pax in E. & P. Pf. 3(5): 82 (1890) & in E.P. IV. 147(3): 105 (1911); Hutch. in F.T.A. 6 (1): 741 (1912); J. Léon. in B.J.B.B. 31: 159 (1961) & in F.C.B. 8: 102 (1962)

Dioecious trees or shrubs. Leaves alternate, distichous, simple, entire to serrate, mucronulate, often unequal at the base, finely pellucid-punctate, petiolate; stipules small, generally persistent. Terminal bud often claw-like. Male flowers in dense many-flowered axillary or supraaxillary fascicles or abbreviated racemes; bracts minute; calyx 5-partite or lobed, slightly imbricate; petals larger than the calyx-lobes, contorted or imbricate in bud, flattened; disc 0; stamens 5, uniseriate, opposite the sepals (in Africa) or 10, biseriate, outer ones opposite the sepals (in Asia), equal; filaments short, broad or thick, ensconced between and free from or adnate to the lobes of the pistillode; anthers with the connective produced or not; staminodes 0; pistillode pentagonal at the base, subcylindric at the apex (in Africa). Female flowers usually in fewer-flowered fascicles than ♂; calyx and petals as in ♂; glands at the base of the sepals; petals caducous; disc 0; ovary 2–5-locular, with 1 ovule per locule, fleshy; styles 2–5, bipartite, laciniate-papillose, spreading. Fruits drupaceous, ovoid or subglobose and smooth in the fresh state, ± smooth, verrucose or muricate when dry; endocarp woody, fairly thin, tubercled or muricate, dark coloured. Seeds broadly ovate, flattened or curved; testa crustaceous, shiny.

10 species, of which 2 are Asiatic, the rest African.

M. puberula *Planch.* in Hook., Ic. Pl. 8, t. 758 (1848), pro parte; Muell. Arg. in DC., Prodr. 15(2): 1041 (1866), pro parte; Pax in E.P. IV. 147 (3): 106 (1911), pro parte; Hutch. in F.T.A. 6(1): 741 (1912), pro parte; V.E. 3(2): 128 (1921); I.T.U., ed. 2: 134 (1952); J. Léon. in B.J.B.B. 31: 167 (1961) & in F.C.B. 8: 104 (1962). Type: Fernando Po, *Vogel* 175 (K, holo.!)

A shrub up to 6 m.; twigs densely pubescent. Leaf-blades elliptic-oblong to ovate-lanceolate or ovate, rarely oblanceolate, 5–15(–20) cm. long, 2–6(–9) cm. wide, acute or subacute, acuminate, often ± unequal at the base (attenuate one side and cuneate the

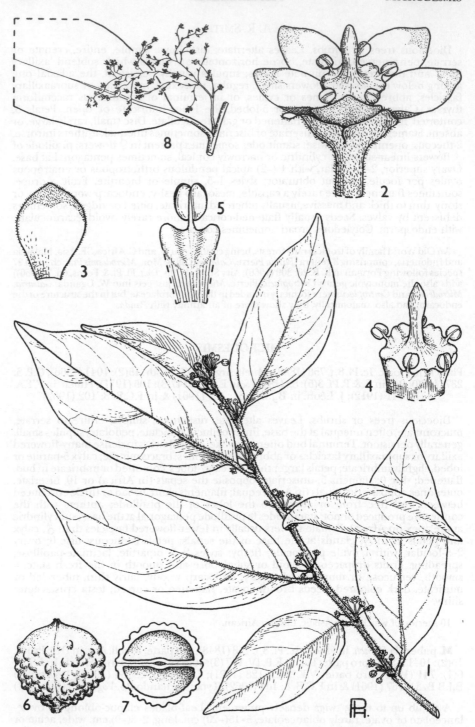

FIG. 108. *MICRODESMIS PUBERULA* — 1, flowering branch, × ⅔; 2, ♂ flower, × 12; 3, petal, × 10; 4, stamens, disc and pistillode, × 12; 5, stamen, × 20; 6, fruit, × 3; 7, same in transverse section, × 3; 8, galled inflorescence, × ⅔. 1, from *Eggeling* 1451; 2–5, from *Zenker* 4329; 6, 7, from *Zenker* 4313; 8, from *Harris* 149. Drawn by Pat Halliday.

other or cuneate one side and rounded the other), finely toothed or subentire, shiny above, pubescent on the midrib above, otherwise glabrous, pubescent on the midrib and main veins beneath, otherwise ± glabrous, lateral veins 6–9 pairs, midrib and main veins somewhat impressed above, prominent beneath, tertiary veins also prominent beneath; petiole 4–12 mm. long, pubescent. Stipules linear or filiform,2–4 mm. long. Male fascicles 5–many-flowered; inflorescence sometimes galled and then pseudo-paniculate. Male flowers: pedicels 5 mm. long, puberulous; calyx 2 mm. long, lobes ovate, 1.5 mm. broad, puberulous without, green; petals subcircular-ovate to ovate-oblong, 3 mm. long, 2–2.3 mm. wide, densely puberulous without and within in the upper part, salmon-pink or pinkish orange, spreading; filaments 1 mm. long and broad, fleshy, obtriangular, ¾ adnate to the pistillode, reddish orange; anthers 0.5 mm. long; connective not produced; pistillode 2–2.5 mm. long, 1.5 mm. diameter below, reddish orange. Female fascicles 1–3(–5)-flowered. Female flowers: pedicels 3–4 mm. long, extending to 5–10 mm. in fruit, puberulous; calyx and petals as in ♂, but sepals with 2–3 minute glands at the base; ovary 2(–3)-locular, ellipsoid, 1 mm. long, pubescent, green; styles 2. Fruits ovoid, 10–12 mm. long, 9–11 mm. in diameter and smooth when fresh, 6–8 mm. long and in diameter and verrucose-muricate when dry, slightly puberulous above, shiny, green at first, later becoming red; mesocarp up to 2.5 mm. thick when fresh, red; endocarp (1–)2(–3)-locular, muricate, 0.5–1 mm. thick. Seeds concavo-convex. Fig. 108.

UGANDA. Bunyoro District: Budongo Forest, Nov. 1932, *Harris* 149! (galled) & Nov. 1933, *Eggeling* 1451! & Dec. 1934, *Eggeling* 1555!
DISTR. U2; S. Nigeria eastwards to Uganda and southwards to N. Angola
HAB. Rain-forest; 1100 m.

SYN. *M. zenkeri* Pax in E. J. 23: 531 (1897); Hutch. in F.T.A. 6(1): 742 (1912), pro parte; F.P.N.A. 1: 472 (1948). Types: Cameroun, Yaoundé, *Zenker* 318 & 354 (B, syn.†)

NOTE. In order that the description and plate could be complete, resort was made to material from Cameroun as well as Uganda.

Addendum

Cleistanthus schlechteri (*Pax*) *Hutch.* var. **pubescens** (*Hutch.*) *J. Léon.* (see Part I, p. 133)

KENYA. Kilifi District: ± 0 km. N. of Gotani, 12 Feb. 1987, *Luke & Robertson* 202
DISTR. Add – **K** 7
HAB. Add – in dry coastal bushland; 280 m.

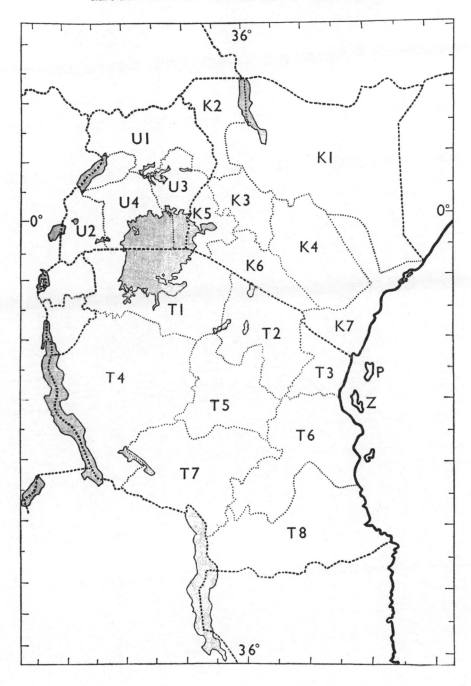

T - #0667 - 101024 - C0 - 244/170/10 - PB - 9789061913382 - Gloss Lamination